Physiological Effects of Noise

Based upon papers presented at an international symposium on the Extra-Auditory Physiological Effects of Audible Sound, held in Boston, Massachusetts, December 28-30, 1969, in conjunction with the annual meeting of the American Association for the Advancement of Science

Edited by

Bruce L. Welch
and
Annemarie S. Welch

Friends of Psychiatric Research, Inc.
Maryland Psychiatric Research Center and
The Johns Hopkins University School of Medicine
Baltimore, Maryland

ℚ PLENUM PRESS • NEW YORK–LONDON • 1970

Organizing Committee for the Symposium

Joseph R. Anticaglia
William F. Geber
Benson E. Ginsburg

Samuel Rosen
Annemarie S. Welch
Bruce L. Welch (Chairman)

Sponsored by
Friends of Psychiatric Research, Inc.
52 Wade Avenue
Baltimore, Maryland 21228

ISBN 978-1-4684-8809-8   ISBN 978-1-4684-8807-4 (eBook)

DOI 10.1007/978-1-4684-8807-4

Library of Congress Catalog Card Number 70-130586

SBN 306-30503-8
© 1970 Plenum Press, New York
Softcover reprint of the hardcover 1st edition 1970

A Division of Plenum Publishing Corporation
227 West 17th Street, New York, N.Y. 10011

United Kingdom edition published by Plenum Press, London
A Division of Plenum Publishing Corporation, Ltd.
Donington House, 30 Norfolk Street, London W.C.2, England

FOREWORD

The remarkable symposium arranged by Bruce L. Welch and
Annemarie S. Welch for the meeting of the American Association
for the Advancement of Science in Boston, Massachusetts at the
end of the year 1969 was devoted to the physiological effects
of audible sound. Dr. Welch and his wife were able to bring
together a distinguished group of scientists from all parts of
the world. It was very remarkable to be able to discuss the
physiological aspects of noise with representative scientists
from Israel, France, Germany, Hungary, Russia, Australia, Canada
and Argentina. Dr. and Mrs. Welch ran the meeting in a delightful
manner and continued to maintain interest and enthusiasm. Now
the results of the conference are available. It is to be hoped
that this volume will find wide interest and attention.

We must differentiate noise from sound. Noise is unpleasant,
unwanted or intolerable sound. On the other hand, even ordinary
sound may at times be unpleasant, simply because we are not conditioned to it.

The general impression that one gets from reading the various
reports on the physiological effects of noise is bad. It's a
pollutant that we can each individually reduce, and maybe we can
have a great enough effect socially so that we can significantly
lower the noise levels which may result in considerable harm to us.

It is interesting that noise as a pollutant has only recently
attracted attention. Noise can be defined in engineering terms
as any disturbance in an integrated equilibrium. We human beings
are in an exquisitely delicate integrated equilibrium. This can be

disturbed in many ways.   The disturbance is usually called stress.

Ordinarily there are physiological factors which tend to pre-
serve the internal steady state, as was emphasized by Claude
Bernard (1813-1878) and by Walter B. Cannon (1871-1945).   Cannon
refers to emotional factors which produce changes in digestion,
blood pressure and other body functions.   It now appears that
noise is one of the stress factors that can produce widespread
disturbances in various biochemical and physiological activities
of the human body.

There is no doubt but that we are being exposed continually
to increasing levels of noise, especially in our cities.   The intensity of noise is usually expressed in terms of "decibels."
The scale is a logarithmic one.   The relative quiet of a wilderness area, punctuated by an occasional cry of a wild bird, or the
rustle of leaves, usually runs around 20-30 decibels.   The inter-
ior of a closed bedroom at midnight in an average city may show a
noise level of around 30 decibels.   The ordinary street sounds of
automobiles and trucks may give a level of around 80 decibels with
respect to pedestrians on the sidewalk.   A heavy truck may run
the noise level to 90 decibels.   The usual kitchen setup with a
TV blaring, a vacuum cleaner running and other appliances operating,
may be in the neighborhood of 100 decibels.   A low-flying jet
airplane may yield around 120 decibels.   A power lawnmower or an
outboard motor will give about the same level of noise.   A rock
and roll band at its peak may run 125 or even higher decibels.
A sharp thunderclap is in the same level.   Sonic booms are different, and involve pressure effects as well as sound levels.

At the ear itself, noise may produce significant reactions if
long continued and intense.   Long continued loud noise itself may
actually result in partial loss of hearing and in interference with
understanding.

The transmission of sound impulses by nerve transmission brings
the sensation to perception in the cortex.   But first these sensory
impulses must pass through the midbrain.   Some of them are directed
towards the cortex for perception, and it is here that noise may
produce emotional disturbances of annoyance, irritability, quarrel-
someness, loss of attention and other conscious factors which are
disturbing.   Meanwhile, however, whether we are conscious of the
unpleasant noise or not, there seems to be a spread into our autonomic nervous systems.   This can produce complicated effects on
the cardiovascular-renal system, on the endocrine system, with the
complications of the chemicals put out by the glands of internal
secretion, and on the reproductive system.

All these matters were explored in considerable detail at the
Boston conference.

Certainly noise is a matter which concerns people all over the world. The conditioning effects of small amounts of noise may be significant. It is to be remembered that I. P. Pavlov (1849-1936) in his remarkable experiments on conditioning, maintained his animals in the famed "towers of silence" which he built especially in the Institute for Experimental Medicine in Leningrad. Here his animals were kept in a superb manner, and protected particularly from the distracting influences of noise.

While we may not be able to protect ourselves by "towers of silence," we should nevertheless be aware of the fact that noise does have complex and harmful effects upon our delicately balanced bodies. We must learn to do everything we can to protect ourselves from noise. It can be done, and it may help all of us.

> Chauncey D. Leake, Ph.D.
> Senior Lecturer
> Pharmacology and
> Experimental Therapeutics
> School of Medicine
> University of California
> San Francisco

PREFACE

Most of the research that has been conducted on the physiological effects of sound during the past two decades has been concerned with its effects upon hearing. It is now clear, however,
that hearing impairment is only one of many important adverse effects that sound, when it becomes noise, may have upon bodily
functions.

This volume is based upon papers that were presented at an
international symposium on the Physiological Effects of Audible
Sound (Extra-Auditory), which was held as a part of the annual
meeting of the American Association for the Advancement of Science
in Boston, Massachusetts, December 28-30, 1969.

It is intended to make apparent areas where present knowledge
is incomplete as well as to present information that is now relatively well established.

It brings together for the first time under a single cover
the major elements of existing information on the effects of
audible sound upon cardiovascular, reproductive, endocrine, and
neurological function. Reports on the studies that have thus far
been conducted on the effects of simulated and actual sonic booms
are also included.

If currently planned supersonic transports fly regular commercial routes over land, approxiamtely 28 million persons in the
continental United States will be routinely exposed each day to
40-50 sonic booms of a higher intensity than those produced by
present military aircraft./ Many millions more will spend their

/Kryter, K.D., 1969. Sonic booms from supersonic transport.
Science 163:359-367.

life being exposed daily to fewer, but nevertheless far too many, disturbing sonic booms.

Motivation for organizing the symposium was provided by the obvious need to encourage an adequate advance evaluation of the probable effects upon health of this dramatic environmental change. No such evaluation, thus far, has been attempted by our government and none is planned.

The symposium was originally organized under the title <u>Assessing the Impact of Technology: Example and Precedent of the SST-Sonic Boom</u>, and its organization was complete in early March, 1969. In addition to technical sessions on the physiological effects of sound, the symposium contained relevant panel discussions on science policy for technology assessment, with emphasis upon the function and malfunction of the advisory system. Due to various pressures, however, control of the major science policy portions of the symposium shifted, and they were recast in more general terms under the title <u>Technology Assessment and Human Values</u>./ The technical sessions that remained form the basis for this volume.//

It is sometimes inordinately difficult to gain recognition for the potentially harmful effects that new technologies may have upon health and to get them properly evaluated in the overall process of "technology assessment." This should not be.

The need to structure science and government in a manner that will assure adequate - and adequately open - advance evaluation of the probable effects of new technologies (both adverse and beneficial), free of pressures from promotional and vested interests, is one of the most pressing needs of our time.

Gratitude is due Mrs. Katherine Embree, Mrs. Elizabeth Borish, The New York Community Trust, Citizens League Against the Sonic Boom, United States Steel Corporation, and Riker Laboratories for meeting the expenses of the symposium participants, and to Mrs. Stephen Lee Magness for typing the manuscripts for publication. Particular appreciation is due Mr. Richard Meachem, Executive Director of Friends of Psychiatric Research for generous and unfailing optimistic support when success seemed so unlikely.

Bruce L. Welch, Ph.D.
Friends of Psychiatric Research, Inc. and
The Johns Hopkins University School of Medicine
Baltimore, Maryland

/Carpenter, R.A. Technology, assessment and human possibilities. Science 166:653, 1969.
//Welch, B.L., 1969. Physiological effects of audible sound. Science 166:533-535.

LIST OF CONTRIBUTORS

Joseph R. Anticaglia, United States Department of Health, Education and Welfare, Public Health Service, Environmental Health Service, Environmental Control Administration, Bureau of Occupational Safety and Health, Cincinnati, Ohio

A. E. Arguelles, Department of Endocrinology and Hormone Laboratory, Hospital Aeronautico, Buenos Aires and Scientific Research Committee of the Argentine

A. Árvay, Department of Obstetrics and Gynecology, University of Debrecen, Hungary

James Bond, Animal Husbandry Research Division, Agricultural Research Service, Beltsville, Maryland

Robert E. Bowman, Regional Primate Research Center, University of Wisconsin, Madison, Wisconsin

Joseph P. Buckley, Department of Pharmacology, University of Pittsburgh School of Pharmacy, Pittsburgh, Pennsylvania

Robert L. Collins, The Jackson Laboratory, Bar Harbor, Maine

Maria V. Disisto, Department of Endocrinology and Hormones Laboratory, Hospital Aeronautico, Buenos Aires and Scientific Research Committee of the Argentine

L. P. Dobrokhotova, Laboratory of the Physiology and Genetics of Behavior, Moscow Lomonosov State University, Moscow, USSR

Gregory B. Fink, Department of Pharmacology, Oregon State
     University, Corvallis, Oregon

D. A. Fless, Laboratory of the Physiology and Genetics of Behavior,
     Moscow Lomonosov State University, Moscow, USSR

Francis M. Forster, Department of Neurology, School of Medicine,
     University of Wisconsin, Madison, Wisconsin

John L. Fuller, The Jackson Laboratory, Bar Harbor, Maine

William F. Geber, Department of Pharmacology, Medical College of
     Georgia, Augusta, Georgia

Jack M. Heinemann, United States Air Force Environmental Health
     Laboratory, Kelly Air Force Base, Texas

Kenneth R. Henry, Regional Primate Research Center, University of
     Wisconsin, Madison, Wisconsin

W. B. Iturrian, Department of Pharmacology, University of Georgia,
     Athens, Georgia

Gerd Jansen, Institut fur Hygiene und Arbeitsmedizin, Essen der
     Rhur-Universitat Bochum, West Germany

Marcus M. Jensen, Department of Microbiology, Brigham Young
     University, Provo, Utah

L. V. Krushinsky, Laboratory of the Physiology and Genetics of
     Behavior, Moscow Lomonosov State University, Moscow, USSR

Karl D. Kryter, Sensory Sciences Research Center, Stanford
     Research Institute, Menlo Park, California

Chauncey D. Leake, School of Medicine, University of California,
     San Francisco, California

Alice G. Lehmann, Laboratoire de Physiologie Acoustique, Domaine
     de Vilvert, Jouy-en-Josas, France

Mary F. Lockett, Department of Pharmacology, The University of
     Western Australia, Nedlands, Australia

Jerome S. Lukas, Sensory Sciences Research Center, Stanford
     Research Institute, Menlo Park, California

M. A. Martinez, Department of Endocrinology and Hormone Laboratory,
     Hospital Aeronautica, Buenos Aires, and Scientific Research
     Committee of the Argentine

L. N. Molodkina, Laboratory of the Physiology and Genetics of
    Behavior, Moscow Lomonosov State University, Moscow, USSR

Charles W. Nixon, Aerospace Medical Research Laboratory, Wright-
    Patterson Air Force Base, Ohio

Eva Pucciarelli, Department of Endocrinology and Hormone
    Laboratory, Hospital Aeronautico, Buenos Aires, and
    Scientific Research Committee of the Argentine

A. F. Rasmussen, Jr., Department of Medical Microbiology and
    Immunology, School of Medicine, University of California,
    Los Angeles, California

L. G. Romanova, Laboratory of the Physiology and Genetics of
    Behavior, Moscow Lomonosov State University, Moscow, USSR

Samuel Rosen, Consultant, The Mount Sinai School of Medicine and
    The New York Eye and Ear Infirmary, New York, New York

A. F. Semiokhina, Laboratory of the Physiology and Genetics of
    Behavior, Moscow Lomonosov State University, Moscow, USSR

Harold H. Smookler, Department of Pharmacology, University of
    Pittsburgh School of Pharmacy, Pittsburgh, Pennsylvania

Lester W. Sontag, The Fells Research Institute, Yellow Springs,
    Ohio

A. P. Steshenko, Laboratory of the Physiology and Genetics of
    Behavior, Moscow Lomonosov State University, Moscow, USSR

Paul Y. Sze, Department of Biobehavioral Sciences, The University
    of Connecticut, Storrs, Connecticut

I. Tamari, Hadassah Medical Centre, Hebrew University, Jerusalem,
    Israel

George J. Thiessen, Division of Physics, National Research Council
    of Canada, Ottawa, Ontario, Canada

Bruce L. Welch, Friends of Psychiatric Research and Department of
    Psychiatry and Behavioral Science, Johns Hopkins University
    School of Medicine, Baltimore, Maryland

Harold L. Williams, Department of Psychiatry and Behavioral
    Sciences, University of Oklahoma Medical Center, Oklahoma
    City, Oklahoma

Z. A. Zorina, Laboratory of the Physiology and Genetics of
    Behavior, Moscow Lomonosov State University, Moscow, USSR

CONTENTS

SUMMARY

INTRODUCTION:
NOISE IN OUR OVERPOLLUTED ENVIRONMENT

JOSEPH R. ANTICAGLIA

U.S. Department of Health, Education and Welfare
Public Health Service, Environmental Health Service
Environmental Control Administration, Bureau of Occu-
pational Safety and Health, 1014 Broadway
Cincinnati, Ohio   45202

The crescendo of noise in our environment is not just an
isolated disturbance or nuisance. It can affect huge numbers
of people, both physically and psychologically. For instance,
50 percent of the machines used in heavy industry may produce
noise at levels that are potentially damaging to a worker's hearing sensitivity. As many as 17 million men, by some estimates,
are exposed to the harmful effects of excessive noise in their
work environment.

Chemicals and gases once thought unique to industry often
escape into our environment. Noise is no exception. The perceived loudness of noise in some cities has more than doubled in
the past two decades. Some impacted communities already have
noise levels that are comparable to those found in industry. Of
course, there is the possibility of daily overland supersonic commercial flights and resulting sonic booms. This sonic insult
could impact an area 25 miles on either side of its flight path
and touch as many as 50 million persons for each transcontinental
crossing of the United States.

From a physicist's standpoint, sound is a random movement of
molecules, which can produce auditory stimulation. He is interested in the "sound source" which liberates energy in a characteristic pattern, in the intensity and frequency of sound and in how
sound waves are propagated. In this limited sense, sound does
not include the sensations it can evoke in the minds of the listeners.

But a psychologist examines an individual's subjective response
to sound in light of some measurable characteristic in the external

1

physical world.   He is interested in identifying those factors that
determine our acceptance or rejection of a sound.   Accordingly, he
may relate the loudness, pitch, or annoyance experienced by individuals to intensity, frequency, or some other physical parameter of
sound.   Here, the human response factors are of paramount importance.   Thus noise - sound without value or unwanted sound - entails
both the objective evaluation of physical sound and the subjective
impression in the listener's mind.

There are innumerable sources of degrading sound or noise, but
whether generated by pneumatic riveters, over-flying aircraft, or
passing trucks, noise can and does adversely affect man.   It can
cause auditory problems by masking speech or by damaging the receptor cells in the Organ of Corti.   It can cause "extra-auditory
effects" involving disturbances of physiologic functions apart
from hearing; and it can hinder task performance and disrupt rest,
relaxation, and sleep.

The extra-auditory effects of special interest to this meeting
involve systemic physiologic responses to sound stimulation.   There
is growing concern about such effects in some academic circles,
although the prevailing view of many "noise experts" in the United
States is that aside from hearing impairment, noise conditions encountered in present day environments can be adapted to without
any ill effects.   While there is evidence to support adaptation
to excessive noise, there are also indications that prolonged exposure to such noise may lead to health problems.   The adaptation
may be achieved but at unacceptable physiological and behavioral
costs.

Along these lines, workers exposed to excessive noise for many
years appear to suffer a greater incidence of neurologic, digestive,
and metabolic disorders compared to the general population.   Exposure to intense noise in heavy industry seems to affect the cardiovascular system by causing angiospastic effects, fluctuations in
blood pressure, and impairment of certain properties of cardiac muscle.

Outside of the industrial environment, residents impacted by
aircraft and highway noises complain that such noise has degrading
effects on their health.   They purport to be suffering from hypertension, undue nervousness, exacerbation of cardiac symptoms, and
assorted mental problems.   There is no factual information to
support these claims, but many questions remain unanswered in this
area.

For example, what are the acute and cumulative health effects,
if any, on workers who are exposed to noise conditions in heavy
industry -- on residents of communities neighboring airports and
expressways?   Can one correctly assume that a worker's harmful
noise exposure is due solely to his job activity?   Or must we now

think in terms of health effects from 24-hour noise exposure, i.e.,
both on and off the job?   Are workers with pre-existing pathological disorders more susceptible to the harmful effects of noise?

With regard to the latter, what stressful effects, if any,
does intrusive community noise conditions have on hospitalized
and convalescent home populations?   Do we have enough information
to unequivocally state that the supersonic transport jet (SST) will
not have any adverse health effects?   Does the social risk out-
weigh the technological benefit?

I acknowledge that environmental noise in the vicinity of
noise-impacted communities has not been clearly identified as
being harmful to the physical and mental health status of such
residents.   However, evidence that might so identify it would
most likely come from long range studies, which have <u>not</u> as yet
been conducted.   Aside from hearing loss, few investigators have
attempted to evaluate the total physical and mental health status
of workers exposed to industrial noise, and nothing has been
accomplished concerning longitudinal studies in this regard.

Of course, one asks where does noise stand on our list of
priorities?   If our goal is the enhancement of the quality of
human life, if we wish to heed our "ecological conscience", we
cannot accept excessive noise as a necessity of progress.   To
do otherwise, possibly reflects unintelligent complacency.   We
need not wait for "Donora"* before we take action.   U Thant,
Secretary General of the United Nations, put it in different terms
when he states " .... It is no longer resources that limit decisions.
It is the decision that makes the resources.   This is the funda-
mental revolutionary change, - perhaps the most revolutionary
that mankind has ever known."

*Deaths in Donora, Pennsylvania, caused by smog during a prolonged
inversion served to dramatically emphasize the seriousness of the
air pollution problem.

ENVIRONMENTAL NOISE, "ADAPTATION" AND PATHOLOGICAL CHANGE

BRUCE L. WELCH

Environmental Neurobiology
Friends of Psychiatric Research and
Department of Psychiatry and Behavioral Science
School of Medicine, Johns Hopkins University
Baltimore, Maryland

The body is physiologically responsive to stimulation of the auditory nerve by sound even during sleep, under anaesthesia and, indeed, even after the cerebral hemispheres have been removed (1).

Sound, either continuous or intermittent, activates subcortical neuronal systems to continually modify the pacing by the brain of cardiovascular, metabolic, endocrine, reproductive and neurological function. These activating neuronal systems are themselves restrained by inhibitory neuronal systems with which they interact, and tonic influences inherent in this interaction determine the level of many peripheral physiological functions.

Bodily functions are different in different stimulus environments. In the truest sense, the individual and his environment are functionally inseparable. They are psychophysiologically linked (2).

The weight of evidence available at this time is that the body does not "adapt" to long-term residence in different stimulus environments by maintaining, or returning to, pre-determined and characteristic "normal" levels of peripheral physiological function. Rather, it responds in a sensitive and continuous manner to its stimulus environment, and different basal levels of function are established and maintained according to the dictates of the prevailing Level of Environmental Stimulation (3).

Pervasive and long-lasting changes are produced even by very short daily exposures to intense stressful stimulation (4).

It is true that a reasonable amount of stimulation is necessary for normal development, and that in some instances the employment of sound in therapeutic treatment has been proposed. It is also true that habituation occurs with exposure to sound, and that this provides some degree of short-lasting protection against subsequent exposure.

Chronic overstimulation, however, has pathological consequences, and a Level of Environmental Stimulation greater than the optimum is clearly harmful to health. It results in the so-called "diseases of adaptation" (5-7). In considering the effects of long-term exposure to audible sound, and particularly to high levels of environmental noise, it is with the pathological processes of adaptation that we must be concerned (8).

REFERENCES

1. Ward, A.A., 1953. Central nervous system effects. BENOX REPORT: An Exploratory Study of the Biological Effects of Noise. The University of Chicago, Office of Naval Research Project NR 144079, Contract N6 ori-020 Task Order 44. pp. 73-80.

2. Mason, J.W., 1968. Organization of psychoendocrine mechanisms. Psychosomatic Medicine 30:565-808.

3. Welch, B.L., 1965. Psychophysiological response to the mean level of environmental stimulation: a theory of environmental integration. In: Medical Aspects of Stress in the Military Climate, U.S. Government Printing Office, Washington, D.C. pp. 39-99.

4. Welch, B.L. and Welch, A.S., 1969. Sustained effects of brief daily stress (fighting) upon brain and adrenal catecholamines and adrenal, spleen and heart weights of mice. Proceedings of the National Academy of Sciences of the United States 64:100-107.

5. Selye, H., 1956. The Stress of Life, McGraw-Hill, New York.

6. Dubos, R., 1965. Man Adapting, Yale University Press, New York.

7. Wolf, S. and H. Goodell, 1968. Harold G. Wolff's Stress and Disease (2nd Ed). C.C. Thomas Publ., Springfield, Illinois.

8. Welch, W.H., 1897. Adaptation in pathological processes. Published in 1937 in Fourth Series of Bibliotheca Americana, Institute of History of Medicine, Vol. 3, Johns Hopkins University Press, Baltimore. pp. 284-310.

AUDIOGENIC STRESS AND SUSCEPTIBILITY TO INFECTION

MARCUS M. JENSEN* and A. F. RASMUSSEN, JR.

Department of Medical Microbiology and
Immunology, School of Medicine
University of California, Los Angeles

Emotional or psychologic stress is often proposed as a factor which may alter the outcome of a host-parasite relationship. Over the past several years, studies in our laboratory in the UCLA Brain Research Institute have dealt with the influences of "psychologic" or CNS-mediated stress on the susceptibility of animals to viral infections. Various aspects of this subject have been reviewed (1,2). In this report, psychologic or nonspecific stress is defined as exposure to any noxious environmental stimulation that mediates responses via the CNS, which in turn stimulate other organs or systems, primarily the pituitary-adrenocortical axis, and is associated with little or no physical trauma. The first experimental stressor systems that we used were avoidance-learning in a shuttle box and confinement in snug wire mesh envelopes (3). Because a slight amount of physical trauma might have been involved with these two systems, a third system was sought. After reports became available indicating the effectiveness of audiogenic stimulation as a stressing agent (4,5), a sound stressing system was added to our studies (6,7).

This report summarizes our research in which sound was used as the inducer of the stress response in mice. Comparison is made with results obtained with avoidance-learning stress. Data is presented on the influences of audiogenic stress on physiologic responses, on changes in susceptibility to both cyto-destructive and oncogenic virus infections, and on host defense mechanisms.

---

*Present address: Department of Microbiology,
Brigham Young University, Provo, Utah

MATERIALS AND METHODS

In most of these studies, 6 to 8 week old Swiss Webster, BRVS, mice were used.

The sound exposure chamber was a heavy cardboard cylinder 40 centimeters in diameter and 3 meters long. Sound was produced by a 35 centimeter loudspeaker installed across one end of the cylinder and driven by a 10-watt audio amplifier. A sine wave signal at 800 cps was produced by a signal oscillator and amplified to about 120 to 123 decibels as measured in the chamber. Mice were placed in open mesh wire cages with 8 x 8 x 20 centimeter individual compartments. The open wire construction allowed unobstructed passage of sound waves to all areas of the cages and the individual compartments kept the mice from huddling. This stress chamber could accommodate 48 mice. During the sound stress period, the open end of the chamber was closed with a 50 centimeter thick fiberglass wedge. Fresh air was supplied to the chamber to maintain it at room temperature. Unless otherwise stated, a 3 hour daily exposure to sound was employed (6).

Avoidance-learning stress was produced by an automatically programmed shuttle box that require the mice to jump over a barrier to the opposite side of a compartment once every 5 minutes on the presentation of a warning buzzer and light. Mice failing to jump received an electric shock. Mice soon learned to avoid the shock and were thus primarily exposed to the stress of anticipation of pain and fear. Daily six hour periods of this stressor were used (3).

RESULTS

Physiological Responses to Audiogenic Stress

We found that variations in leukocyte counts were a useful index for following immediate and transitory responses of mice exposed to stressful stimuli (8). Blood samples were collected from tail veins and were assayed by standard procedures for total and differential leukocyte counts. A typical leukocytic response to a 3-hour exposure to sound (Fig. 1) consisted of a marked leukopenia developing shortly after the onset of stress. After about 90 minutes the leukocyte count began to increase and returned to the control level at about 90 minutes after termination of the stress period. The count continued to increase for an additional 2-3 hours, until a marked leukocytosis was present. By 6 hours after the stress period the leukocyte counts had returned to normal. The same basic response was the same on each daily stress exposure and has been followed for over 30 days. The variations in leukocyte counts were not observed in stressed-adrenalectomized mice.

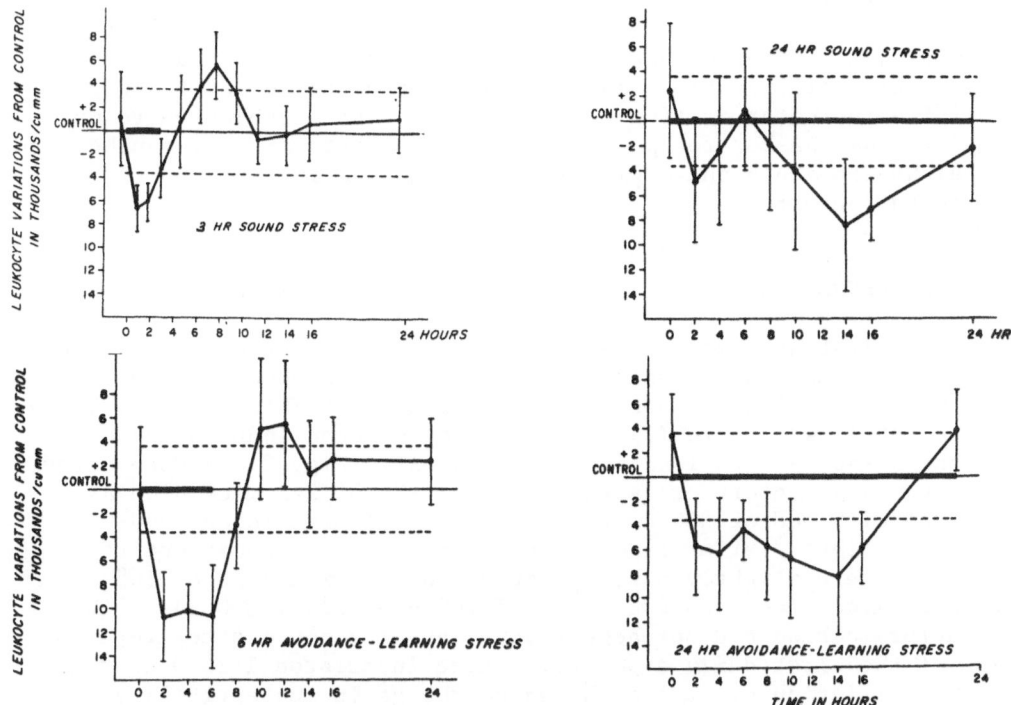

Figure 1. Leukocyte variations during sound and avoidance-learning
stress. Control values, center line: S.D., broken lines. Stress
period is indicated by the heavy bar. Mean for stressed mice,
solid line; vertical lines show S.D. at different sampling intervals.

The biphasic leukocytic response was also induced by avoidance-
learning stress (Fig. 1). A basic difference was the continued
maintenance of leukopenia during the entire avoidance stress period,
while leukopenia in sound stressed mice lasted for a shorter time
with counts beginning to return towards normal even in the presence
of continuous sound exposure. Mice exposed to a continuous 24
hour period of sound responded somewhat differently from mice
stressed by avoidance-learning for 24 hours (Fig. 1). A transitory
inhibition of the stress induced leukopenia was seen very prominently after 6 to 8 hours of sound stress before a more sustained leukopenia was induced. This transitory inhibition of leukopenia was
very minimal in mice subjected to 24 hours of avoidance-learning.

During the first 4 daily exposures to sound, a progressive
hypertrophy of the adrenal glands was induced (6). A significant
atrophy of the liver, spleen or thymus was not seen in sound
stressed mice (6), whereas atrophy of these organs was induced in
mice subjected to avoidance-learning or confinement stress (9).
A significantly increased resistance to passive anaphylaxis was
induced by sound (6) which indicated stimulation of the pituitary

adrenal axis (10).

Dr. H.E. Weimer of our department has shown the appearance of
a serum globulin fraction, called alpha-Two acute phase, to be a
sensitive indicator of tissue damage (11).   Rats subjected to our
sound stress chamber for periods of 4 and 8 hours did not synthesize
the alpha-Two acute phase globulin, thus indicating that no tissue
damage had occurred.

INFLUENCES ON SUSCEPTIBILITY TO ACUTE VIRAL INFECTIONS

The first and most successful experimental system we employed
to study influences on susceptibility was one in which we challenged
mice with approximately one LD_{50} of vesicular stomatitis virus (VSV)
by the intranasal route (7).   Exposures to sound induced marked,
but transient, alterations in susceptibility (Fig. 2).   When virus
was given just before exposure to sound, 188 of 249 mice died as
compared with 108 of 249 controls (p<.001).   Mice inoculated with
virus immediately after exposure to sound on days subsequent to
the first were slightly more resistant to the virus; 60 of 165
stressed mice died compared to 77 of 160 controls (p<.003).   This
refractory period did not appear until about 2 hours after termination of the first daily exposure.   Mice inoculated 1 to 3 hours .
before initiation of stress showed no change in susceptibility.

Adrenalectomized mice that were inoculated on the second day
of stress showed increased susceptibility both before (36 deaths
of 55, p<.001) and after (28 deaths of 53, p<.05) the stress period
(Fig. 2) when compared to adrenalectomized controls (18 deaths of
55).

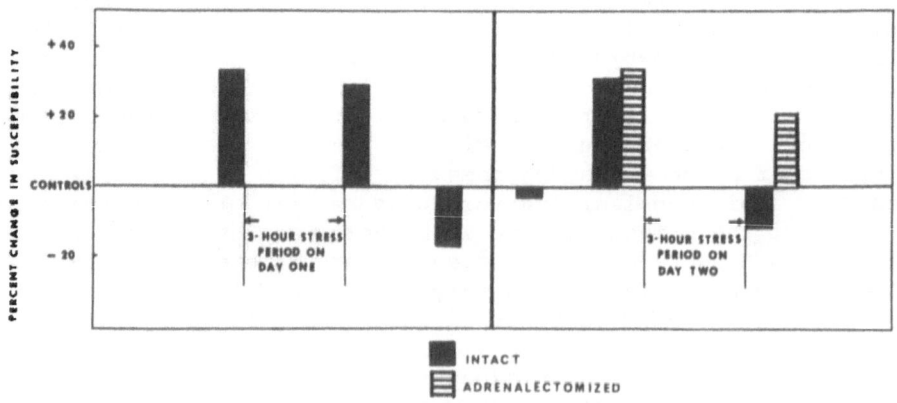

Figure 2.   Changes in susceptibility of mice inoculated intranasally with vesicular stomatitis virus at various times in relation
to 3-hour periods of sound stress.   Vertical bars represent deviation in death rates from nonstressed controls.

Table 1.   Tumor incidence in young Swiss Webster
mice inoculated with polyoma virus and subjected
to 6 days sound stress followed by 14 days avoidance-learning stress.

Age at inoculation	Group	Number of mice	Tumor Incidence (%) at	
			14 wk	32 wk
14 days	Stressed	21	38*	86*
	Controls	21	5	57
19 days	Stressed	22	0	52*
	Controls	21	0	23

*p<.05 to p<.01 compared to respective controls.

Table 2.   Influence of sound stress on splenomegaly
in mice inoculated with Rauscher leukemia virus.

Group	Number of mice	Days of stress before virus	after virus	Spleen (mg)*	p**
I Stress	238	10	15	310 ± 119	0.001
Control	197	–	–	405 ± 129	
II Stress	42	10	0+	240 ± 75	0.26
Control	42	–	–	247 ± 76	
III Stress	49	0	10	223 ± 44	0.01
Control	49	–	–	247 ± 41	

* Mean ± S.D.
** Compared to own controls.
+ Mice sacrificed 12 days after inoculation.

Mice exposed to sound stress were also challenged with intra-peritoneal inoculations of coxsackie B-1 and herpes simplex virus-es. No sharp alterations in susceptibility were noted; mice inocu-lated immediately before the stress period were slightly, but not significantly more susceptible than controls. Exposure to sound had no observable influence upon susceptibility to intranasally inoculated influenza virus (12).

The same biphasic alteration in susceptibility observed in mice inoculated intranasally with VSV when subjected to sound stress, was not seen in VSV infected mice subjected to periods of avoidance-learning stress. No reproducible patterns of induced changes in susceptibility to VSV could be obtained in avoidance-learning stressed mice, although repeated experiments were con-ducted (12).

Influences on Susceptibility to Tumor Viruses

An initial series of experiments, demonstrated that newborn mice inoculated with polyoma virus and exposed to sound stress starting one month later resulted in no marked differences in tumorigenesis as compared to unstressed infected mice (13). Later studies, carried out in our laboratory by Dr. S.S. Chang, demonstrated that 14 to 30 day old mice when inoculated intra-cardially with polyoma virus developed high incidences of tumors. This lead to a re-study of the effect of stress on susceptibility to polyoma induced tumors (14). In these experiments young mice were inoculated with 2.8×10^5 pfu of polyoma virus and then sub-jected to 6 daily exposures to sound stress followed by 14 days of avoidance-learning stress. Sound stress was used in these studies as the young mice were unable to negotiate the avoidance-learning procedure. Mice were examined periodically for palpatable tumors and were sacrificed after 32 weeks and the total incidence of tumors determined. The values obtained with Swiss Webster mice are listed in Table 1; they show a significantly higher rate of tumorigenesis in the stressed mice. Similar studies with other strains of mice of both high and low natural susceptibility con-sistently demonstrated either reduced latent periods or a higher incidence of tumors in stressed mice.

Next, the influence of stress on the course of Rauscher murine leukemia virus infection was studied. Initially mice which were subjected to daily exposures to sound were infected i.v. with a moderate dose of virus and periodic blood samples were examined. No differences were seen between stressed and nonstressed mice as measured by the induction of lymphatic leukemia. However, it was noted that virus induced splenomegaly was retarded in the stressed mice. This observation was then studied in detail (15) and the results are summarized in Tables 2 and 3. Mice were first subjected

Table 3. Influence of stress on splenomegaly in adrenalectomized and intact mice 13 days after inoculation
with rauscher leukemia virus.

Group	Number of mice	Spleen (mg)*	p<
Adrenalectomized stressed	21	290 ± 62	0.38
Adrenalectomized controls	20	323 ± 116	
Intact stressed	23	175 ± 54	0.001
Intact controls	24	269 ± 106	

* Mean ± S.D.
** Compared to own controls.

to stress for 10 days, then injected i.v. with a dose of virus that
induced palpable spleens in 12 to 14 days.   An additional 15 days
of stress was applied before the mice were sacrificed and their
spleens weighed.   The stressed mice had significantly smaller
spleens than the controls.   When no post infection stress was
applied, no retardation in spleen size was seen.   Stress applied
only after infection induced a moderate retardation of splenomegaly.
Similar results were obtained with avoidance-learning stress.

To determine if the stress-induced hyperactivity of the adrenal
glands caused the retardation of splenomegaly, mice were first subjected to stress (in this case avoidance-learning) for 5 days, then
adrenalectomized and rested 1 day.   Mice were then inoculated with
the virus and stressed for an additional 13 days before the spleens
were weighed.   Control mice received the same inoculation and surgical procedure.   Intact stressed and control mice were simultaneously tested.   A slight, but statistically insignificant, retardation was seen in spleen weights of the stressed-adrenalectomized
mice, while a marked difference was seen between the intact groups
(Table 3).

Influences on Host Defense Mechanisms

We were unable to measure any impairment in the production of
circulating antibodies in mice subjected to avoidance-learning
stress (16).   These studies were not extended to sound stressed
mice.

The inflammatory response as measured by granuloma formation against subcutaneously-implanted cotton pellets, was significantly decreased in sound stressed mice (17).   Sound stress was applied for 5 days before and 14 days after implantation of a single small dental cotton pellet.   The granulomas were excised on the 14th day, weighed and examined histologically.   The mean granuloma weight from 56 stressed mice was 37.4 mg with a $\pm$ 8.7 mg standard deviation compared to 53.8 $\pm$ 12.4 mg for granulomas from 57 control mice ($p < 0.001$).   Marked differences were seen on histological examination (Fig. 3).   Granulomas from stressed mice had narrow capsules with only slight cellular infiltration into the cotton pellet, while control mice developed granulomas with wide capsules which were poorly differentiated from the dense cellular infiltration.

Sound and other forms of stress have been shown to retard interferon production.   Much less interferon was produced in the strains of mice inoculated intranasally with VSV just before exposure to sound stress (18).   While other studies using primarily avoidance-learning stress have shown a single exposure to stress to induce a transient suppression of interferon production which was stimulated by i.v. injection of Newcastle Disease Virus (19).

No differences were noted in the rate of clearance of virus particles (T-2 coliphage) from the blood of sound stressed and nonstressed mice, regardless of the time of viral inoculation in relation to the stress period (12).   Some evidence suggested an impaired clearance of bacteriophage from the peritoneal cavity of mice during the sound stress period; at the termination of the stress period the normal clearance rate resumed (20).

 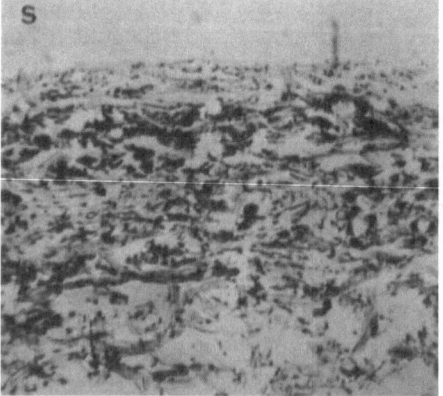

Figure 3.   Histological appearance of granulomas, showing capsule and marginal area, when removed from Control (C) and Stressed (S) mice after 14 days.

DISCUSSION

Based on our observation (6,8) and those of others (4,5) high
intensity audiogenic stimulation qualifies as a psychologic or non-
specific stressor, i.e., it induces a CNS response which in turn
causes hyperactivity of the pituitary-adrenocortical axis and possibly other responses, yet produces no direct physical trauma (11).
Variations in total leukocyte counts reflected immediate and transitory responses of mice to audiogenic stress and appeared to be
primarily associated with adrenocortical secretions (6,8). This
association is further suggested when the biphasic adrenocortical
response measured by Henkin and Knigge (21) in rats during continuous sound stress (characterized by an initial period of hyper-
secretion followed by a period of normal or depressed output with
an eventual return to a high rate of secretion) is inversely correlated with the biphasic pattern of leukocyte counts in the mice
exposed to continuous sound for 24 hours (Fig. 1).

The basic differences between sound and avoidance-learning
stress, as measured by leukocyte variations, was the sustained
suppression obtained with the avoidance system. This may have
been due to the intermittent nature of the avoidance stressor, with
a 5 second warning stimulus and response period interspaced with
5 minute waiting periods, while the sound stimulus was continuous.
From both leukocytic responses and from direct behavioral observations, the mice appeared to be unresponsive after about 1 to $1\frac{1}{2}$
hours of a 3-hour stress period, and perhaps we could have obtained
the same results with a less than 3 hour exposure period. Short
bursts of sound at 5 minute intervals might have induced the more
sustained response seen with the avoidance system (Fig. 1); this
was not tried. Based on organ weight changes (6), the sound system we used was not as stressful to mice as the avoidance-learning
system (9). We found no evidence of hearing loss in the sound
stressed mice.

The susceptibility changes to viral infections induced by stress
can be divided into two general categories. The first, referred to
as accumulative, is characterized by increased susceptibility after
several weeks of daily stress exposures. Such stress exposure, in
these examples avoidance-learning, has predisposed mice to greater
susceptibility to such viral infections as herpes simplex (3),
poliomyelitis (22), and coxsackie B (23). The accumulative response
has best been measured by such changes in the host as adrenal hypertrophy and atrophy of the thymicolymphatic tissues (9) and body
weight changes (24). The second type of response is exemplified
by the findings with vesicular stomatitis virus and exposure to
sound stress (7) in which a transitory and in some cases a biphasic
response was observed. It was characterized by periods during
daily stress when mice may be either more, equally, or less susceptible to viral challenge than control mice. While the adreno-

cortical response to a stress period is transitory, the transitory
increase in susceptibility to vesicular stomatitis virus was independent of adrenal function.   Likewise, a transitory impairment
of interferon production in response to the daily stress exposure
was not associated with adrenal secretions (19).   Evidence indicates
that multiple factors may be associated with the physiologic mechanisms by which sound or other forms of nonspecific stress influence
host resistance to infectious diseases; certainly not all mechanisms
are a result of adrenal hyperactivity.   The transitory and even
multi-directional responses of animals to stressful stimuli make
studies of this type difficult.   Such responses may increase
resistance to one infection and decrease it to another, depending
on the pathogenesis and the temporal relationships involved.

The influence of sound stress on the response of mice to
virus-induced cancers further illustrated the diverse influences
that might be observed (25).   On the one hand, a definite stress-
induced suppression of the progression of Rauscher virus leukemia
was seen (15), while on the other hand, an increased tumoriogenesis
was observed in stressed mice infected with polyoma virus (14).
The Rauscher virus-infected mice were given a fairly large inoculum
which probably resulted in the infection and malignant conversion
of many cells.   The influence of stress appeared to be a simple
retardation of cellular proliferation during the early erythroid
phases of the disease, and might be compared to the anti-inflammatory
influence of stress on nonmalignant tissues (17,26).   There was no
evidence that stress had any influence on malignant conversion at
the cellular level in the Rauscher virus-infected mice.   In the
pathogenesis of the polyoma infection, there is first a virulent
cycle of cellular infection with viral replication and cellular
destruction, followed by infection of different cells in which no
complete viral replication or cellular destruction occurs.   However, some of the latter cells undergo malignant conversion.   The
first phase is highly efficient and is during the asymtomatic stage
of the disease.   The malignant conversion is relatively rare and
requires as much or more than a million times as many infectious
units of virus per cell as does the virulent cycle.   For stress
to be effective in increasing the number of tumors, it must be
applied early in the infection (13,14).   If the response to stress
impaired the host's ability to remove or neutralize the virus during
the early virulent cycles, more viruses would then be available and
would result in more malignant conversions and hence a higher incidence of tumors.

The state of our knowledge regarding the influence of stress
on host defense mechanisms is fragmentary.   While our preliminary
efforts indicated no influence of stress on circulating antibodies
(16), more recent studies have indicated a reduction in antibody
response in stressed animals (27).   Stress has been shown to have
an influence on tissue immunity by delaying the rejection of skin

grafts (28).   Impairment of interferon production has been demonstrated in stressed mice (18,19).   Evidence was not found for impaired clearance of viruses from the blood of sound stressed mice
(12); however, evidence has been presented for impaired circulatory
clearance in other stress systems (25).   Impaired clearance of
virus from the peritoneal cavity (20) and from muscle tissues (16)
has been noted in stressed mice.

SUMMARY

Mice exposed to sound intensities of about 120 db for 3 hours
each day showed evidence of increased stimulation of the pituitary-
adrenocortical axis.   Each daily exposure to sound was characterized
by a leukopenia, followed by a leukocytosis after the stress period.
Mice inoculated intranasally with vesicular stomatitis virus just
before exposure to sound were more susceptible to the infection,
while mice inoculated after the exposure were more resistant than
non-exposed mice.   Mice infected with polyoma virus and subjected
to sound stress developed more tumors than nonstressed mice, while
sound stress slightly suppressed the progression of Rauscher virus
leukemia.   The inflammatory and interferon responses were impaired
in mice exposed to sound.

ACKNOWLEDGEMENT

This research was supported by USPHS Mental Health Training
Grant 5 T1-MH-6415.

REFERENCES

1. Friedman, S.B. and L.A. Glasgow, 1966.   Psychologic factors
   and resistance to infectious disease.   Pediatric Clinic
   North America 13:315-335.

2. Wolf, S. and H. Goodell, 1968.   Harold G. Wolff's Stress and
   Disease.   Thomas, Springfield, Illinois.

3. Rasmussen, A.F., Jr., J.T. Marsh and N.Q. Brill, 1957.
   Increased susceptibility to herpes simplex in mice subjected
   to avoidance-learning stress or restraint.   Proceeding
   Society Experimental Biology and Medicine 96:183-189.

4. Anthony, A., E. Ackerman and J.A. Lloyd, 1959.   Noise stress
   in laboratory rodents.   I. Behavioral and endocrine response
   of mice, rats, and guinea pigs.   Journal Acoustical Society
   America 31:1430-1437.

5. Sackler, A.M., A.S. Weltman, M. Bradshaw and P. Jurtshuk, Jr., 1959. Endocrine changes due to auditory stress. Acta Endocrinologica 31:405-418.

6. Jensen, M.M., and A. F. Rasmussen, Jr., 1963. Stress and susceptibility to viral infections. I. Response of adrenals, liver, thymus, spleen and peripheral leukocyte counts to sound stress. Journal of Immunology 93:17-20.

7. Jensen, M.M., and A.F. Rasmussen, Jr., 1963. Stress and susceptibility to viral infections. II. Sound stress and susceptibility to vesicular stomatitis virus. Journal of Immunology 90:21-23.

8. Jensen, M.M., 1969. Changes in leukocyte counts associated with various stressors. Journal Reticuloendothelial Society 6:457-465.

9. Marsh, J.T. and A.F. Rasmussen, Jr., 1960. Response of adrenals, thymus, spleen and leucocytes to shuttle box and confinement stress. Proceedings Society Experimental Biology and Medicine 104:180-183.

10. Treadwell, P.E. and A.F. Rasmussen, Jr., 1961. Role of the adrenals in stress induced resistance to anaphylactic shock. Journal of Immunology 87:492-497.

11. Weimer, H.E., and D.C. Benjamin, 1965. Immunochemical detection of an acute-phase protein in rat serum. American Journal of Physiology 209:736-744.

12. Jensen, M.M. Unpublished data.

13. Rasmussen, A.F., W.H. Hildemann and M. Sellers, 1963. Malignancy of polyoma virus infection in mice in relation to stress. Journal National Cancer Institute 30:101-112.

14. Chang, S.S. and A.F. Rasmussen, Jr., 1964. Effects of stress on susceptibility of mice to polyoma virus infection. Bacteriological Proceedings 64:134 (abstract).

15. Jensen, M.M., 1968. The influence of stress on murine leukemia virus infection. Proceedings Society Experimental Biology and Medicine 127:610-614.

16. Yamada, A., M.M. Jensen and A.F. Rasmussen, Jr., 1964. Stress and susceptibility to viral infections. III. Antibody response and viral retention during avoidance-learning stress. Proceedings Society Experimental Biology and Medicine 116:677-680.

17. Funk, G.A., and M.M. Jensen, 1967.   Influence of stress on
    granuloma formation.   Proceedings Society of Experimental
    Biology and Medicine   124:653-655.

18. Chang, S.S. and A.F. Rasmussen, Jr., 1965.   Stress-induced
    suppression of interferon production in virus infected mice.
    Nature 205:623-624.

19. Jensen, M.M., 1968.   Transitory impairment of interferon pro-
    duction in stressed mice.   Journal of Infectious Diseases
    118:230-234.

20. Funk, G.A., 1965.   The Influence of Stress on Host Defense
    Mechanisms.   Masters Thesis, Univ. of Calif., Los Angeles.

21. Henkin, R.I., and K.M. Knigge, 1963.   Effect of sound on the
    hypothalamic-pituitary-adrenal axis.   American Journal of
    Physiology 204:710-714.

22. Johnsson, T. and A.F. Rasmussen, Jr., 1965.   Emotional stress
    and susceptibility to poliomyelitis virus infection in mice.
    Archive fur die gesonte Virusforschung 17:392-397.

23. Johnsson, T., J.F. Lavender, E. Hultin and A.F. Rasmussen, Jr.,
    1962.   The influence of avoidance-learning stress on resist-
    ance to coxsackie B virus in mice.   Journal of Immunology
    91:569-575.

24. Friedman, S.B. and R. Ader, 1965.   Parameters relevant to the
    experimental production of "stress" in the mouse.   Psychoso-
    matic Medicine   27:27-30.

25. Rasmussen, A.F., Jr., 1969.   Emotions and Immunity.   Annals
    New York Academy of Sciences   164 (2):458-461.

26. Smith, L.W., N. Molomut and B. Gottfried, 1960.   Effect of
    subconvulsive audiogenic stress in mice on turpentine induced
    inflammation.   Proceedings of Society Experimental Biology
    and Medicine 103:370-372.

27. Solomon, G. F., 1969.   Emotions, stress, the central nervous
    system, and immunity.   Annals New York Academy of Sciences
    164(2):335-343.

28. Wistar, R.T. and W.H. Hildemann, 1960.   Effect of stress on
    skin transplantation immunity in mice.   Science 131:159.

EFFECTS OF SOUND ON ENDOCRINE FUNCTION AND ELECTROLYTE EXCRETION

MARY F. LOCKETT

Department of Pharmacology
The University of Western Australia
Nedlands, W.A.   6009

EFFECTS OF SOUND ON ADENOHYPOPHYSEAL FUNCTION

Loud sounds, intense light, immobilization, anxiety, forced exercise, surgery, cold and many other stressful agents increase the secretion of corticotrophin (ACTH) from the adenohypophysis. In each case, the mechanism by which the secretion of ACTH is accelerated is neurohumoral (1) and is mediated through the central nervous system (2).   The resulting elevation in plasma concentrations of ACTH causes an increase in the secretion of adrenal corticoids; the additional corticoid secreted is that characteristic of the particular species under stress (4).   Thus, loud sounds raise plasma concentrations of corticosterone in the rat (5) and of 17-hydroxycorticosterone in man (6) and in monkeys (7).   High concentrations of ACTH also increase the rate of secretion of aldosterone (8).

In 1963 Henkin and Knigge (5) exposed rats to 220 cycles/sec at 130 decibels (d.b.) for periods up to 48 hours and studied the resultant changes in adrenal secretion of corticosterone.   This intense sound doubled the adrenal output of corticosterone in 30 min and trebled it in 60 min.   The same secretion rate attained after one hour of exposure was maintained for 11.5 hours but, although the stimulus was continued, at 12 hours the adrenal output of corticosterone fell rapidly to normal or subnormal levels, then rose again to the former high rate which was maintained from 24 hours through 48 hours of high intensity sound.   This suppression of adrenal corticoid secretion during responses to stressful agents is well known: it is caused by an inhibition of ACTH release, since the adrenal remains fully responsive to exogenous ACTH (9).   Neither the stores of ACTH in the adenohypophysis

21

nor of corticotrophin releasing factor of the hypothalamus (10, 11, 12) are exhausted, for application of a second stressful agent during suppression of ACTH release by the first immediately re-starts ACTH secretion (13, 9, 5).   It is also difficult to attri-bute this sudden inhibition of ACTH secretion, which is prolonged over hours, to a steroid feed back modification of neurohumoral transmission since the amounts of exogenous steroid necessary to abolish responses to external stress (14, 15, 16) produce plasma concentrations markedly in excess (17) of the physiological range. The probable source of this suppression of ACTH release therefore lies within the central nervous system.   Stimulation of the hippo-campus causes a prolonged inhibition of ACTH secretion (17, 18, 19, 20) and the destruction of the hippocampus raises plasma corti-coid (21, 22, 13).   Conversely stimulation (24) and destruction (25) of the amygdaloid nucleus respectively enhance and depress the secretion of ACTH.   Stimulation of reticular components of the tegmentum in the mid brain (22, 26) and of its projections into the dorsal longitudinal fasciculus (27) enhances adrenal cortical secretion.   Conversely, a "tonic inhibitory effect" of the reti-cular formation on the hypothalamic-pituitary mechanism controlling ACTH release has also been documented (5).   Nauta and Kuypers (1958) (28) describe reticular projections from the medial caudal mid brain to the hypothalamus and limbic system as travelling in two main bundles, the dorsal longitudinal fasciculus of Schutz and the mammillary peduncle.   Limbic and reticular activating systems are conceived as constituting a single two-way major system formed by multisynaptic relays situated, for the main part, in the hypo-thalamus (29, 30).   Mason (1958) (7) also considers such a two-way system.   In Mason's system the hippocampus exerts an inhibitory influence on the hypothalamus and reticular formation proportionate to the excitation it receives from them via the ascending fibres. Hence, hippocampal negative feed back is considered reciprocally related to activity in the lower centres.

SOUND, ENDOCRINES AND HYPERTENSION

It has often been suggested that a relationship exists in man between psychological stresses and essential hypertension.   Psycho-genic hypertensions can be generated in laboratory animals during conditioned avoidance studies and by exposure of rats to air blasts for 5 to 10 min. on 5 days a week for 33 weeks (31, 32).   Audio-genic stimulation can also generate an hypertension which persists for several months after exposures to intense sounds have ceased (33).   Watzman and Buckley(34) have studied the changes in adrenal and plasma corticosterone and in urinary catecholamines associated with the development of hypertension by rats subjected to repeated but variable stressors (sound, light, cage rocking) for 16 weeks. Initially the urinary catecholamines and the plasma and adrenal concentrations of corticosterone rose: these values had however

returned to normal by the 8th week.   Hypertension had however
reached full development at 8 weeks and persisted throughout and
beyond the 8th - 16th week of stress.   During this latter part of
the experiment corticosterone levels again rose to high levels but
the urinary catecholamine output remained within normal limits.
It is, of course, tempting to speculate on the part that may have
been played by the early overactivity in the sympathetic nervous
system.   Hyperactivity in this system may be expected to enhance
the rate of secretion of renin from the juxtaglomerular apparatus
in the kidney (35).   The enzyme renin then interacts with its a_2
globulin substrate in the blood stream to form the decapeptide
angiotensin I.   The angiotensin I is rapidly converted, mainly by
the lung, to the strongly vasopressor octapeptide angiotensin II
(36).   Hypertension can also be produced by prolonged administration of high salt diets, with or without unilateral nephrectomy
and by the injection of large and repeated doses of the salt-
retaining mineralocorticoid deoxycorticosterone.   The mechanism
of genesis of these salt hypertensions is unfortunately not under-
stood.   It is of interest, in this connection, that the hypertension
of adrenal regeneration is associated with a supranormal production
and secretion of deoxycorticosterone by the regenerating glands,
in vivo, but not in vitro (37).

SOUND AND THYROID FUNCTION

Loud noises (38) and many of the stresses which increase the
secretion of ACTH reduce the rate of secretion of the thyrotrophic
hormone from the adenohypophysis (39, 40).   Overall, there appears
to be an inverse relationship between the rates of secretion of
ACTH and of TSH by laboratory animals subjected to acute stresses.
These observations may reflect reciprocal availability of hypothalamic releasing factors (41).   No successful demonstration has
however been made of stress induced change in the thyroid function
of man (42, 43).

THE INFLUENCE OF SOUND ON NEUROHYPOPHYSEAL FUNCTION

In 1964 there was nothing in the literature known to us which
indicated a response of the neurohypophysis to sound.   Our interest
was however awakened to this possibility by a series of chance observations.   Thunderstorms were found to increase the rates of the
urinary excretion of water, sodium (Na^+) and potassium (K^+) by normal but not by neurohypophysectomized rats, and the changes in urinary excretion produced by these storms resembled those seen in the
first hour after the subcutaneous injection of 4-8 m Units oxytocin
per rat (44).   Female rats weighing 160-200 gm were therefore
accustomed to stomach tubes and handling for use in experiments
designed to determine the mechanism of the diuresis and natriuresis

caused by thunder.

METHODS

Every experiment was designed as a series of cross-over tests in which each animal served as its own control. Tests constituting a single experiment were carried out at intervals of 3 days and at a fixed time of day, since the rate of excretion of Na+ falls throughout the day. Each test began with a 2 hr. period in which no solid food was available. An oral water load equivalent to 2.5% body weight was given at the end of this period, immediately before each animal was put into a separate cage for the collection of all the urine which entered the bladder in the following hour. This collection period was extended to 2 hours for adrenalectomized and totally hypophysectomized rats and for all rats receiving high doses of vasopressin. Concentrations of Na+ and K+ were determined by flame photometry. Urinary excretion rates were expressed as ml. or u-equiv./100 gm body weight/hr and are throughout presented as mean values ± the standard error of the mean. The significance of differences between means has been determined by t tests within groups.

Sound Effects. Thunderclaps, each lasting 3-4 sec were recorded on tape and were played back, 2 claps at an interval of 1 min every 5 min for 20 min starting 30 min after withdrawal of solid food. A second 20 min session began 15 min after administration of the water load. The sound produced had an intensity of 98-100 d.b. amongst the rat cages and a predominant frequency range of 50-200 cycles/sec. An audiogenerator was used to produce a pure frequency of 150 cycles/sec at 100 d.b. This stimulus was applied for 2 min every 15 min throughout a period which started 10 min after water loading and terminated at the end of the urine collection. Exposures to 20 kcycles/sec at 100 d.b. were given for 2 sec and were applied 15 min before, during and 15, 30 and 45 min after the water load (45).

Bioassays. Total urinary epinephrine was extracted from acidified rat urines by absorption onto and elution from aluminium oxide according to the method of Euler and Orwen (46). This method gave 80-85% recoveries in our hands. Eluates containing epinephrine were stored overnight at pH 4.0 and 4°C and were brought to pH 6.0-6.4 immediately before they were assayed on the oestrus uterus of the rat by the method of Gaddum and Lembeck (47). Thioglycollate-labile (48) oxytocic activity in plasma from heart blood was assayed in vitro on cubes of lactating mammary glands from rats by the method of van Dongen and Hays (49). A linear relationship between log dose and log ejection time was found for concentrations of 1×10^{-10} to 1×10^{-2} Units oxytocin per ml, and the index of precision, was 0.56.

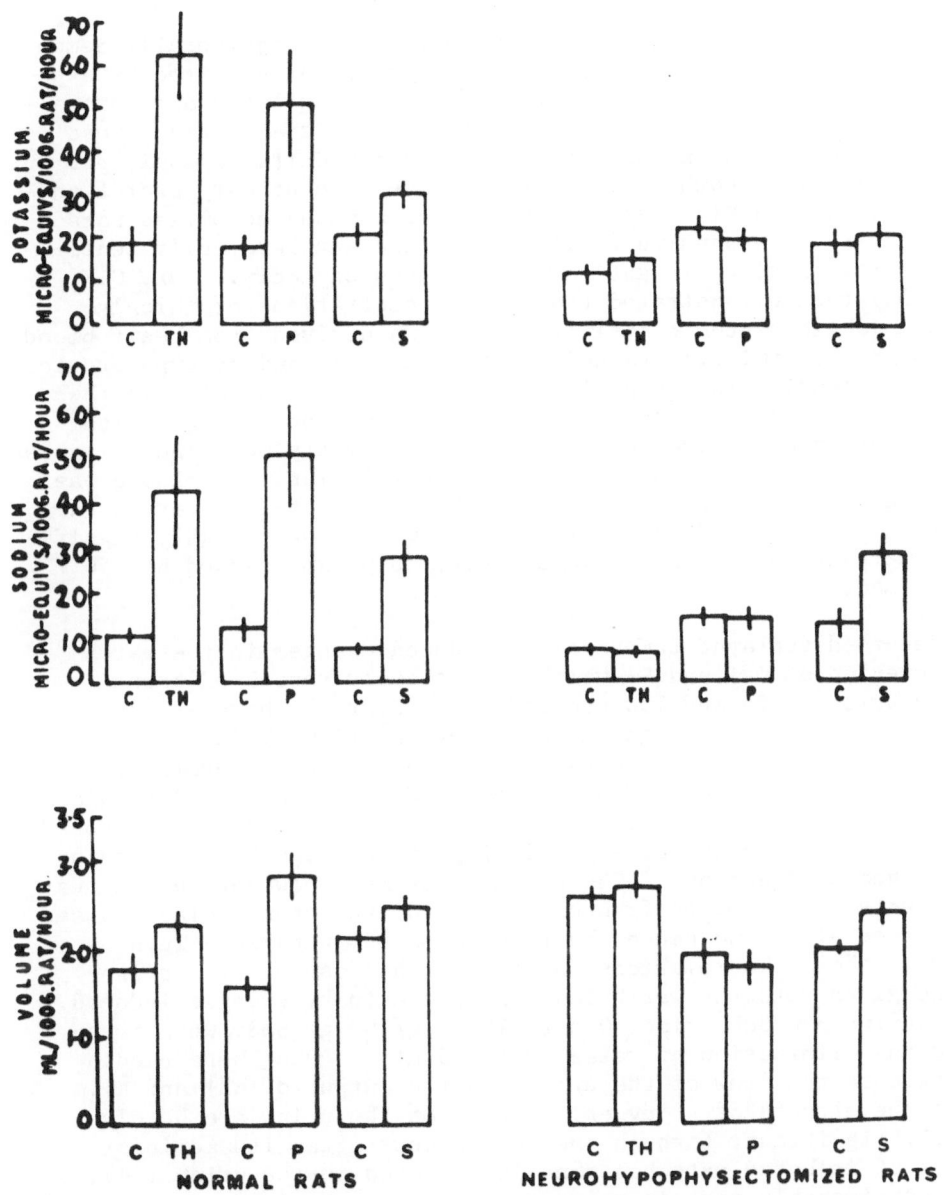

Figure 1. Sound-induced changes in the urinary excretion of K^+ Na^+ and water in normal and neurohypophysectomized rats. The heights of the open rectangles show mean responses for a group of 12 rats, the inset bars the standard errors of these means. C = control values. TH, P and S, respectively, are the effects of replayed thunderclaps, 150 cycles/sec and 20 kcycles/sec.

Experimental Observations

Collaborative work with Dr. C.W. Ogle (45) unequivocally demonstrated that the changes in the urinary excretion elicited by exposure to 150 cycles/sec in normal, adrenal demedullated and adrenalectomized salt-maintained rats were indistinguishable from those
caused by the subcutaneous injection of 4-8 m Units oxytocin per
rat (Figure 1).   Both treatments increased the urinary excretion
of water, Na+ and K+.   By contrast, neurohypophysectomized rats
were totally unresponsive to 150 cycles/sec but were fully sensitive to the natriuretic action of exogenous oxytocin.   Dr. C.W.
Ogle (50) then demonstrated that the concentrations of thioglycollate-labile oxytocic activity in plasma derived from heart blood
were somewhat similarly raised by s.c. oxytocin and by exposure to
150 cycles/sec.   The comparison of blood levels was made in the
middle of the one hour period of urine collection through which
the urinary changes induced by the two treatments were indistinguishable.   Hence, by reason of the experimental plan, the plasma was
collected 30 min after s.c. oxytocin but immediately after an exposure to 150 cycles/sec.   Figure 2 shows the very highly significant increase in oxytocin-like activity in plasma caused by 150
cycles/sec.

Recorded replayed thunderclaps induced changes in the rates
of excretion of Na+ and K+ which very closely resembled those
caused by oxytocin and 150 cycles/sec.   Overall, however, the
thunderclaps caused a significantly smaller ($P < 0.1$) increase in
urine volume despite an equivalent increase in Na+ excretion.
Neither replayed thunderclaps nor 150 cycles/sec appeared to alarm
the rats: indeed, the animals scarcely seemed to notice the sounds
and continued uninterruptedly sniffing at the cage bars, preening
themselves or sleeping.   The possibility did, however, exist that
components of the wider frequency range of the thunderclaps caused
unobserved alarm and hence the release of sympathetic amines.
Figure 3 shows the relationship between the dose of s.c. epinephrine given with the water load and the urinary changes induced.
Epinephrine hydrochloride, 0.5 to 10.0 ug/100 gm body weight reduced the elimination of water, Na+ and K+.   40 ug were needed
to increase the flow of the urine and the output of Na+ and K+.
Since the effects of 40 ug epinephrine on the urine are barely
($P < 0.1$) significant through the second hour after its administration
and 4% of injected catecholamine is excreted in the urine (51), a
very considerable rise in urinary epinephrine would be expected
if the diuresis, natriuresis and kaluresis induced by thunderclaps were attributable to the release of epinephrine.   Neither
150 cycles/sec (52) nor replayed thunderclaps had significant
effect on urinary epinephrine.   Moreover, 5 ug epinephrine s.c./
100 gm body weight completely antagonized the urinary changes induced by 5 m Units oxytocin in normal (Figure 4) and adrenalectomized

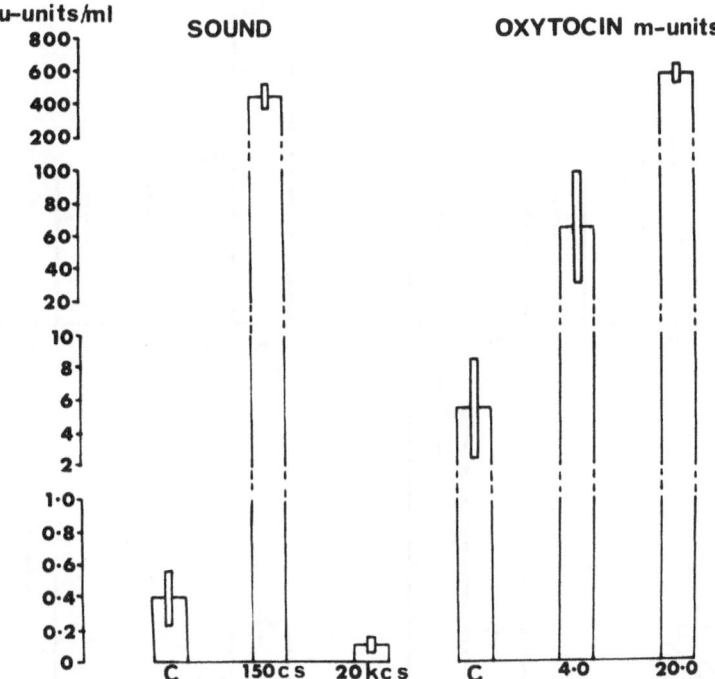

Figure 2. Effect of sounds and subcutaneous oxytocin upon plasma oxytocic activity. Heart blood withdrawn at the midpoint of urine collection. Rectangles as in Figure 1. Twelve rats/group.

rats, in rats with denervated kidneys and in rats which had been pretreated with propylthiouracil. A comparable inhibition by epinephrine of milk-ejection by oxytocin has recently been described (53). By contrast, small amounts of vasopressin, 0.1 m Unit per rat, selectively reduced the elimination of water during an oxytocin-induced diuresis and natriuresis (Figure 4).

Clearly then, under the conditions of these experiments, exposure of rats to 150 cycles/sec induces selective release of oxytocin from the neurohypophysis. Exposure to replayed thunderclaps with a dominant frequency range of 50–200 cycles/sec also liberates oxytocin but probably only a small amount of vasopressin that has only marginal functional significance. Selective release of neurohypophysial hormones is a well-known phenomenon that is dependent both on the type of stimulus and on the species. For the most part, however, oxytocin and vasopressin are released simultaneously and the quantity of oxytocin released exceeds that of vasopressin (54). Suckling or milking releases oxytocin and vasopressin in the ratio of approximately 100:1 (55, 56) but hemorrhage releases vasopressin unaccompanied by any significant amount of oxytocin (57). The concept that the two hormones are synthesized

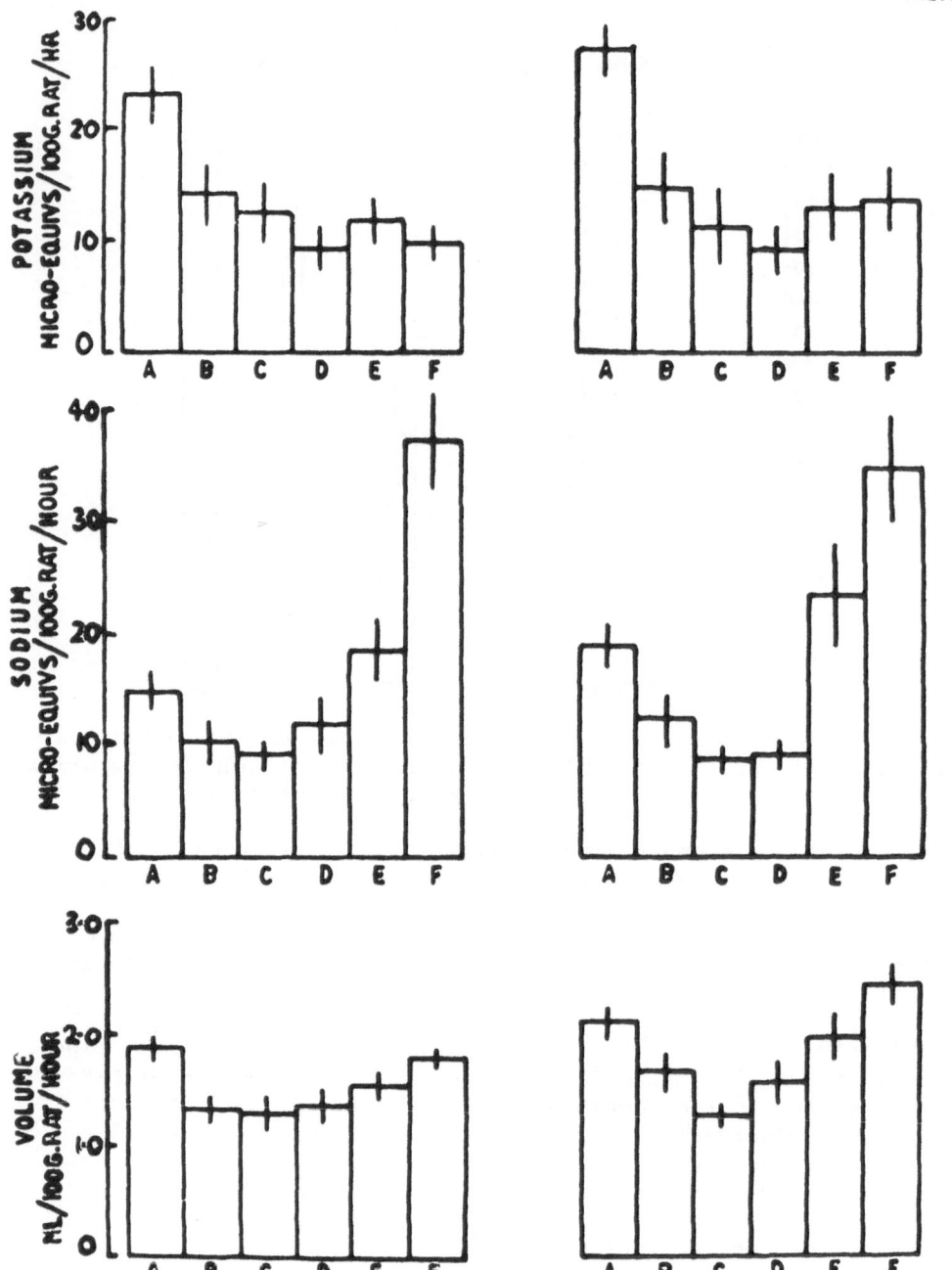

Figure 3. Dose effect relationships for the effects of subcutaneous
1-epinephrine on the excretion of water and electrolytes by normal
rats. Doses in ug per 100 gm of body weight: A = 0, B = 2.5, C = 5.0,
D = 10.0, E = 20 and F = 40. Rectangles as in Figure 1.

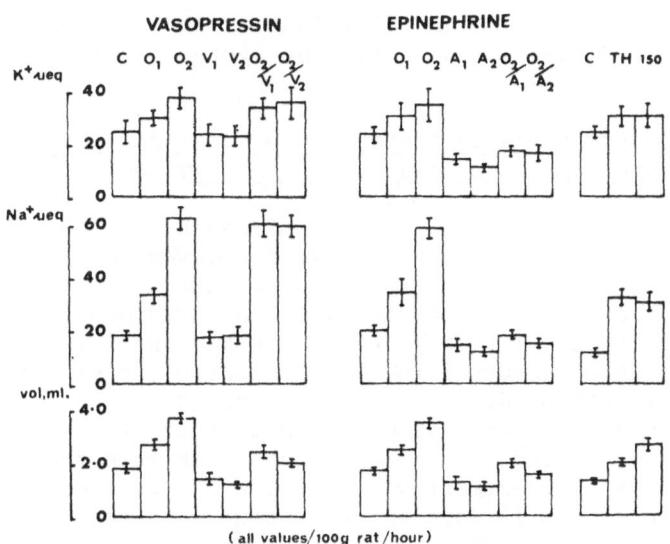

Figure 4. Inhibition of oxytocin-induced diuresis and electrolyte excretion by vasopressin and epinephrine in normal rats. C = control values; O_1 and O_2 = oxytocin, 4 and 8 mU s.c.; V_1 and V_2 = 0.05 and 0.1 m units vasopressin s.c./100 gm body weight; A_1 and A_2 = 2.5 and 5.0 ug (-) epinephrine s.c./100 gm body weight. TH and 150 = replayed thunderclaps and 150 cycles/sec respectively. Rectangles as in Figure 1. Sound intensity was 100 dB.

in different neurones in the hypothalamus arose from a series of observations. First, the relative concentrations of oxytocin and vasopressin vary in different parts of the hypothalamus (58). Second, lesions of the paraventricular nuclei decrease the stores of oxytocin in the posterior lobe (59). Third, stimulation of the corticomedial amygdala in the cat (60) selectively released vaso-pressin without detectable quantities of oxytocin. The very beautiful work of Aulsebrook and Holland (61) has demonstrated the selective release of oxytocin in lactating rabbits by stimu-lation in the prelimbic cortex, portions of the nucleus accumbens, the hippocampal rudiment, the diagonal band and the periventricular system rostral to the paraventricular nucleus. The paraventricular nucleus was related to the release solely of oxytocin: the supra-optic nucleus was not exclusively related to the release of vaso-pressin. Extensive studies on the central inhibition of oxytocin release (62) have led to the concept that overall integration of somatovisceral input occurs within the limbic system, with which all pathways facilitatory to the release of oxytocin have connections: hence, the limbic system can regulate oxytocin release through fornical connections with the transitional periventricular grey. Cross (63) has suggested that a diffuse afferent pathway which includes the leminiscal and reticular system is involved in the

suckling reflex. The cochlear divisions of the auditory nerves
are intimately related to the reticular system in the region of
the trapezoid body (64).

Preliminary observations on adaptation of neurohypophysial function
to recurrent and continuous exposure to 150 cycles/sec.

     Rats, housed in a sound-proofed semi-anechoic room, well ven-
tilated and temperature controlled, were exposed to 150 cycles/sec
at 98-100 db for 30 min every 4 hours, day and night. Natriuresis
and kaluresis rose during the first week of treatment to plateau
levels reached by the 10th day and maintained until the 16th or
17th day when normal control levels of electrolyte excretion were
rather suddenly restored. Accompanying changes in water excretion
were less marked. Figure 5 shows the magnitude of these effects.
The excretion of sodium and potassium has· been more than doubled
by fourteen days of treatment with 150 cycles/sec. Moreover, the
electrolyte losses at 14 days exceed those caused by a first single
exposure to 150 cycles/sec.

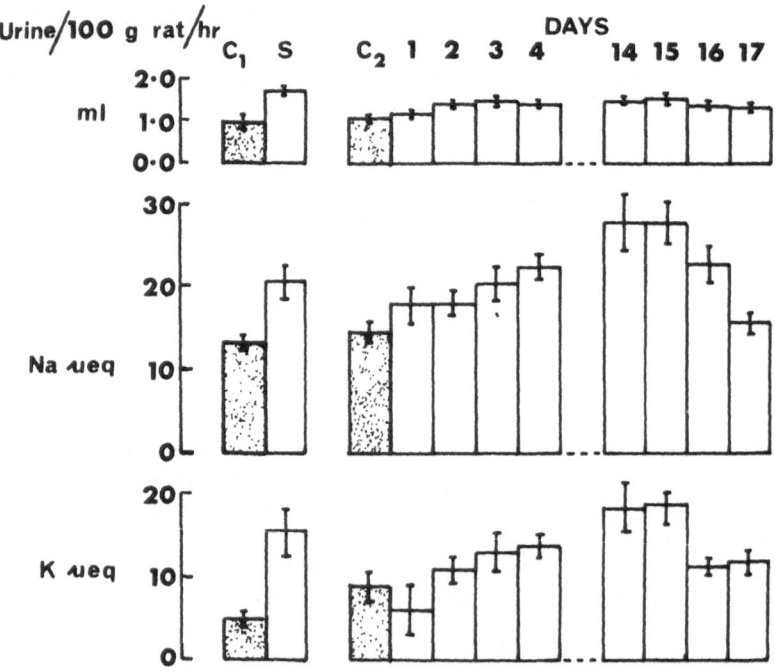

Figure 5. Effects of intermittent long-term exposure of rats to
150 cycles/sec. 12 rats were housed in a sound-proofed semi-
anechoic chamber and subjected to a single 30 min exposure (S)
to 150 cycles/sec, 98-100 db or to 30 min exposure every 4 hours
day and night for 1-17 days. Control values (silent) are stippled.
Rectangles as in Figure 1.

This intermittent exposure of rats to 150 cycles/sec was not without influence on the concentrations of corticosterone in the peripheral blood. Values found on the fourth day for exposed animals were 19.82 = 1.23 (12) ug/100 ml and for similarly treated but unexposed rats 24.39 = 1.48 (12) ug/100 ml (despite adrenal hypertrophy in the exposed rats). On the seventeenth day of treatment the blood levels of corticosterone were 18.02 ug/100 ml. Values found on the fourth day for the weights of the paired adrenals were 35.63 ± 1.09 (12) mg and on the seventeenth day, 32.98 ± 0.78 (18) mg. The paired adrenal weights for similarly treated but unexposed animals were 30.06 ± 0.87 ug (12) mg.

The changes in urinary excretion of water and electrolytes produced by exposure to 20 kcycles/sec.

Exposure to 20 kcycles/sec provoked diuresis and natriuresis in intact and neurohypophysectomized rats (Figure 1). The rate of excretion of K^+ also increased, but less markedly, so that the Na:K ratio in the urine rose significantly in approximately two-thirds of all experiments. Adrenalectomized and adrenal-demedullated rats failed to respond to 20 kcycles/sec by an increase in the rate of excretion of electrolytes, and a highly significant antidiuresis ($P < 0.001$) replaced the diuresis which this frequency produced in intact animals (Figure 6). Responses to 20 kcycles/sec were not modified by previous denervation of the kidneys. Additionally, exposure to 20 kcycles/sec markedly increased the urinary excretion of epinephrine by normal rats (Figure 7) and raised the thioglycollate-labile antidiuretic activity excreted in the urine of adrenalectomized rats from 0.5 to 24.8 ± 5.5 u-units/rat/hour.

The diuretic natriuretic effect of 20 kcycles/sec in normal and in neurohypophysectomized rats was clearly attributable to the release of epinephrine since this response to 20 kcycles/sec was accompanied by a very highly significant rise in urinary epinephrine and was abolished by demedullation of the adrenals. Since the antidiuresis produced by 20 kcycles/sec in adrenal-demedullated rats was accompanied by a large increase in thioglycollate-labile antidiuretic activity in the urine, the urinary changes induced in adrenal-demedullated rats by 20 kcycles/sec are attributed to the release of ADH. Because it is known that raised plasma concentrations of epinephrine inhibit the secretion of ADH (65, 66, 67), the ADH component of the response to 20 kcycles/sec should be minimal in normal rats and should proceed uninhibited in adrenal-demedullated animals. The minor part played by ADH release in the response of intact rats to 20 kcycles/sec is uncovered by the data presented in Figure 8. Significant correlation between the rates of water and Na^+ excretion was shown by 25 normal rats during the diuretic natriuretic action of 5 to 10 ug (-) epinephrine, s.c., per 100 gm rat. These same animals responded by diuresis and natriuresis to

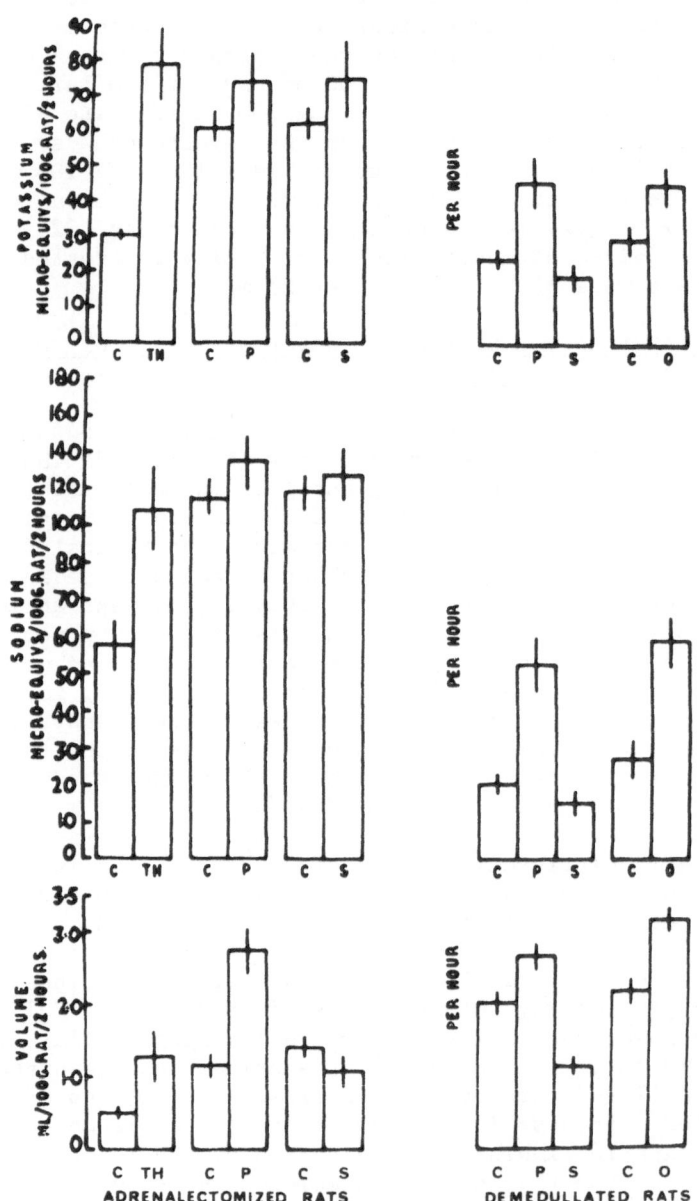

Figure 6. Effects of sound on the excretion of water and electrolytes by adrenalectomized salt-maintained rats and by rats with
demedullated adrenals. C = control values; TH, P and S, respectively,
are the effects of replayed thunderclaps, and of sound at 150 cycles/
sec and at 20 kcycles/sec (100 dB). O = 4 m U oxytocin, s.c.
Rectangles as in Figure 1.

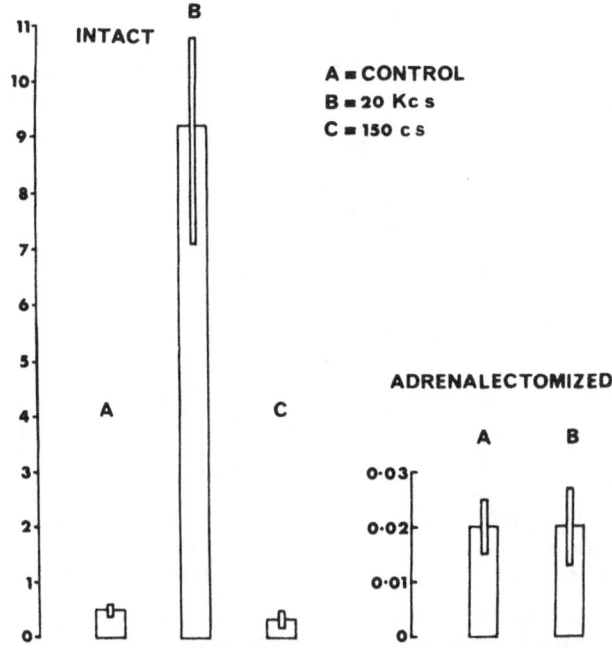

Figure 7. Effects of sound on urinary epinephrine in normal and in adrenalectomized rats.   Ordinates, ng (–) epinephrine/100 gm rat/hour.

20 kcycles/sec but no correlation was demonstrable between the rates of water and of Na$^+$ excretion, almost certainly by reason of the excretion of ADH in amounts which varied from one animal to the next.   Moreover, at equivalent rates of Na$^+$ excretion, the urine excreted on exposure to 20 kcycles/sec was significantly (P<0.05) more concentrated than the urine excreted during the action of epinephrine.

A further point of interest arising from these experiments is the apparent release of large amounts of epinephrine from the adrenal medulla by 20 kcycles/sec and the accompanying evidence that the sympathetic denervation of the kidneys did not modify the urinary changes induced by 20 kcycles/sec.   Selective secretion of epinephrine from the adrenal medulla can be induced by hypo-thalamic and by pontine stimulation (68, 69, 70, 71, 72).   Since the sites where such responses can be evoked seem few, the peculiar nature of the response to 20 kcycles/sec may, in the future, assist in the location of those nerve centres and pathways involved in its genesis.*

───────

*In discussion, the point was made that rats do not hear as well at 150 cyc/s as at 20 Kcyc/s and that the difference in observed effect at these frequencies might be due to differences in intensity of nervous stimulation.

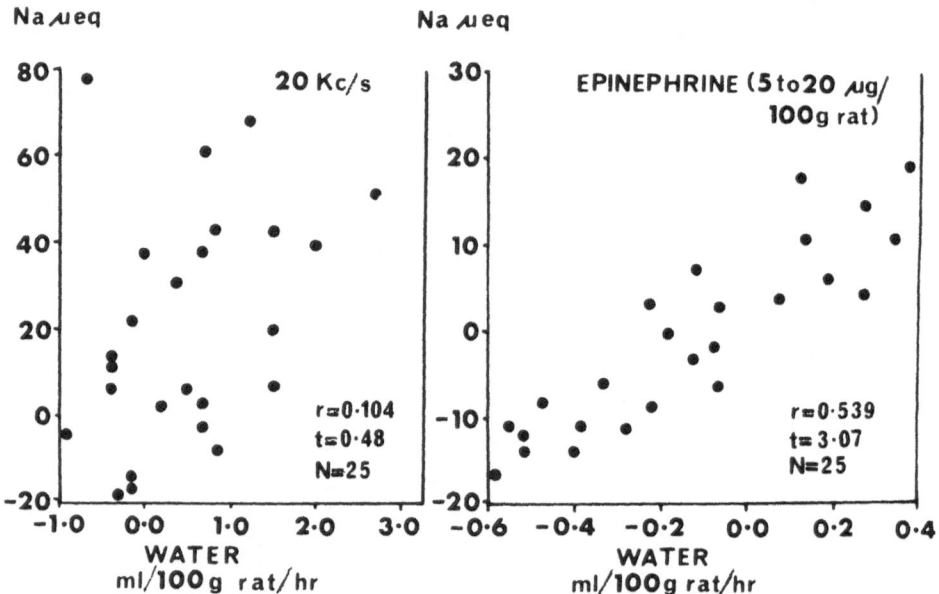

Figure 8.   Changes in water output are not correlated with changes
in sodium output during responses to 20 kcycles/sec but are correlated during responses to (−) epinephrine s.c. in normal rats.
Ordinates: Rates of sodium (Na^+) excretion in u equiv./100 gm
rat/hour.

The relationship of the changes in the excretion of electrolytes caused by 150 cycles/sec and 20 kcycles/sec to audiogenic seizures.

     The audiogenic seizure is the most widely studied effect of
sound in animals, probably because these seizures bear a considerable resemblance to human epilepsy, and within a species those most
susceptible to these seizures are those which suffer from spontaneous
epileptiform attacks (73).   The species susceptible to audiogenic
seizures are mice, rats, rabbits, chickens and dogs.   Cats and
monkeys are not normally susceptible but can be made so by pretreatment with thiosemicarbazide (74) or methionine sulfoxine (75).
These drugs also lower the threshold for seizures in susceptibles.
Susceptibility to epileptiform attacks in response to sound is
commonly found only in certain strains or substrains of a species
dubbed susceptible.   Certainly in mice (76) and in rats (77)
susceptibility is dependent on several genes: hence its genetic
origin is complex, and in the absence of extensive inbreeding and
narrow selection individuals of high and low susceptibility to
these seizures will be found within a single stock.

     The intensity of sound needed to produce these seizures lies

between 90 and 134 d.b. and the effective frequency band is described as extending from 4 to 80 kcycles/sec according to species.
Effective intensities undoubtedly vary with species, and at least
as far as rats and mice are concerned, the percentage of seizures
at differing frequencies is directly related to the audiogram curve
of the species (73).

The pattern of the audiogenic seizure is characteristic for
a species. In the rat there is an alerting, followed by a frozen
state which typically progresses through episodic running to a
convulsion in which clonic movements precede a tonic phase.
Stupor often follows. Death may occur from asphyxia during the
tonic phase of the convulsion or from cerebral hemorrhage (73).

The rats used for experiments on the urinary changes caused
by thunder and by exposure to 150 cycles/sec and 20 kcycles/sec
came from a stock in which susceptibility to audiogenic seizures
is not rare. Approximately 10% of the females of this stock
respond to 20 kcycles/sec at 100 d.b. for 3.5 sec either by compulsive running, or by compulsive running terminating in a convulsion, or by a convulsion. The less susceptible animals all
respond to this sound by the immediate assumption of an alert
posture in which they appear to freeze until the sound stops.
Relaxation in the ensuing silence has a latency of 2 to 12 sec
varying between individuals. All the animals whether susceptible
or not susceptible to seizure respond to the sound by release of
epinephrine. Hence 20 kcycles/sec is either alarming or causes
aural pain in rats. Since individual rats respond quite regularly
to 20 kcycles/sec, two or three times a week by changes in urinary
electrolytes which do not diminish with successive exposures to
this sound, it seems probable that the release of epinephrine is a
consequence of aural pain. Repetition of the stimulus over weeks
might be expected to yield diminishing alarm reactions since these
animals are capable of learning. Our data therefore suggests
that the involvement of the sympathetic nervous system in audiogenic fits is possibly no more than a coincidental effect of aural
pain. Genetic factors, resulting in biochemical deviations from
the norm determine the susceptibility to seizure.

By contrast, no audiogenic seizures have yet been produced by
exposure of rats to 150 cycles/sec, and this frequency at 100 d.b.
elicits no change in behaviour. It is therefore improbable that
the release of oxytocin by 150 cycles/sec in our stock of rats is
related to the genesis or occurrence of audiogenic seizures.

REFERENCES

L. De Groot, J., and G.W. Harris, 1950. Hypothalamic control of
the anterior pituitary and blood lymphocytes. Journal of

Physiology (London) 111:335-346.

2. McCann, S.M., 1953. Effect of hypothalamic lesions on the
   adrenal cortical response to stress in the rat.   American
   Journal of Physiology 175.:13-20.

3. Fortier, C., 1963. Hypothalamic control of the anterior pituitary. In Comparative Endocrinology, U.S. Von Euler and H. Heller
   (Eds.) Chap.1, Vol.1. Academic Press, New York, pp. 1-24.

4. Bush, I.E., 1953. Species differences in adrenocortical secretion. Journal of Endocrinology 9:95-100.

5. Henkin, R.I. and K.M. Knigge, 1963. Effect of sound on the
   pituitary adrenal axis. American Journal of Physiology 204:710-714.

6. Persky, H., R.R. Grinker, D.A. Hamburg, M. Sabshin, S.J. Korchin,
   H. Basowitz and J.A. Chevalier, 1956. Adrenal cortical function
   in anxious human subjects; plasma level and urinary excretion of
   hydrocortisone. Archives of Neurology and Psychiatry 76:549-558.

7. Mason, J.W., 1958. Plasma 17-hydroxycorticosteroid response to
   hypothalamic stimulation in the conscious rhesus monkey.
   Endocrinology 63:403-411.

8. David, J.O., 1967. The regulation of aldosterone secretion.
   In The Adrenal Cortex A.B. Eisenstein (Ed.)pp. 203-247.
   Little, Brown, Boston.

9. Brodish, A. and C.N.H. Long, 1956. Changes in blood ACTH under
   various conditions studied by cross-circulation technique.
   Endocrinology, 59:666-676.

10. Saffran, M. and A.V. Schally, 1955. The release of cortico-
    trophin by anterior pituitary tissue in vitro. Canadian
    Journal of Biochemistry and Physiology 33:408-415.

11. Guillemin, R. and A.V. Schally, 1959. Re-evaluation of a
    technique of pituitary incubation in vitro as an assay for
    corticotrophin-releasing factor. Endocrinology 65:555-562.

12. Guillemin, R., A.V. Schally, H.S. Lipscomb, R.N. Andersen and
    J.M. Long, 1962. On the presence in the hog hypothalamus of
    8-corticotrophin-releasing factor, a - and 8-melanocyte
    stimulating hormones, adrenocorticotrophin, lysine-vasopressin
    and oxytocin. Endocrinology 70:471-477.

13. Gemzell, C.A., D.C. Van Dyke, C.A. Tobias and M.M. Evans, 1951.
    Increase in the formation and secretion of ACTH following adrenal-
    ectomy. Endocrinology 49:325-336.

14. Abelson, D. and D.N. Baron, 1952. The effect of cortisone
    acetate on adrenal ascorbic acid depletion following stress.
    Lancet 2:663-664.

15. Forgacs, P. and L. Hajdu, 1953. Adrenocorticotrophic hormon
    (ACTH) elualasztas gatlasa cortisonnal. Kiserletes Orvostudomany
    5:444-448.

16. Schwartz, N.B. and A. Kling, 1960. Stress-induced adrenal
    ascorbic acid depletion in the cat. Endocrinology 66:308-310.

17. Smelik, P.C., 1963. Failure to inhibit corticotrophin secretion
    by experimentally induced increases in corticoid levels. Acta
    Endocrinologica 44:36-46.

18. Endroczi, E. and K. Lissak, 1961. The role of the rhinencephalon
    in the activation of the hypophyseo-adrenocorticogonad system
    and in the formation of emotional and sexual behaviour.
    Problemy Endokrinologii i Gormonoterapii. (Moscow) 4:18-25.

19. Endroczi, E. and K. Lissak, 1962. Interrelations between
    paleocortical activity and adrenocortical function. Acta
    Physiologica Academiae Scientiorum Hungaricae 21:257-263.

20. Porter, R.W. 1954. The central nervous system and stress in-
    duced eosinopenia. Recent Progress in Hormone Research 10:1-18.

21. Endroczi, E., K. Lissak, C. Szep and A. Tigyi, 1954. Examina-
    tions of the pituitary-adrenal-thyroid system after ablation
    of neocortical and rhinencephalic structures. Acta Physiologica
    Academiae Scientiorum Hungaricae 6:19-31.

22. Endroczi, E. and K. Lissak, 1960. The role of the mesencephalon,
    diencephalon, and archicortex in the activation and inhibition
    of the pituitary-adrenocortical system. Acta Physiologica
    Academiae Scientiorum Hungaricae 17:39-51.

23. Knigge, K.M., 1960. Neuroendocrine mechanisms influencing ACTH
    and TSH secretion and their role in cold acclimatization.
    Federation Proceedings 19:45-51.

24. Mason, J.W., 1959. Plasma 17-hydroxycorticosteroid levels
    during electrical stimulation of the amygdaloid complex in
    conscious monkeys. American Journal of Physiology. 196:44-48.

25. Mason, J.W., W.J.H. Nauta, J.V. Brady and J.A. Robinson, 1959.
    Limbic system influences on the pituitary-adrenal cortical
    system. Proceedings of the Endocrine Society 41:29.

26. Okinaka, S., H. Ibayashi, K. Motohashi, T. Fumita, S. Yoshida

and N. Ohsawa, 1960. Effect of electrical stimulation of the limbic system on pituitary-adrenal function: posterior orbital surface. Endocrinology, 67:319-325.

27. Mason, J.W., 1958. The central nervous system regulation of ACTH secretion. In Reticular Formation of the Brain, H.H. Jaspar, L.D. Proctor, A.S. Knighton, W.C. Noshaw and R.T. Costello. (Eds), Little, Brown, Boston, pp. 645-662.

28. Nauta, W.J.H. and H.G.J.M. Kuypers, 1958. Some ascending pathways in the brain stem reticular formation. In Reticular Formation of the Brain, H.H. Jaspar, L.D. Proctor, A.S. Knighton, W.C. Noshay and R. T. Costello. (Eds), Little, Brown, Boston, pp. 3-30.

29. Nauta, W.J.H., 1961. Limbic system and hypothalamus. Physiological Reviews 40:102-104.

30. Nauta, W.J.H., 1963. Central nervous organization and the endocrine motor system. In Advances in Neurology. A.V. Nalbandov. (Ed), University of Illinois Press, Urbana. pp. 5-21.

31. Medoff, H.S. and A.M. Bongiovanni, 1945. Audiogenic stimulation and blood pressure. American Journal of Physiology 143:300-305.

32. Farris, E.J., E.H. Yeakel and H.S. Medoff, 1945. Development of hypertension in emotional animals. American Journal of Physiology 144:331-333.

33. Smirk, F.R., 1949. Pathogenesis of essential hypertension. British Medical Journal 1 :791-799.

34. Rosecrans, J.A., N. Watzman and J.P. Buckley, 1966. The production of hypertension in male albino rats subjected to experimental stress. Biochemical Pharmacology 15:1707-1718.

35. Vander, A.J., 1967. Control of renin release. Physiological Reviews 47:359-382.

36. Peart, W.S., 1965. The renin-angiotensin system. Pharmacological Reviews 17:143-182.

37. Rapp, J.P., 1969. Deoxycorticosterone production in adrenal regeneration hypertension, in vitro and in vivo comparison. Endocrinology, 84:1409-1420.

38. Brown-Grant, K. and G. Perthes, 1960. The response of the thyroid gland of the guineapig to stress. Journal of Physiology (London) 151:40-50.

39. Brown-Grant, K., G.W. Harris and S. Reichlin, 1954. The effect of emotional and physical stress on thyroid activity in the rabbit. Journal of Physiology (London) 126:29-40.

40. Harris, G.W., 1955. Neural control of the pituitary gland. Arnold. London.

41. Brown-Grant, K., 1966. The control of TSH secretion. In The Pituitary Gland. G.W. Harris and B.T. Donovan (Eds). Vol. II, Butterworths, London, pp. 235-261.

42. Shipley, R.A. and F.H. MacIntyre, 1954. Effect of stress, TSH and ACTH on the level of hormonal I^{131} of serum. Journal of Clinical Endocrinology and Metabolism 14:309-317.

43. Engstrom, W.W. and B. Markhardt, 1955. The effect of serious illness and surgical stress on the circulating thyroid hormone. Journal of Clinical Endocrinology and Metabolism 15:953-963.

44. Lees, P. and M.F. Lockett, 1964. The influence of hypophysectomy and of adrenalectomy on the urinary changes induced by oxytocin in rats. Journal of Physiology (London) 171:403-410.

45. Ogle, C.W. and M.F. Lockett, 1966. The release of neuro-hypophyseal hormone by sound. Journal of Endocrinology 36: 281-290.

46. Euler, U.S. von and I. Orwen, 1955. Preparation of extracts and organs for estimation of free and conjugated noradrenaline and adrenaline. Acta Physiologica Scandinavica 33:(Suppl. 118) 1-9.

47. Gaddum, J.H. and F. Lembeck, 1949. The assay of substances from the adrenal medulla. British Journal of Pharmacology and Chemotherapy 4:401-408.

48. Bisset, G.W., 1961. The assay of oxytocin and vasopressin in blood and the mechanism of inactivation of these hormones by thioglycollate. In Oxytocin, R. Caldeyro-Barcia and H. Heller (Eds) Pergamon Press, London, pp. 380-398.

49. Van Dongen, C.G. and R.L. Hays, 1966. A sensitive in vitro assay for oxytocin. Endocrinology 78:1-6.

50. Ogle, C.W., 1967. Low frequency sound and oxytocic activity of plasma, in rats. Nature (London) 214:1112-1113.

51. Euler, U.S. von, R. Luft and T. Sundin, 1954. Excretion of urinary adrenaline in normals following intravenous infusion. Acta Physiologica Scandinavica 30:249-259.

52. Ogle, C.W. and M.F. Lockett, 1968. The urinary changes induced in rats by high pitched sound (20 kycycles/sec) Journal of Endocrinology 42:253-260.

53. Bisset, G.W., B.J. Clark and G.P. Lewis, 1967. The mechanism of the inhibitory action of adrenaline on the mammary gland. British Journal of Pharmacology and Chemotherapy 31:550-559.

54. Kleeman, C.R. and R.E. Cutler, 1963. The neurohypophysis. Annual Review of Physiology 25:385-432.

55. Peeters, G. and R. Coussens, 1950. The influence of milking on the lactating cow. Archives Internationales de Pharmaco-dynamie et de Therapie 84:209-220.

56. Cross, B.A., 1951. Suckling antidiureses in rabbits. Journal of Physiology (London) 114:447-453.

57. Ginsburg, M. and M.W. Smith, 1959. The fate of oxytocin in male and female rats. British Journal of Pharmacology and Chemo-therapy 14:327-333.

58. Lederis, K., 1961. Vasopressin and oxytocin in the mammalian hypothalamus. General and Comparative Endocrinology 1:80-89.

59. Olivecrona, H., 1957. Paraventricular nucleus and pituitary gland. Acta Physiologica Scandinavica Suppl. 136:1-178.

60. Slotnick, B.M. and A.B. Rothballer, 1964. Vasopressin release following stimulation of limbic forebrain structures in the cat. Federation Proceedings 23:150.

61. Aulsebrook, L.H. and R.C. Holland, 1969. Central regulation of oxytocin release. American Journal of Physiology 216:818-829.

62. Aulsebrook, L.H. and R.C. Holland, 1969. Central inhibition of oxytocin release. American Journal of Physiology 216:830-842.

63. Cross, B.A., 1961. Neural control of oxytocin secretion. In Oxytocin R. Caldeyro-Barcia and H. Heller (Eds), Pergamon Press, London, pp. 24-47.

64. Zeman, W. and J.R.M. Innes, 1963. In Craigie's Neuroanatomy of the Rat. New York Academic Press, pp. 61-77.

65. O'Connor, W.J. and E.B. Verney, 1942. The effect of removal of the posterior lobe of the pituitary on the inhibition of water diuresis by emotional stress. Quarterly Journal of Experimental Physiology and Cognate Medical Sciences 31:393-408.

66. Duke, H.N. and M. Pickford, 1951. Observations on the action
    of acetylcholine and adrenaline on the hypothalamus. Journal
    of Physiology (London) 114:325-332.

67. Pickford, M., 1960. The release of oxytocin and vasopressin in
    Polypeptides which affect smooth muscle. M. Schachter (Ed.)
    Pergamon Press, London, p. 42.

68. Brauner, F., F. Brucke, F. Kaindl and A. Neumayer, 1951.
    Quantitative Bestimmungen uber die Sekretion des Nebennieren-
    markes bei elektrischer Hypothalamus Reizung. Archives Inter-
    nationales de Pharmacodynamie et de therapie 85:419-430.

69. Brucke, F., F. Kaindl and H. Mayer, 1952. Uber die Veranderung
    in der Zusammensetzung des Nebennierenmarkinkretes bei elektrischer
    Reizung des Hypothalamus. Archives Internationales de Pharmaco-
    dynamie et de therapie 88:407-412.

70. Grant, R., P. Lindgren, A. Rosen and Uonas B., 1958. The
    release of catechols from the adrenal medulla on activation
    of the sympathetic vasodilator nerves of the skeletal muscles
    in the cat, by hypothalamic stimulation. Acta Physiologica
    Scandinavica 43:135-154.

71. Iremoto, T., 1955. Studies on the central mechanism for the
    adrenaline secretion - Accelerating mechanism for adrenaline
    secretion in the pons. Folia Endocrinologica Japonica 31:268-270.

72. Folkow, B. and U.S. von Euler, 1954. Selective activation of
    noradrenaline and adrenaline producing cells in the suprarenal
    gland of the cat by hypothalamic stimulation. Circulation
    Research 2:191-195.

73. Lehman, A. and R.G. Busnel, 1963. A study of the audiogenic
    seizure. In Acoustic Behavior of Animals. R.G. Busnel (Ed),
    Elsevier Publishing Company, pp. 244-262.

74. Wada, J.A. and T. Asakura, 1969. Susceptibility to audiogenic
    seizure induced by thiosemicarbazide. Experimental Neurology
    24:19-37.

75. Wada, J.A. and M. Ikada, 1966. The susceptibility to auditory
    stimuli of animals treated with methionine sulfoxime. Experi-
    mental Neurology 15:157-165.

76. K. Schlesinger, R.C. Elston and W. Boggan, 1966. The genetics
    of sound induced seizure in inbred mice. Genetics 54:95-103.

77. Krushinski, L.V., 1959. Genetic investigations in experimental
    pathophysiology of higher nervous activity. Bulletin Moskov.
    Obstrch. Biol. 64:105-117 (Read in translated abstract only).

ENDOCRINE AND METABOLIC EFFECTS OF NOISE IN NORMAL, HYPERTENSIVE

AND PSYCHOTIC SUBJECTS

A. E. ARGUELLES, M. A. MARTINEZ, EVA PUCCIARELLI
and MARIA V. DISISTO
Department of Endocrinology and Hormone Laboratory
Hospital Aeronautico, Buenos Aires and
Scientific Research Committee of the Argentine (J.I.C.E.F.A.)

In a previous paper (1) we reported an increased cortical adrenal activity in the human, exerted by certain levels of audio-stimulation. The original experiments suggested that the release of increased levels of glucocorticoid in response to noise is mediated through ACTH released from the pituitary.

It is well known that a number of stressors can elicit an increased release not only of corticoids but of sympatho-adrenomedullary hormones. These hormones can cause a mobilization of free fatty acids from adipose tissue, with elevation of circulating triglycerides and cholesterol, both constituents of plasma lipoproteins. Chronically elevated blood lipid and adrenal hormones can contribute to degenerative changes in the arterial and myocardial tissue (2,3).

The aim of the present study was to investigate whether abnormal responses to sound exposure were reflected in the catecholamines and corticoid hormones pattern of patients with abnormal circulatory conditions such as cardiac infarction and hypertension as well as those of some psychiatric patients.

MATERIALS AND METHODS

The different subjects studied were assembled in four groups. The first was of twelve patients with myocardial infarction in the chronic state (more than 3 months after the acute myocardial damage).

Twelve subjects were hypertensive with mean diastolic pressure between 100 and 130 mm Hg. Aldosterone secreting adrenomas and

renovascular hypertension had been excluded in all of them.
Another group of twelve psychotic patients was also studied.
Seven of them were schizophrenics, four were alcoholic psychotics
and one was a melancholic depression case.

Five normal controls underwent the same tests and assays.
In order to overcome any emotional tension induced by the blood
extraction, the subjects were allowed to rest for at least half
an hour.   They were explicitly told to refrain from smoking and
from taking alcoholic and caffeine containing beverages on the
morning of the study and any kind of drug during the previous 24
hours.   Unfortunately this could not be done with the psychiatric
patients, whose drug treatments could not be interrupted.   This
was made of large doses of Chlorpromazine (200 to 2,000 mg).
Plethysmographic measurements of forearm blood pressure and heart
rate were recorded during exposure to a pure noise of 2,000 cycles/
second from a tape recorder, which was amplified up to a level of
90 dB.   We chose this frequency and intensity considering that
higher ones would not represent the characteristics of sound found
normally in ordinary conditions of urban day life in large cities.
All the subjects were placed in a sound isolated booth and had
been checked for audiogram abnormalities with the exception of
the psychotic patients that were studied in an ordinary more-or-
less silent room adjoining their ward.   Unfortunately these
patients could not be audiographically explored.

In previous experiments (1,4) the subjects were exposed via
earphones to pure sound emitted by an audiometer valve at 1000,
5000 and 10,000 cycles per second.   However, since noise can be
transmitted through other parts of skeletal and soft tissue, we
felt that a more realistic approach to the study of the effects
of noise could be obtained with the use of a tape-recorder.
Noise levels obtained with repeated playing of the magnetophonic
tapes were frequently checked by a specialized sound technical
expert.

The subjects were comfortably seated and most of them (except
the psychotic cases) read books or magazines throughout the noise
period.

Fasting blood samples were drawn into heparinized syringes
immediately before starting and at the end of the noise period.

At six o'clock the subjects emptied their bladders and drank
250 ml of water; exactly three hours later they emptied their
bladders again.   Immediately afterwards the experiment started,
and from that moment on another three hours' urine sample was
collected.

Determination of plasma corticoid-steroids were done with

Mattingly's method (5) using an Aminco-Bowman spectro-photo-fluorometer. Adrenaline and nor-adrenaline, where assayed with Crout's method (6), also with fluorescence measured with Aminco-Bowman equipment.

Urinary 17-OH corticoids were determined with Metcalf's method (7) and 17 Ketosteroids with Ricca's method (8).

Determination of plasma cholesterol was done with Abell et al., method (9).

RESULTS

Normal Controls (Table 1)

In normal controls, urinary epinephrine and norepinephrine were significantly increased by sound exposure. Under the conditions of these experiments, blood pressure was not affected. Figure 1 shows comparative basal urinary catecholamine levels in controls and in patients with essential hypertension, chronic cardiac infarction, schizophrenia and in alcoholic psychotics. The following sections describe their response to sound.

Table 1. Catecholamines in resting condition and with sound stimulation in normal controls. *

Patient	Urinary Epinephrine ug/hr		Urinary Norepinephrine ug/hr		Blood Pressure mm Hg	
	b	s	b	s	b	s
HO	0.36	1.0	0.9	1.8	125/80	125/80
PE	0.4	0.76	1.3	2.1	110/75	110/70
AE	0.2	0.46	0.61	1.2	130/80	130/80
LR	0.36	0.51	1.6	1.2	120/80	120/80
AU	0.48	0.42	1.08	1.24	120/80	120/80
Mean	0.36	0.63	1.09	1.50	120/80	120/80
SD	0.08	0.22	0.38	0.42	–	–
p<	0.02		0.05		–	–

*b = baseline; s = stimulated by sound.

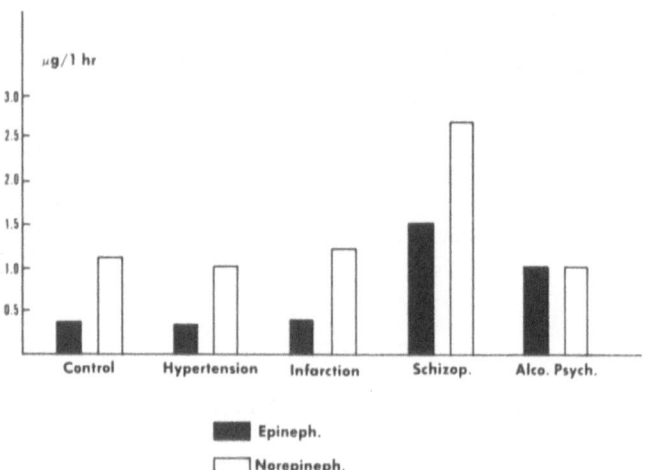

Figure 1. Mean basal levels of urinary catecholamines in controls, essential hypertensives, chronic myocardial infarction, schizophrenia and alcoholic psychotics.

Patients with Myocardial Infarction (Table 2)

Plasma cholesterol. Under basal resting conditions, plasma cholesterol in cardiac infarction patients was somewhat higher

Table 2. Chronic cardiac infarction patients. Response to sound of plasma cortisol and cholesterol and urinary catecholamines.*

Patient	Plasma Cortisol ug%		Plasma Cholesterol ug%		Urinary Epinephrine ug/hr.		Urinary Norepinephrine ug/hr.	
	b	s	b	s	b	s	b	s
1 PH	24	7.6	217	230	0.36	0.53	1.3	1.2
2 FP	32	31	290	290	0.46	0.68	0.86	1.11
3 OA	18	22	209	240	0.43	0.54	0.9	2.53
4 VR	36	44	365	320	0.39	0.42	1.27	2.29
5 DE	21	60	260	260	0.5	0.59	1.6	2.42
6 MC	24	-	295	340	0.14	0.18	0.88	1.79
7 GO	18	15.5	268	285	0.45	0.78	1.12	1.42
8 GM	34	22	245	250	0.41	0.67	1.24	2.01
9 GJ	12	-	260	255	0.33	0.73	0.82	1.4
10 GA	29	22	215	290	0.43	0.74	1.5	1.8
11 LC	19	19	270	300	0.41	0.46	1.08	4.16
12 GE	-	18.4	-	250	0.68	0.34	2.20	5.03
Mean	24.2	26.1	263	275	0.41	0.55	1.23	2.26
SD	7.74	15.2	44	37	0.1	0.17	0.3	1.1
p<	-	-	0.40		0.001		0.001	
% Change					+34%		+83%	

*b = baseline; s = stimulated by sound.

Table 3. Hypertensive patients. Modification of plasma cholesterol and corticoids, urinary catecholamines and blood pressure with sound.*

Patient	Plasma Cortisol ug%		Plasma Cholesterol ug%		Urinary Epinephrine ug/hr.		Urinary Norepinephrine ug/hr.		Blood Pressure mm Hg	
	b	s	b	s	b	s	b	s	b	s
1 CE	25	20	210	230	0.16	0.29	0.38	1.38	140/110	150/110
2 AA	43	15	355	320	0.18	0.30	0.60	0.72	140/100	140/100
3 UR	14	15	285	220	0.34	0.23	0.36	1.5	160/100	160/100
4 CA	26	15	295	370	0.14	0.67	0.57	3.4	170/105	200/130
5 AE	12	15.6	200	240	0.18	0.65	1.72	1.95	160/90	170/100
6 RN	28.4	22.7	320	260	0.35	0.61	1.23	1.55	170/110	200/110
7 TE	10	14.2	210	195	0.35	0.19	0.64	1.09	160/100	160/120
8 SL	46	51	275	360	0.16	1.26	1.7	2.46	160/120	190/130
9 RI	24	22.7	295	260	0.41	0.55	1.6	1.8	150/95	150/95
10 RE	18.2	10	210	265	1.05	0.53	1.6	1.32	170/110	170/120
11 CE	18.9	9.1	250	255	0.38	0.77	0.89	1.69	160/100	175/110
Mean	24.1	19.1	264	270	0.33	0.55	1.02	1.80	160/100	170/110
SD	11.7	11.4	52	65	0.24	0.3	0.54	0.64	-	-
$p <$	-		0.8		0.05		0.001		0.01/0.01	

*b = baseline; s = stimulated by sound.

than in normals.   Some degree of elevation in cholesterol levels
with sound exposure was observed in seven out of twelve patients,
but the overall increase was not significant.

Plasma 17-hydroxycorticosteroids.   Under basal resting conditions,
plasma 17-hydroxycorticosteriods were somewhat higher than normal.
In response to sound they increased in only four of the twelve
patients tested, but in two of these extremely high levels were
reached at the end of the test.   Some patients that responded to
sound with a reduction of plasma corticoid values had abnormally
high levels under resting conditions immediately prior to the tests.

Epinephrine. The average resting level of urinary epinephrine
excretion was in the upper end of the normal range (0.41 ± 0.1
ug/hr, S.D.).   Following sound exposure, significant increases
in epinephrine were observed (p $<$ 0.001), with responses occurring
in all subjects but one; the mean after sound exposure was 0.55±
0.26 ug/hr, S.D.

Norepinephrine. Under resting conditions, urinary norepinephrine
values were either in the upper normal range or abnormally high
(four of twelve cases); the resting average was 1.23 ± 0.3 ug/hr,
S.D.   Sound caused large increases in most patients.   In the only
patient that did not show an elevation of epinephrine, there was a
marked elevation of norepinephrine.   Two patients showed extremely
high norepinephrine levels, in the range of pheochromocytoma
patients (cases 11 and 12, Table 2) when stimulated with sound.

The response of plasma corticoids and urinary catecholamines
to sound in chronic cardiac infarction patients is summarized in
Figure 2.

Patients with Essential Hypertension (Table 3)

Plasma 17-hydroxycorticosteroids.   Plasma 17-hydroxycorticoid
concentrations, in most patients, were well above upper normal
levels under resting conditions (24.1 ± 11.7 ug/100 ml, S.D.),
as if these patients were suffering a chronic stress condition.
With sound exposure, these values decreased in seven out of eleven
patients, and the levels at the end of the sound exposure were
significantly lower than in resting conditions (19.1 ± 11.4,S.D.,
p $<$.05).

Epinephrine.   All patients showed resting epinephrine values
in the normal range.   During the sound period, there was a significant increase in the output of epinephrine.   The mean value
of the epinephrine excretion during the sound experiments was 0.55
ug/hr, as compared with 0.33 ug/hr under resting conditions (p $<$ 0.05).

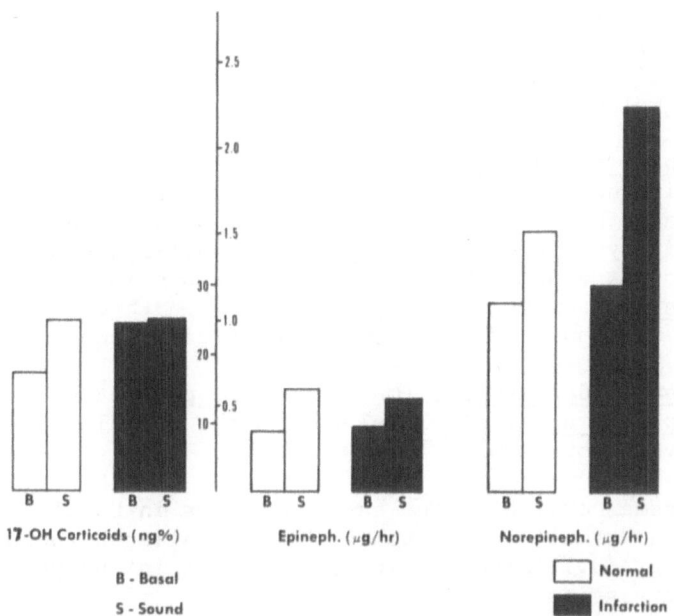

Figure 2. Response of urinary catecholamines and plasma corticoids to sound in chronic cardiac infarction patients.

Norepinephrine. Norepinephrine was well above the normal upper limit in four of the eleven patients under resting conditions. Increases in post-sound urinary norepinephrine were observed in all cases (p < .001). One patient (#4) showed a six-fold increase in norepinephrine and more than a five-fold increase in epinephrine.

Blood pressure. Blood pressure, either diastolic, systolic or both, increased to some degree in eight of twelve labile hypertensive patients. The increase in the mean diastolic and systolic pressures over the pre-noise period was significant (p < .01).

Patients with some Mental Diseases (Table 4)

1. Schizophrenic patients. In three patients where the plasma 17-OH corticoids were studied in resting conditions, two were in the high normal range and the other was elevated. Noise caused an important increase in one, no change in the second and the one with the abnormally high basal value showed a diminished level.

Epinephrine excretion in basal conditions was non-detectable in one, normal in another, and markedly elevated in the other two. With the sound stimulation the responses were not significant, with slight reduction of the mean value.

Norepinephrine was significantly elevated in resting conditions
in three out of five cases, and in three of them (aggressive) the
sound stimulation caused increases that in one patient reached an
extremely high level (5.9 ug/hour).

2. Alcoholic psychotics. The plasma 17-OH corticoids were
practically in the normal range in the three patients in which it
was studied and the sound test caused an important elevation in one
patient. Urinary epinephrine in resting conditions showed normal
levels in two cases and markedly elevated values in two others.
Norepinephrine in resting state was very low in three out of four
patients and only one patient showed significant response.

3. Endogenous depression. Only one male patient was studied.
His resting plasma corticoid level was in the normal range and was
markedly elevated by sound.

Table 4. Plasma corticoids and catecholamines in basal conditions
and with sound stimulation in schizophrenic patients taking chlorpromazine (3 gr./d.), alcoholic psychotics and in one patient with
endogenous depression.*

Patient	Plasma cortisol ug%		Urinary epinephrine ug/hr.		Urinary norepinephrine ug/hr.	
	b	s	b	s	b	s
Schizophrenia						
FA	–	–	1.0	0.7	4.6	0.45
QI	–	–	0.45	0.9	1.5	1.3
SA	19	33	0.0	1.1	2.7	3.1
NR	23	23	3.05	0.3	3.0	5.9
RL	40	26	3.1	3.8	1.4	1.8
Mean	27.3	27.3	1.52	1.36	2.68	2.51
SD	–	–	1.4	1.4	1.2	2.3
p<	–	–	0.8		0.6	
Endogenous depression						
ZT	15	27	1.5	0.4	2.9	0.6
Alcoholic dementia						
ME	–	–	2.1	1.8	0.85	1.2
ER	12	12	0.3	0.52	0.58	2.5
GR	11	22	1.5	2.3	0.5	0.1
LU	23	15	0.47	0.43	2.8	1.3
Mean	15.3	16.3	1.1	1.27	1.1	1.27
SD	–	–	0.8	0.9	1.1	0.9
p<		–	0.7		0.4	

*b = baseline; s = stimulated by sound.

Epinephrine and norepinephrine were high in basal conditions but instead of increasing with the sound experiment a marked reduction of both catecolamines was registered.

4. Anxiety neurosis. These patients had been previously studied and reported (4), and are shown here for comparison purposes. They were stimulated with 10,000 cyc/sec. noise for 1 hour and their basal mean plasma corticoids were 13.4 ± 4.1 ug / 100 ml, with a highly significant elevation with sound stimulation of 23.1 ± 2.4 ug/100 ml (p<0.01).*

DISCUSSION

The present study has confirmed that marked endocrine disturbances can be caused by sound stimulation as it has been reported by several authors (10), (11), (12), (1), (13).

These effects have been particularly intense when highpitched sound (10 or 20 kcyc/sec.) was used, causing release of antidiuretic hormone (ADH) and epinephrine in the rat (14) and of cortisol in the normal and neurotic human (4).

In the present study the first striking finding has been the fact that the sound stimulation in some of our patients with myocardial infarction did not cause a plasma corticoid response at all, although highly significant responses of epinephrine and particularly norepinephrine were observed, and in two cases extremely high levels of plasma 17-OH corticoids were reached with noise exposure. It is difficult to explain the decrease of some of the plasma 17-OH corticoids after sound exposure in this group of patients as well as in the hypertension cases. Circadian variations are never so marked even in the absence of the slightest corticoadrenal stimulation. As the resting levels of plasma corticoids were very high and equivalent with those found in very stressed people, we feel that some of our patients with coronary thrombosis and hypertension are easily emotional stressed people who arrived at the hospital with very increased pituitary-corticoadrenal activity caused by high emotional tension, and that lack of pain and other discomfort throughout the experiments reassured them, relaxing their emotional tensions and letting them concentrate on amusing reading during the noise tests with subsequent normalization of the corticoadrenal function.

An interesting practical finding was the increases of blood pressure caused by the noise stimulation in almost all the labile hypertensive patients, which correlates fairly well with the catechol modifications.

_____
*For "normals", Dr. Arguelles' non-stimulated and stimulated plasma corticoid values, respectively, are about 11.5 and 18.8 (4).

The important epinephrine and norepinephrine responses to sound stimulation in normal subjects, myocardial infarction and hypertension patients were highly significant, but myocardial infarction patients showed significantly larger responses of norepinephrine than did the other groups, with much less pronounced epinephrine responses than in the controls.

According to Elmadjian (15) this larger norepinephrine output with sound stress would indicate an aggressive tendency. Other authors have mentioned aggressiveness as a psychological trait of the myocardial infarction prone individuals.

Wallace (16) measured the 24 hour catecholamine excretion of patients with acute myocardial infarction and found a four-fold increase, being highest in those with heart block. In uncomplicated cases, the catecholamine excretion remained elevated during a ten day period with a step-wise decrease each day. Although our cases had been uncomplicated during the acute period and had been studied after a prolonged convalescence the intense response of epinephrine and norepinephrine to a continuous moderatedly high level of noise suggests that they react easily to sound stress with large catecholamine discharges. It is well known that catecholamines can increase myocardial oxygen consumption and that prolonged adrenergic stimulation may be harmful to the myocardium (2). Intense sound stimulation as produced by truck, bus and automobile traffic and found in aircraft or surface transportation according to these findings and by increasing catecholamine levels could therefore aggravate the hypoxia of the already infarcted myocardium, thereby predisposing to arrhythmias and failure.

Release of catecholamines elicited by sound may increase the mobilization of free fatty acid from adipose tissue and the formation of triglyceride and cholesterol integrating the lipoproteins. Levi (3) mentions that chronically elevated blood lipids plus the direct cardiotoxic effects of epinephrine and norepinephrine can contribute to degenerative changes in the myocardium.

Holmberg et al. (17) found an increase in plasma norepinephrine in three labile hypertensives exposed to tape recorded industrial noise of 100 dB combined with intermittent light for 6 min., although these subjects reacted in a more pronounced manner to the emotional stress of mental arithmetic. They also found moderate rises in systolic and diastolic pressures.

Our studies with a larger group of hypertensive patients agree with the forenamed authors and prove that plain noise at ordinary urban levels can make hypertensives react more strongly than other people, with blood pressure and adrenergic activity increase.

The schizophrenics presented high mean levels of resting plasma corticoids, epinephrine and norepinephrine, and although there was no increase of the mean values with noise, there were very high levels of norepinephrine excretion induced by the experiment in two out of the five cases.

As one of the characteristics of some forms of schizophrenia (paranoic schizophrenics) is their frequent dangerous aggressive outbursts, noise stimulation might be responsible, perhaps by turning over large amounts of norepinephrine, for starting aggressive and even homicidal episodes in these unfortunate patients. Considering that the psychotic patients of this study while stimulated with sound were under large dose chlorpromazine, their responses may have been much more pronounced had they been off drug treatment.

We have not found a significant difference in basal epinephrine excretion between the alcoholics and control group as reported by Holmberg et al. (17), who did not detect any traces of plasma epinephrine or norepinephrine in some of his chronic alcoholic cases. Mean norepinephrine excretion was reduced in some of the alcoholic psychotics of this study but some good responses to the sound experiment were registered.

In the only endogenous depression case studied, a large plasma corticoid response to noise was noticed, but the normal epinephrine and norepinephrine basal levels fell after noise stress as if his adrenergic system had become exhausted.

Epinephrine and norepinephrine had very significant increases after noise exposure of the control subjects. This proves that noise may be a powerful stressing agent in normal people and capable of causing serious disturbances in some cardiovascular and psychotic patients.*

SUMMARY

Exposure of normal human subjects, cardiocirculatory and psychotic patients to sound (2,000 cyc/sec) for 30 min. with an intensity of 90 decibels caused increased activity of the cortico-adrenal and adrenergic functions.

Plasma 17-OH corticoids, urinary epinephrine and norepinephrine were studied both in basal conditions and after sound stimulation.

Twelve patients with chronic myocardial infarction showed very high plasma 17-OH corticoid resting levels and some of them did not respond behaviorally to sound as normal people, but the rate of urinary excretion of norepinephrine was higher than in the other groups.

---

*Differences in urinary hormone excretion may be partly due to differences in volume of urine, since urine volume of creatinine was not reported (Editor).

This supports the hypothesis that cardiac infarction patients react to audiogenic stress with increased release of norepinephrine because of a more active aggressive psychological personality. Significant epinephrine and norepinephrine responses to sound stimulation were also present in normal subjects and hypertensives.

Systolic and diastolic pressure rose significantly with the sound experiments in many labile hypertensive patients, correlating with increased catecholamine excretion.

Some schizophrenic patients with noise exposure showed very high levels of urine norepinephrine although they were under large doses of chlorpromazine. These high excretion rates of norepineph- rine caused by sound stimulation are similar to those that have been registered in psychotics during extreme aggressive episodes.

Mean norepinephrine excretion was reduced in alcoholic psychotics, but they responded to sound stress as the controls. Noise proved to be an important stressing agent capable of causing disturbances in cardiovascular and psychotic subjects.

We are thankful to Mr. Jorge Loyenicou, president of AMPLITONE S.A. ARGENTINA, for preparation and checking of the noise recording.

REFERENCES

1. Arguelles, A.E., D. Ibeas, J. Pomes Ottone and M. Chekherdemian, 1962. Pituitary-adrenal stimulation by sound of different frequencies. Journal of Clinical Endocrinology 22:846-852.

2. Selye, H., 1961. The Pluricausal Cardiopathies. C. C. Thomas, Publisher, Springfield, Illinois.

3. Levi, L., 1968. Symatho-adrenalmedullary and related bio- chemical reactions during experimentally induced emotional stress. In: E. Michel (Ed.), Endocrinology and Human Be- havior, Oxford University Press, New York. p. 200.

4. Arguelles, A.E.,1967. Endocrine response to auditory stress of normal and psychotic subjects. In: E. Bajusz (Ed.) An Introduction to Clinical Neuroendocrinology, S. Karger, Publisher, Basel, pp. 121-132.

5. Mattingly, D., 1962. A simple fluorometric method for the estimation of free 17-hydroxycorticoids in human plasma. Journal of Clinical Pathology 15:374-379

6. Crout, R., 1961. Catecholamines in urine. In: D. Seligson

(Ed.) <u>Standard Methods of Clinical Chemistry</u>, Vol. 3, Academic Press, New York, pp. 62–80.

7. Metcalf, M. G., 1963.   A rapid method for measuring 17-hydroxy-corticosteroids in urine.   Journal of Endocrinology 26:415–423.

8. Ricca, A., 1957.   Valoracion de los 17-ceto-esteroides totales neutros.   Revista de la Asociacion Bioquimica Argentina 22: 11–17.

9. Abell, L., B. B. Levy, B. B. Brodie and F. B. Kendall, 1952. A simplified method for the estimation of total cholesterol in serum and demonstration of its specificity.   Journal of Biological Chemistry 195:357–366.

10. Biro, J., V. Szokolai and A. G. Kovach, 1959.   Sound stimulation of the pituitary-adrenal cortical system.   Acta Endocrinologica 31:542–552.

11. Zondek, B. and I. Tamari, 1960.   Effect of audiogenic stimu-lation on genital function and reproduction.   American Journal of Obstetrics and Gynecology 80:1041–1048.

12. Grognot, P., H. Bioteau and A. P. Gibert, 1959.   Variation dutaux de potassium plasmatique au cours de l'exposition aux bruits.   Comm. au 1er. Colloque internat. sur le bruit, Paris. pp. 16–21.

13. DeWied, D., 1964.   Site of blocking action of dexamethasone on stress induced pituitary ACTH release.   Journal of Endo-crinology 29:29–37.

14. Ogle, C. W. and M. Lockett, 1968.   The urinary changes induced in rats by high pitched sound.   Journal of Endocrinology 42: 253–260.

15. Elmadjian, F., J. M. Hope and E. T. Lamson, 1957.   Excretion of epinephrine and norepinephrine in various emotional states. Journal of Clinical Endocrinology and Metabolism 17:608–620.

16. Wallace, A. G., 1968.   Catecholamine metabolism in patients with acute myocardial infarction.   In: <u>Acute myocardial in-farction</u>. In: D.G. Julian and M. F. Oliver (Eds.), Williams & Wilkens, Baltimore. pp. 237–242.

17. Holmberg, G., L. Levi, A. Mathe, A. Rosen and H. Scott, 1967. Catecholamines and the effects of adrenergic receptor blockade on cardiovascular and mental reaction during stress.   Fors-varsmedicin 3 suppl. 2:201–210.

NOISE, HEARING AND CARDIOVASCULAR FUNCTION

SAMUEL ROSEN*

Consultant, The Mount Sinai School of Medicine, New York,
and The New York Eye & Ear Infirmary

It is known that loud noise causes a number of reactions in
the human body which the recipient cannot control, in addition to
the psychic shock. The blood vessels constrict, the skin pales,
the pupils dilate, the eyes close, one winces, holds the breath,
and the voluntary and involuntary muscles tense. Gastric secretion
diminishes and the diastolic pressure increases. Adrenalin is
suddenly injected into the blood stream, which increases neuro-
muscular tension, nervousness, irritability and anxiety.

These changes occur via the vegetative nervous system. A very
important function of this system is the regulation of the changing
caliber of the blood vessels. Lehmann and Tamm (1) and Jansen (2)
found that a short or prolonged noise caused vaso-constriction of
precapillary blood vessels which persisted for the duration of the
noise and longer. After five minutes of noise the constriction of
the blood vessels begins to disappear but may persist for 25 minutes
before disappearing completely.

Peripheral vaso-constriction was recorded plethysmographically
by a cuff on the end of the finger. This vaso-constriction occurs
independent of annoyance or any other emotion and is present with
each noise stimulus and at all intervals. A meaningless white
noise at 90 db SPL produced the same degree of vaso-constriction
in persons long accustomed to such noise as it did in those un-
accustomed to it. During exposure to strong light the eyes can be
protected by closing the lids but the ears have no lids and are

---

*Address: 101 East Seventy-Third Street, New York, New York 10021 always vulnerable.

Rosen et al.(3,4) studied the hearing of an isolated primitive black tribe in southeast Sudan, the Mabaans. They live in an at-mosphere of virtual silence -- 35-40 db on the "C" scale. They have neither guns nor drums and use a five-string lyre for their songs and dances. They are mainly vegetarians although they catch a few fish with spears. They never have elevated blood pressure and coronary heart disease is unknown. They live long under mini-mal stress. They have better hearing in the high frequencies with aging than appears in any similar studies in modern western civili-zation. Their hearing, compared with that in several studies in western cultures done by Glorig et al.(5) was better with aging at 4000 cps. This was also found to be true in the higher frequencies, using 14 KC as an example, comparing the Mabaans with subjects of the same age in New York, Dusseldorf, and Cairo.

Of course, the Mabaans have learned to "listen" for survival in the jungle and their cardiovascular tree is exceptionally healthy, supplying excellent nutrition to the auditory nerve. When some of these tribespeople reach and live in noisy Khartoum under conditions of stress and diet changes they develop hypertension and coronary heart disease accompanied by some loss of hearing.

Jansen, et al.(6) compared the vaso-constriction of the Mabaans to that of the Dortmunders in Germany, both exposed to identical loud noise stimuli of 90 db pure tones and 90 db white noise. The vaso-constriction was much greater in the Mabaans of all ages, and it also disappeared much quicker in the Mabaans (3 and 4). A similar degree of vaso-constriction occurs during sleep as well as in wakefulness. The Mabaans seem to have greater elasticity of all their blood vessels, which may account for their slower aging process.

Auditory fatigue tests were done on the Mabaans at 106 db. In comparing the threshold shift, they showed significantly less fatigue than subjects tested by Glorig (7) in the United States. The threshold shifts in 50-year-old Mabaans was less than that in Glorig's 28-year-olds, showing that the Mabaans' recovery rate from auditory fatigue is much faster (5.)

Exposure to white noise of 90 db or more just once, or even a few times, will not damage the hearing. However, when an individual is exposed to such noise for many years then a significant cochlear hearing loss will result. If there is already present somatic disease like atherosclerosis or coronary heart disease such noise exposure could endanger health and aggravate the pathology by adding insult to injury. This could happen with even less intense noise and shorter exposure time.

Recently Lawrence, Gonzalez and Hawkins (8) demonstrated

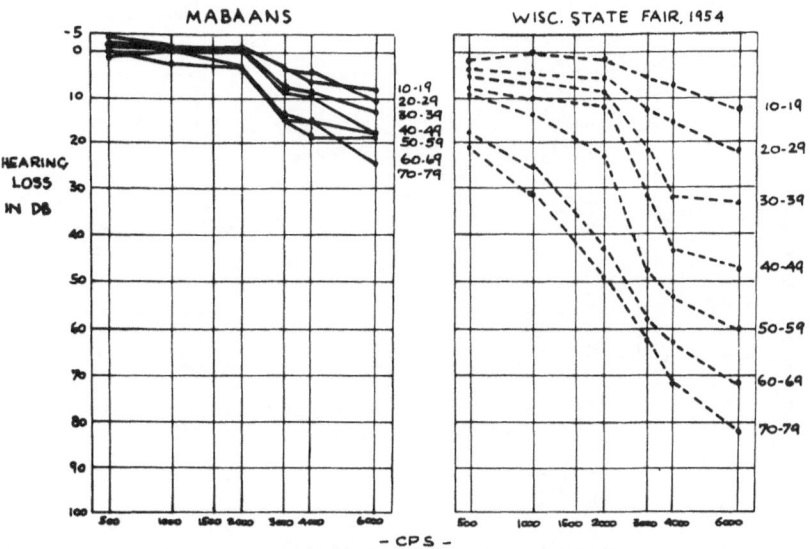

Figure 1. Comparison of median decade audiograms for men. Mabaans and Wisconsin State Fair, 1954.

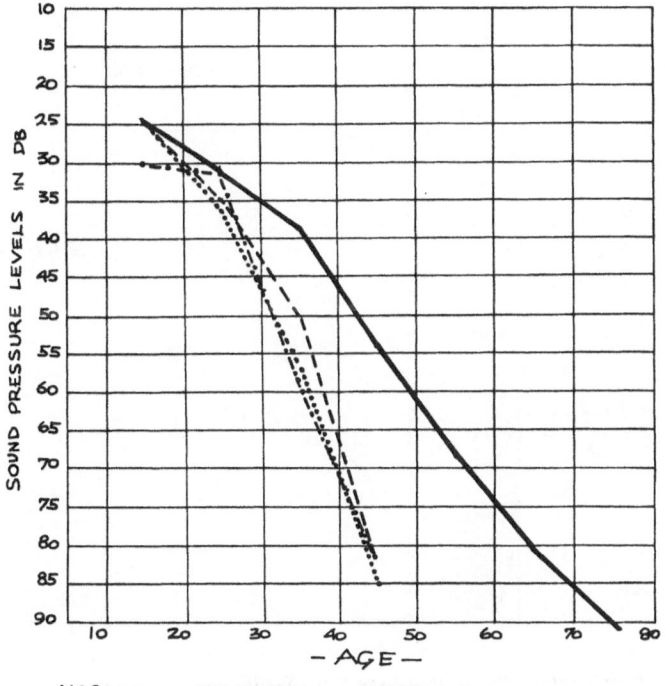

Figure 2. Comparison of decade audiograms. Mabaans, New York, Dusseldorf and Cairo. 14 KC. Median values plotted.

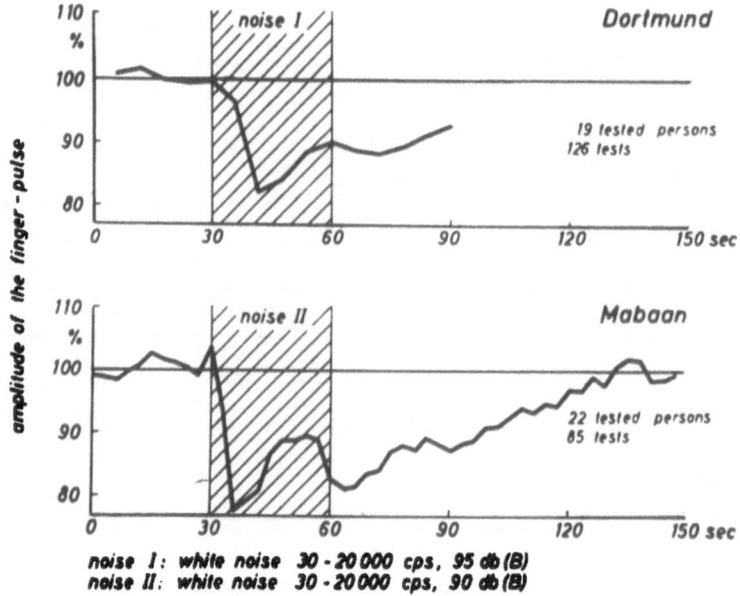

Figure 3.   Influence of octave-band noise on adults.   Mabaans
and Dortmund.

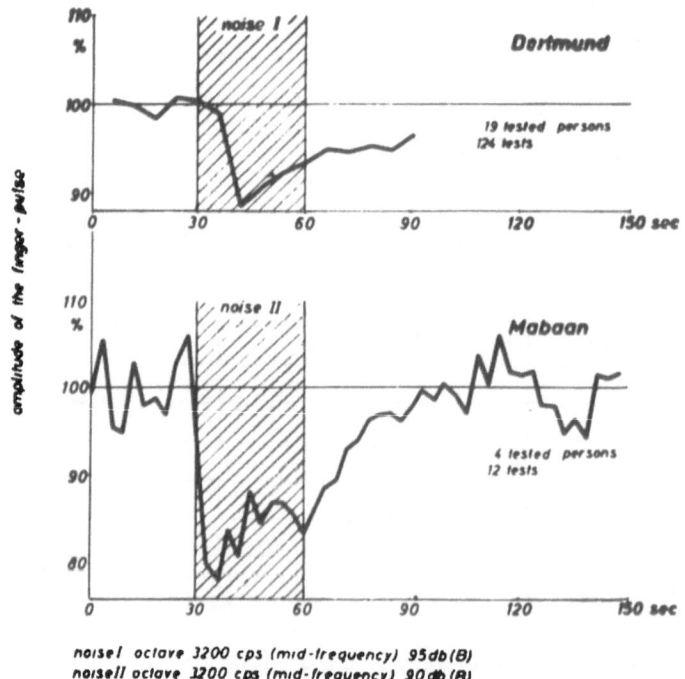

Figure 4.   Vaso-constrictive effect of noise on adult Mabaans and
adults in Dortmund, Germany.

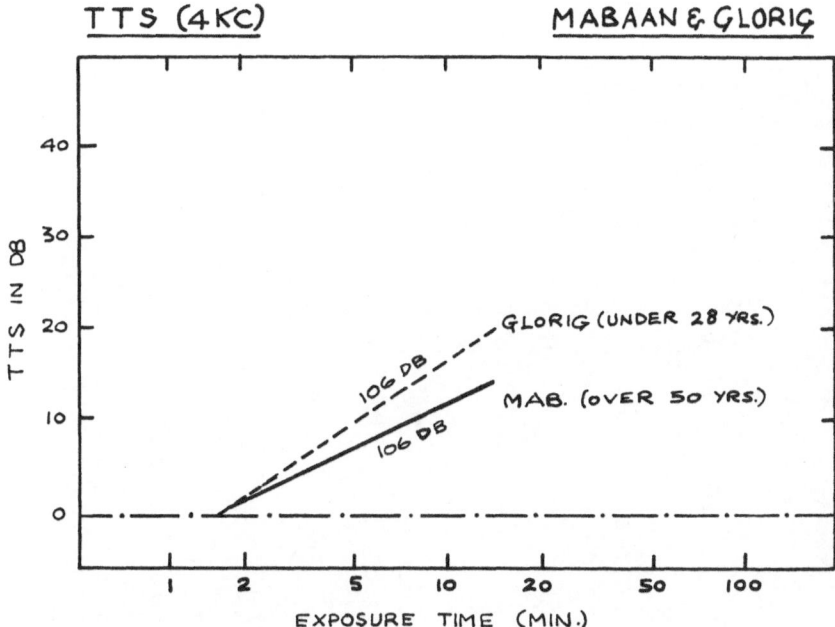

Figure 5. Auditory fatigue tests. Mabaans and Glorig.

histologically in animal experiments that noise causes vaso-
constriction of the cochlear vessels, particularly the spiral
vessels underlying the basilar membrane and the stria vascularis,
accompanied by a significant reduction in the flow and number of
red blood cells reaching the Organ of Corti. One can see the
difference in circulation between the non-noise-exposed and noise-
exposed animals in Figures 6 and 7. Obviously this vaso-constriction
of the cochlear vessels causes ischemia, or anemia, and the sensory
cells show structural changes and loss of function. The above
experiments tend to confirm the findings of vaso-constriction
demonstrated plethysmographically among the Mabaan tribe.

     Additionally, it has been shown by Friedman et al.(9) recently
that rabbits exposed to 102 db of white noise for 10 weeks showed
a much higher level of blood cholesterol than non-noise-exposed
animals although diet was identical in both groups. Noise-exposed
animals developed a greater degree of aortic atherosclerosis with
higher cholesterol content than control animals not exposed to
noise (Figure 8). Also, cholesterol deposits in the iris were more
severe and extensive in the noise-exposed than in the non-noise-
exposed animals (Figure 9). There was a rough correlation between
the degree of lipid infiltration in the iris observed during life
and the extent of aortic atherosclerosis seen at autopsy.

     All of these studies seem to emphasize the noxious effects of

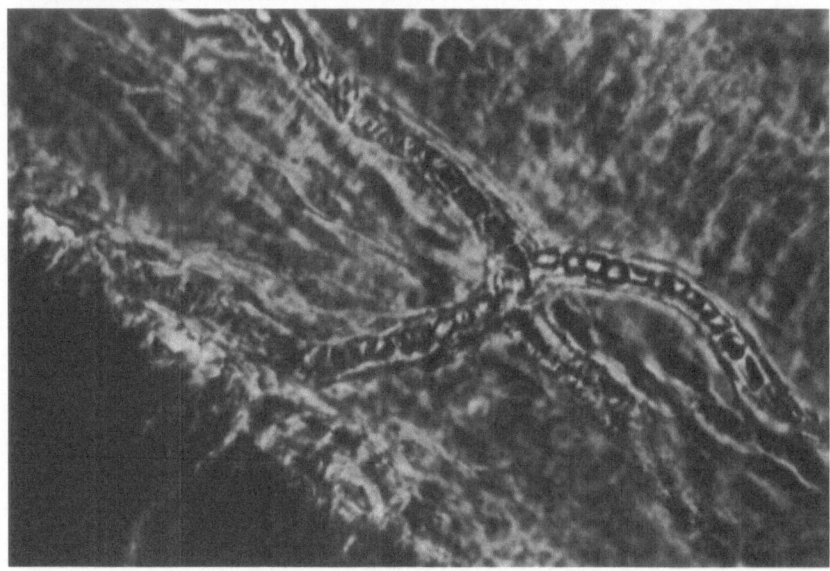

Figure 6.   Cochlear circulation in non-noise-exposed guinea pigs.

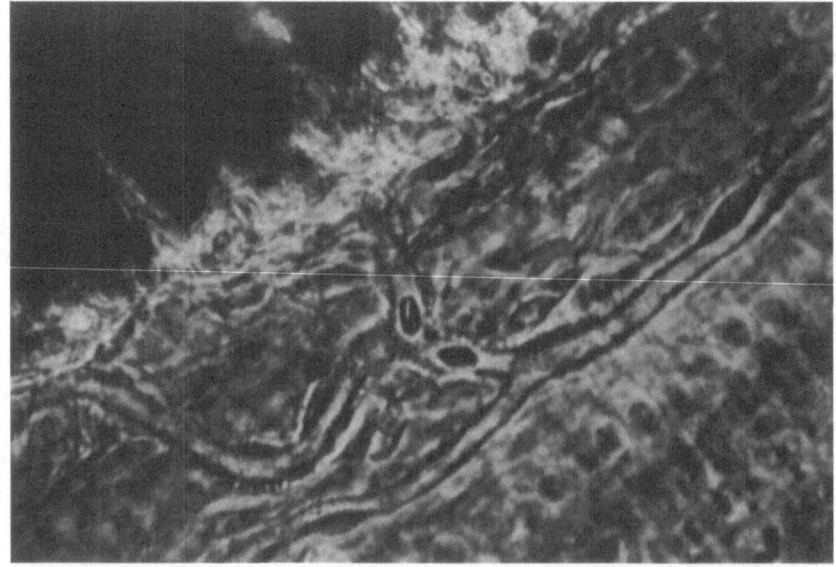

Figure 7.   Cochlear circulation in noise-exposed guinea pigs.

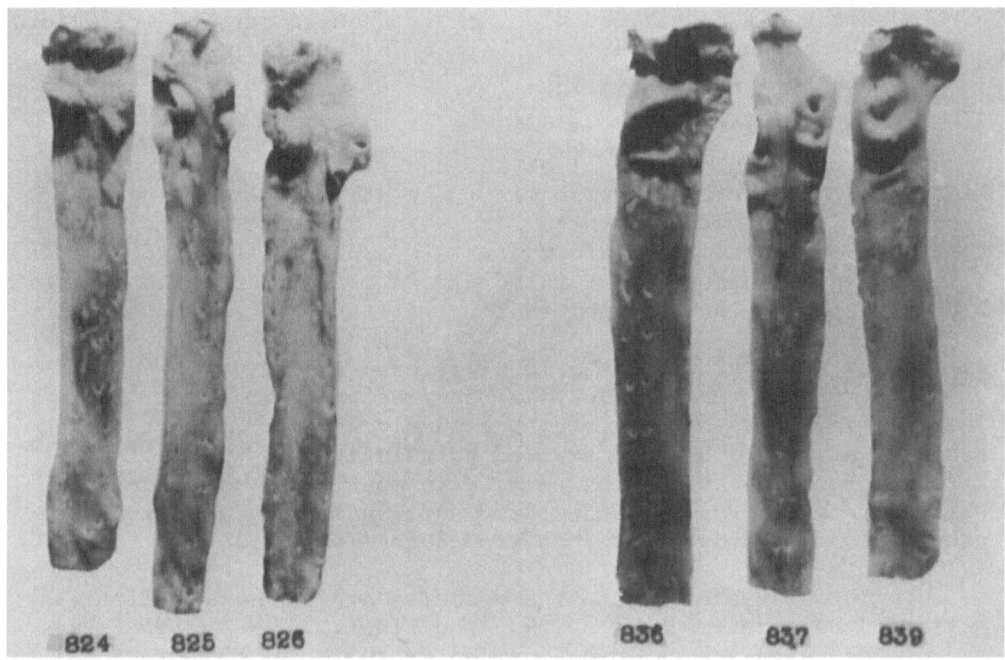

Figure 8. Aortic atherosclerosis in noise-exposed and non-noise-exposed rabbits.

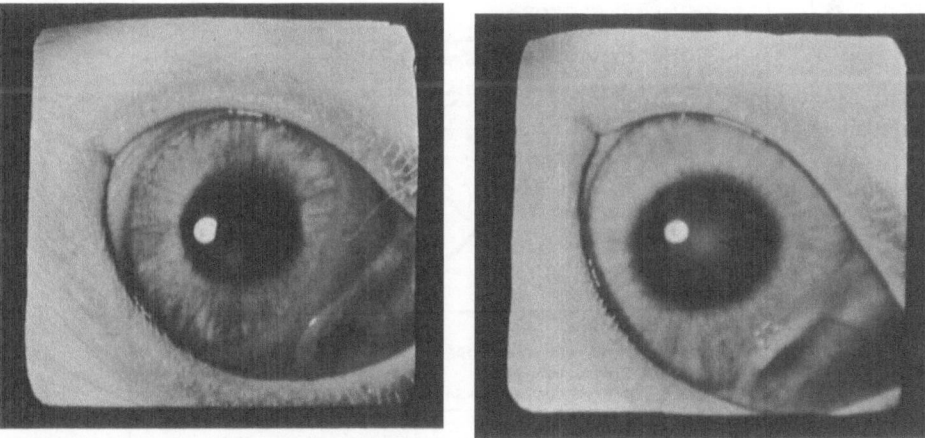

Figure 9. Lipid infiltration in the iris in noise-exposed and non-noise-exposed rabbits.

noise in man and animals.   However, it is difficult to assess the
effect of any one factor, such as noise, when so many other factors
in our environment may enhance or diminish the effect of noise.
This concept of multiple factors may be illustrated schematically
in Figure 10.

    How are we to assess all of the contributing factors already
mentioned?   The silent environment, the habit of listening, the
excellent cardiovascular system furnishing ample nourishment to
the inner ear -- all are important.   To separate these factors is
like trying to restore a scrambled egg into a single white and yolk
placed neatly in the original shell.

    Among still other populations we found similarities and dis-
crepancies in hearing studies, as reported by Rosen et al.(10).
The Finnish people have hearing not nearly as good as that of the
Mabaans, although they are exposed to relatively little noise --
at least much less noise than exists in most urban areas in the
United States.   But the Finns, as we do, suffer from excessive
atherosclerosis and coronary heart disease (11).

    In Yugoslavia and Crete the peasants are exposed to little
noise and have better hearing than the Finns.   But the Yugoslavs
and Cretans have a much lower incidence of atherosclerosis and
coronary heart disease.   The Maaban study suggests that their
hearing acuity, metabolic activity and excellent nutrition of the
inner ear might provide some resistance to noise trauma or any
other noxious factor.

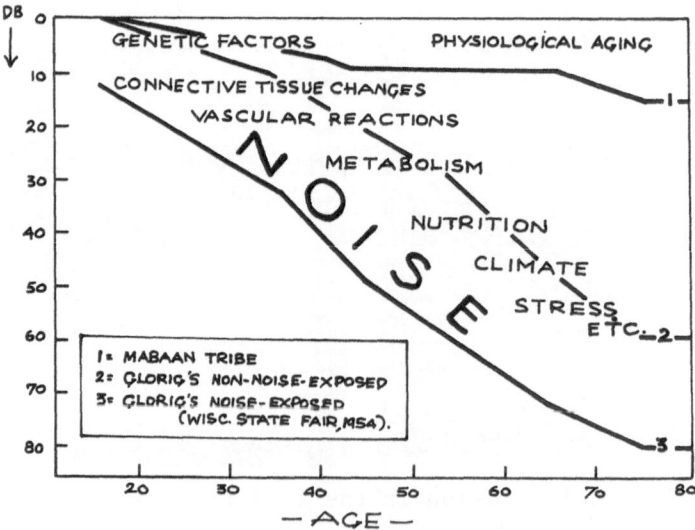

Figure 10. Schematic concept of presbycusis.   6000 cps.

The healthier an organism is the better it can resist injury from any cause, including noise. We now have millions with heart disease, high blood pressure, and emotional illness who need protection from the additional stress of noise. Rest, relaxation and peaceful sleep are necessary to all, especially to those already tense or ill. Innumerable noises invade the daily lives of great masses of people, yet nobody becomes indifferent to them. Even though such noise is not likely to damage the hearing it does inflict stress, tension, and sometimes intolerable nervous strain. People become irritable, unsociable, and more quarrelsome at work and at home.

There seems to be little doubt that noise pollution is a health hazard. The physiological effects of noise in the human, other than auditory, have not been given sufficient attention by scientists. It is our hope that further research and study will help initiate a world-wide movement for effective control of noise pollution.

REFERENCES

1. Lehmann, G. and J. Tamm, 1956. Uber Veränderungen der Kreislaufdynamik des ruhenden Menschen unter Einwirkung von Geräuschen. Internationale Zeitschrift fur Angewande Physiologie einschliesslich Arbeitsphysiologie (Berlin) 16:217-227.

2. Jansen, G. and P.-y Rey, 1962. Der Einfluß der Bandbreite eines Geräusches auf die Stärke vegetativer Reaktionen. Internationale Zeitschrift fur Angewande Physiologie einschliesslich Arbeitsphysiologie (Berlin) 19:209-217.

3. Rosen, S., M. Bergman, D. Plester, A. El-Mofty and M.H. Satti, 1962. Presbycusis study of a relatively noise-free population in the Sudan. Annals of Otology, Rhinology and Laryngology. 71:727-743.

4. Rosen, S., D. Plester, A. El-Mofty and H.V. Rosen. 1964. High frequency audiometry in presbycusis: a comparative study of the Mabaan Tribe in the Sudan with urban populations. Archives of Otolaryngology. 79:18-32.

5. Glorig, A. 1957. Some medical implications of the 1954 Wisconsin state fair hearing survey. American Academy of Opththalmology and Otolaryngology Transactions. 61:160-171.

6. Jansen, G., S. Rosen, J. Schulze, D. Plester and A. El-Mofty, 1964. Vegetative reactions to auditory stimuli: comparative studies of subjects in Dortmund, Germany, and the Mabaan Tribe

in the Sudan.   Transactions of the American Academy of Opthal-
mology and Otolaryngology. 68:445-455.

7. Glorig, A. and J. Nixon, 1960.   Distribution of hearing loss
    in various populations.   Annals of Otology, Rhinology and
    Laryngology. 71:727-743.

8. Lawrence, M., G. Gonzalez and J. E. Hawkins, 1967.   Some
    physiological factors in noise-induced hearing loss.   American
    Industrial Hygiene Association Journal. 28:425-430.

9. Friedman, M., S.O. Byers and A.E. Brown, 1967.   Plasma lipid
    responses of rats and rabbits to an auditory stimulus.
    American Journal of Physiology.   212:1174-1178.

10. Rosen, S., D. Plester, A. El-Mofty and H. V. Rosen, 1964.
    Relation of hearing loss to cardiovascular disease.   Trans-
    actions of the American Academy of Opthalmology and Otolaryn-
    gology. 68:433-444.

11. Rosen, S. and P. Olin, 1965.   Hearing loss and coronary
    heart disease.   Archives of Otolaryngology. 82:236-243.

RELATION BETWEEN TEMPORARY THRESHOLD SHIFT AND

PERIPHERAL CIRCULATORY EFFECTS OF SOUND

GERD JANSEN

Institut fur Hygiene und Arbeitsmedizin

Essen der Rhur-Universitat Bochum, West Germany

It has long been known that there are relations between hearing and the function of various organs of the human body (1,2,4,9,
10). Lawrence (8) discussed the possibilities of inducing temporary threshold shift (TTS). His results show that TTS occurs
not only with noise but also with changes in the nervous processes
within the cochlear nerve that are influenced by other factors than
noise. From other authors, too, we know relationships between
the hearing organ and some other extraaural functions. But when
looking up the literature for quantitative relations of TTS and
certain physiological functions we failed in finding exact results.

We know the expected number of decibels of TTS caused by a
certain intensity of noise (13). On the other hand we also know
the degree of circulatory system changes (vegetative reactions =
VR). Our problem to be solved is to find the quantitative relation
between TTS and VR.

METHODS

1. For our experiments we used a white noise of 105 dB(A)
applied by loud-speakers during a period of 8 min. After this
we determined the TTS at 4000 cps 2 min after the end of the noise
period.

2. As indicator of the VR we used the changes of the blood
volume at the finger tip, as measured with a strain gauge (finger-
pulse-amplitude - FPA).

3. Noise induces a higher degree of sympathicotonus in the

vegetative nervous system. Hildebrandt (3,4) has described a new method to pick up the initial value of vegetative tonus by using pulse rate and breathing frequency. The quotient of these two parameters (PAQ) is, in most cases, 4 for a lying person; if sitting or erect the PAQ is somewhat greater, for example 5. Within a group of tested persons there are always existing differences in PAQ. Persons with a PAQ of 7 or 8 are regarded as of high sympathicotonus, whereas persons with parasympathicotonus have a PAQ of 2.5 to 5.

ORDER OF TESTS

The scheme of the range of tests is described in Figure 1. In the beginning of the whole test period we proved the hearing of our persons by an audiogram. Only persons with a normal hearing were examined. We tested 50 persons with a normal hearing. After the first audiogram the persons had a few minutes rest, after which we began recording the FPAs with the subject sitting in a comfortable chair in a sound proofed room. We regarded the first 10 min as a reference period or quiet period (q). Then we applied a white noise of 90 dB(A) for half a minute. After the following 2 min of the second q-period the white noise of 105 dB(A) was applied for 8 minutes. After this we determined the TTS at 1, 2, 4, 6, 8 kcps, but TTS at 4 kcps was determined exactly 2 min past the end of noise application.

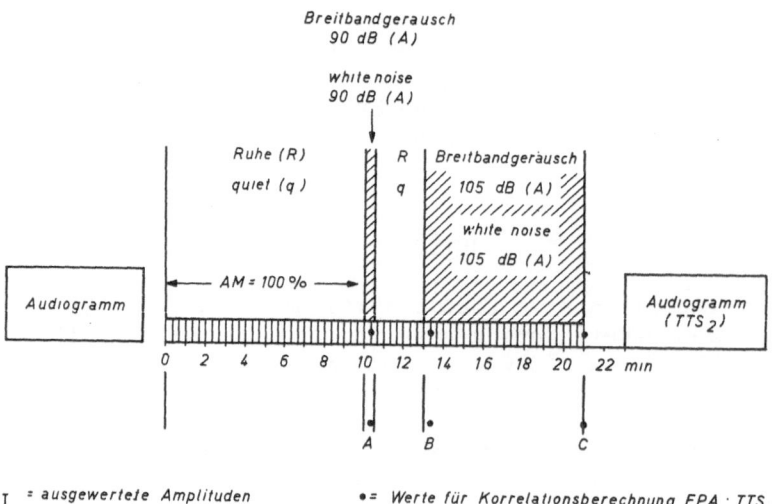

Figure 1. Anordnung der versuchsreihe beziehung zwischen finger – puls – amplitude und TTS   (Order of tests relation between finger – pulse – amplitude and TTS).

EVALUATION OF TESTS

The audiograms were evaluated by taking the difference between the pre- and post-test values (TTS$_2$). The pulse-amplitudes
were measured every 20 sec from the beginning to the end of the
registration. For the assessment of noise effects on VR we used
the points A (20 sec past beginning of white noise 90 dB(A), B
(20 sec past beginning of white noise 105 dB(A) and C (end of
noise period) as shown in Figure 1.

The pulse rate was calculated from the recorded pulse amplitude, the breathing rate from the registered curves picked up by
strain gauge within the first 10 min (quiet-period). The quotient
PAQ was found by dividing pulse rate by breathing rate within the
first 10 min.

RESULTS

The audiograms before the tests showed that we had chosen 50
normal hearing persons. The TTS in these 50 persons showed an
average value of 19.7 dB, which is in accordance with the predictable TTS of 105 dB for 8 min. After a separation of TTS$_2$ at 4000
cps we found in the better hearing ear an average TTS of 16.3 dB
and of 23.1 dB in the ear with greater TTS. For further calculations we used the smaller and the greater TTS as well. This
is shown in Figure 2 (TTS : FPA (b)) where the regression equations
and the correlation-coefficients are given.

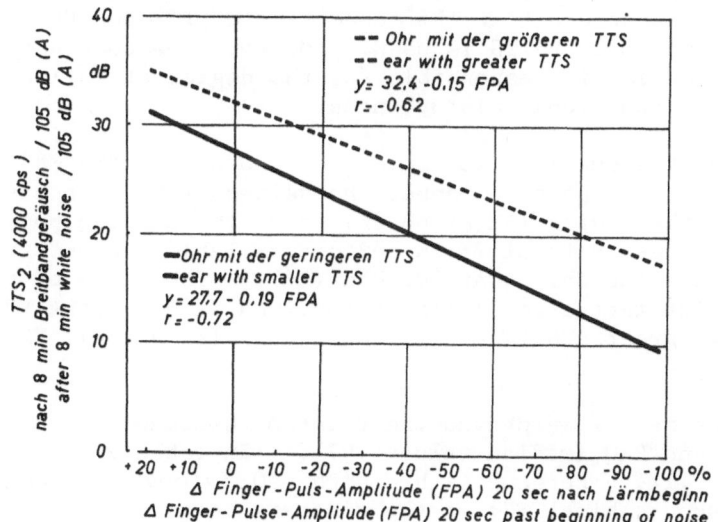

Figure 2. Beziehung zwischen finger - puls - amplitude und TTS
(Relation between finger - pulse - amplitude and TTS).

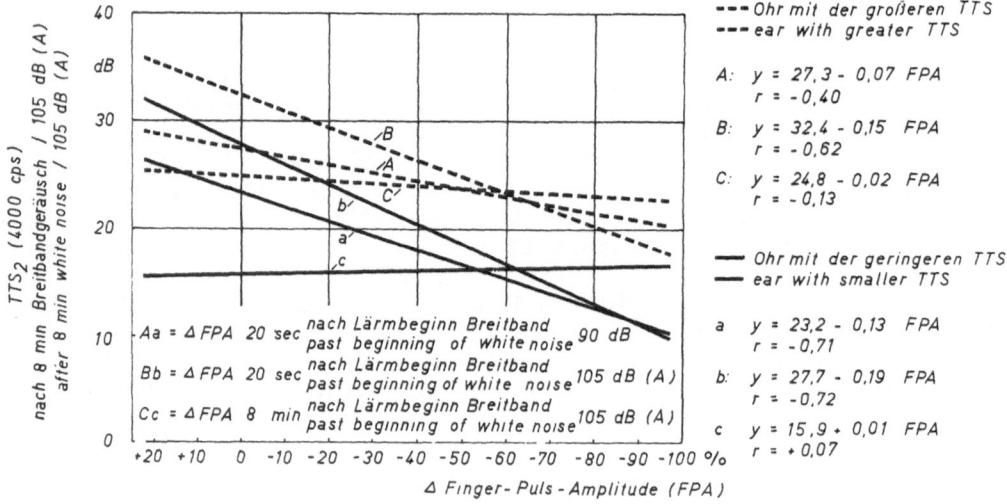

Figure 3. Finger - puls - amplitude und TTS.

In the two noise periods we saw the typical influence of noise on the pulse-amplitudes. The amplitudes were decreased at 90 dB after a short latency of 2 to 3 sec beyond commencement of the noise, and there was a significantly greater influence in the second noise burst (105 dB, 8 min). As shown in Figure 3, there is a correlation between TTS_2 and FPA in A and in B whereas in TTS_2 and point C there is no correlation. The equations and correlations were calculated for the smaller TTS and the points mentioned above (minuscles) and for the greater TTS and the points mentioned above (maiuscles). We conclude that the connection of TTS and FPA is closer if only the smaller TTS is used. Moreover we conclude that the better hearing ear is responsible for the degree of influence on the vegetative functional disturbances.

We split up the smaller TTSs into two groups from the median value (16.3 dB). In the same manner the values of VR were divided above and below the median (AM of 68.8%), e.g. the average pulse-amplitude was decreased to 31.2% in point B. Table 1 gives the results when using the Chi-Square-method (X^2). The value of p = 10.6% gives a hint that a small TTS is combined with a great VR in one person, v.v. great TTS is always connected with a small VR in another person.

In the same manner we proved the relation between TTS (median value 16.3 dB) and PAQ (median value 4.97). In Table 2 the X^2 is calculated: p < 0.1%. Persons with a high PAQ (sympaticotonus). in most cases, show a high TTS; on the contrary persons with a low PAQ (parasympathicotonus), in most cases, show a small TTS.

Table 1. Relation between smaller
TTS (4000 cps) and VR (FPA in B).

	VR AM	VR AM	AM
TTS AM	8	19	27
TTS AM	12	11	23
	20	30	50

Table 2. Relation between smaller
TTS (4000 cps) and PAQ.

	PAQ AM	PAQ AM	AM
TTS AM	23	4	27
TTS AM	9	14	23
	32	18	50

DISCUSSION

As mentioned in the introduction, noise affects the sympathetic part of the vegetative nervous system. In former experiments with short noise bursts and long noise periods during the natural sleep in the night we could prove this fact of vegetative disturbances(5); within the different stages of sleep we never saw an increase of pulse-amplitudes as it is sometimes seen in an awakened person on a high level of sympathicotonus. Wilder (14) describes the importance of the law of initial value in the vegetative nervous system. Persons in a high sympathicotonus with a high level of vasoconstriction often react with a very small or no decrease in FPA and sometimes they react with a decrease in FPA (vasodilatation).

Moreover, we saw in the sleep, as a parasympathicotropic phase, VR occurring at intensities of 12 to 15 dB below the intensity that produced VR in awakeness. This means that noise is more effective in a parasympathic phase concerning the pulse-amplitudes. On the other hand, this is an additional proof for noise as a sympathicotropic influence.

In a number of test series we got the same results when recording the dilatation of pupil area. We never saw a constriction of the pupil by noise. The question of a conditioned reflex with noise and light may not be discussed at this place. The basic reaction caused by noise is a sympathetic one.

Other experiments with a vasodilatative remedy (450 mg xanthinol-nicotinat = Complamin) resulted in a partial compensation of noise effects (6). With the same doses we saw a significant but very small positive influence on the audiograms in quiet phase and after a broad band masking of 110 dB for 15 min at 4000 cps.

We have to decide now if the TTS in our experiments is caused by a vascular or by a nervous influence. As we have heard from Lawrence (8), nervous changes may cause or influence TTS. A similar conclusion may be drawn from experiments of Plester (11). He examined the influence of drugs affecting nervous and vasal functions. He worked with Hydergin and Gynergen. He saw a significant influence on adaptation with respect to hearing fatigue and a more rapid recovery of hearing with Hydergin. This effect he attributed to the sympathicolytic and not the vasoactive influence of Hydergin. After the application of Gynergen, which is sympathicolytic and vasoconstrictive in the periphery, he saw a positive influence on adaptation and recovery of hearing functions. These results have led Plester (11) to the conclusion that the inner ear functions are regulated by the vegetative nervous system in quiet periods. This equilibration is changed into sympathicotropic direction by noise. Tanner (12) also dealt with the problem of the vegetative regulated functions in the inner ear. His experiments resulted in the conclusion that a higher level of parasympathicotonus is much more positive for the inner ear. By inhibition of the sympathetic effect a better circulation in the inner ear is occurring. These investigations and others make it possible to draw the conclusion that persons with a high sympathicotonus are much more susceptible to negative noise influences than persons with a high parasympathicotonus.

CONCLUSIONS

Comparing the experiments of Plester (11), Tanner (12) and others with the results of our own, it seems to be proved that TTS and VR are functions influenced by the vegetative system. This is the reason that there must be a connection between the degree of changes of these two functions (TTS and FPA). This statement implies that the limits for hearing loss and vegetative injuries have to be found around the same noise intensities. The limits for extraaural noise effects (VR) established in former experiments (11) show a critical curve of 94 dB (for narrow bands) and of 88 dB (for broad bands) at 1000 cps with a decline of 1 dB/octave. The negative effects of noise cannot be principally different upon TTS and VR. This means that a man who will not have a hearing loss from high intensity noise is, nevertheless, highly endangered by the extraaural influences of high intensity noise.

The recommendation of the various national standards associations for allowable noise exposure only refer to TTS with respect to hearing loss; but these limits are valid for VR as well.

SUMMARY

With 50 normal hearing persons we examined temporary threshold
shift (TTS), finger-pulse-amplitude (FPA) which is an indicator of
vegetative reactions (VR) and the initial value of vegetative nervous system by calculating the quotient of pulse rate and breathing
frequency (PAQ).   There was existing a correlation between FPA and
TTS_2 at 4000 cps and between PAQ and TTS_2 at 4000 cps as well.
Great TTS and small FPA occurs in persons with a high PAQ.   On
the other hand persons with small TTS showed a great FPA and belong
to the group with small PAQ.   From these and many other experiments
we have to conclude that TTS and FPA are influenced by the nervous
system, and that noise changes the vegetative reactions in the direction of sympathicotonus.   This means that limits for hearing loss
and for extraaural injury are close connected and that persons
suffering from high noise may be endangered either by hearing loss
or by extraaural injury.

REFERENCES

1. Grandjean, E., 1960.   Physiologische und psychologische
   Wirkungen des Lärms. Mensch und Umwelt, Documenta Geigy
   4:13-42.

2. Grandjean, E., 1969.   Effects of noise on man. In: W. D.
   Ward and J. E. Fricke (Eds.), Noise as a Public Health Hazard,
   American Speech and Hearing Association, Washington, D.C.,
   Report 4:99-102.

3. Hildebrandt, G., 1965.   Die Koordination von Puls- und Atem-
   frequenz bei der Arbeit.   Internationale Zeitschrift für
   Angewandte Physiologie einschliesslich Arbeitsphysiologie
   (Berlin) 21:27-48.

4. Jansen, G., 1967.   Zur nervösen Belastung durch Lärm.
   Beihefte zum Zentralblatt für Arbeitsmedicin und Arbeitsschutz
   9:1-75.

5. Jansen, G., 1969.   Effects of noise on physiological state.
   In: W. D. Ward and J. E. Fricke (Eds.), Noise as a Public
   Health Hazard, American Speech and Hearing Association, Wash-
   ington, D.C., Report 4:99-102.

6. Jansen, G., W. Klosterkotter und R. Reineke, 1969.   Experimentelle Untersuchungen zur Kompensation lärmbedingter Gefäßreaktionen.
   Arbeitsmedicin, Sozialmedicin, Arbeitshygiene 29:303-330.

7. Jansen, G. and J. Schulze, 1964.   Beispiele von Schlafstörungen
   durch Gerausche.   Klinische Wochenschrift (Berlin) 42:132-134.

8. Lawrence, M., 1960.   Some physiological correlates of noise in-
    duced hearing loss.   Industrial Medicine and Surgery 29: 445-452

9. Lehmann, G. und J. Tamm, 1956.   Die Beeinflussung vegetativer
    Funktionen des Menschen durch Geräusche.   Forschungsbericht des
    Wirtschafts und Verkehrsministeriums Nordrhein Westfalen 257:
    1-37.

10. Maugeri, S.   Effetti biologici del rumore industriale.   Fors-
    chungsbericht 350 der EGKS, Hohe Behörde Luxemburg, Patologia
    da Rumore.   II Progresso Medico 17 (13):423-429.

11. Plester, D., 1953.   Der Einfluß vegetativ wirksamer Pharmaka
    auf die Adaptation bzw.   Hörermüdung.   Archiv für Ohren-Nasen-
    und Kehlkopf Heilkunde 162:473-478.

12. Tanner, K., 1955.   Über Hörermüdung und akustisches Trauma
    und deren Beeinflussung durch vegetativ wirksame Pharmaka.
    Acta Oto-Laryngologia 45:65-81.

13. Ward, W.D., 1966.   Temporary threshold shift in males and
    females.   Journal of the Acoustical Society of America 40:
    478-485.

14. Wilder, J., 1967.   The law of initial value.   In: J. Wilder
    (Ed.), Stimulus and response, John Wright & Sons, ltd., Bristol.
    pp. 1-91.

CARDIOVASCULAR AND BIOCHEMICAL EFFECTS OF CHRONIC INTERMITTENT NEUROGENIC STIMULATION

JOSEPH P. BUCKLEY and HAROLD H. SMOOKLER

Department of Pharmacology
University of Pittsburgh School of Pharmacy
Pittsburgh, Pennsylvania

Investigators have called attention to the unresolved psychologic conflicts which give rise to chronic emotional tension which they believe are specific etiological factors in the development of such diseases as hypertension, coronary vascular disease and gastrointestinal ulcers in man (1-3). Selye (4-6) has postulated that organisms subjected to alarming stimuli will respond in a given manner, which he termed "the general adaptation syndrome" or "stress syndrome." Barry and Buckley (7) have reviewed in detail the effects of exposing animals to stress upon their behavioral performance as well as physiological, biochemical, and endocrinological function. Various types of stimuli can act as the systemic stressors and produce all three stages of the general adaptation syndrome. Audiogenic stimuli have been found to induce cardiovascular changes (8,9), changes in adrenal weights and adrenal ascorbic acid (10) and death (11). Herrington and Nelbach (12) exposed rats to a complex type of stimulation including audiogenic and cage vibration and observed changes in the experimental animals similar to those produced by thyroid overdosage. Hudak and Buckley (13) modified this procedure and subjected rats to a combination of audiogenic stimuli, visual stimuli, and motion. These animals became hypertensive, irritable, and exhibited excessive salivation, lacrimation, urination, and defecation. Buckley et al. (14) demonstrated that daily doses of chlorpromazine (4 mg/kg) and reserpine (0.1 mg/kg) failed to prevent the development of the hypertensive state and reported that the lethal effects of the environmental stressors were markedly potentiated by these tranquilizers. Rosecrans et al. (15) found that there was a maximal elevation in urinary catecholamines in rats following a single 4-hour exposure to the combined audiogenic, visual, and motion stress; however, the levels returned to control levels within the fourth week of continued exposure to the stressors. In addition, both adrenal and plasma corticosterone levels markedly increased following a single exposure but returned to control levels by the eighth week of stress.

The present studies were undertaken to obtain a more detailed understanding of the time course of certain biochemical processes that occur during prolonged exposure to intermittent neurogenic stimulation and to correlate these effects with certain physiological alterations of the cardiovascular system.

MATERIALS AND METHODS

The stress chambers utilized were three semi-soundproof rooms (Industrial Acoustics Co., New York City) having inner dimensions of 4 ft. x 4 ft. x 6 ft. with one 12-inch speaker installed in the ceiling and four 6-inch speakers on the walls at cage height. One hundred and fifty-watt spotlights were installed in each corner and adjusted so that the intensity of the light at the cage site was approximately 140-foot candles, and the cages were placed on an Eberbach shaker.

The following stressors were utilized in this study: (1) Audiogenic stimulation was produced by tape recordings of noxious sounds (compressed air blasts, bells, buzzers, and tuning fork impulses for one-half minute periods at 5-minute intervals). The intensity of stimulation at the center of the cage was approximately 100 decibels. (2) Flashing lights were on for one-quarter second and off for three-quarter second with two lights on simultaneously, so that there was one-half second of complete darkness. (3) Motion stimuli were produced by oscillation of the cage at a rate of 140 per minute.

A control unit was utilized so that a variable 4-hour sequence of stressors applied singly or in various combinations could be automatically programmed into the three stress rooms, and the animals were subjected to the stressors three days per week on a randomized schedule between Monday and Saturday of each week. Each study utilized between 120 and 260 male albino Wistar and Sprague-Dawley rats weighing between 60 and 100 grams when received in our laboratory. The animals were housed under standard laboratory conditions for a period of four weeks prior to being subjected to the stress program. The animals were numbered, and, during this period, were recorded periodically and their systolic blood pressures were determined on at least six different occasions using a photoelectric tensometer (Metro Scientific, Inc.). Most of the studies used six groups of rats with 20 rats per group as follows: Group I, stressed-drug treatment #1; Group II, stressed-drug treatment #2; Group III, stressed-nontreated; Group IV, nonstressed-drug treatment #1, Group V, nonstressed-drug treatment #2; Group VI, nonstressed-nontreated.

Chemical Assays

Brain Catecholamines.   Brain norepinephrine (NE) and dopamine (DA) were determined according to the method of Brodie et al.   (16) with the samples being oxidized according to the method of Chang (17) and fluorescence being measured on an Aminco-Bowman spectrophotofluorometer.

Serum Free Fatty Acids (FFA).   Serum free fatty acids were determined colorimetrically according to the method of Novak (18).

Serum Corticosterone.   Serum corticosterone levels were estimated fluorometrically using the method of Guillemin et al.   (19) to extract the corticosterone from the serum and fluorescence developed using the reagent described by Zenker and Bernstein (20).

RESULTS

Male albino Sprague-Dawley rats weighing approximately 60 grams when received by our laboratory consistently developed hypertensive states when subjected to chronic intermittent neurogenic stimulation. Preliminary studies indicated that cardiovascular pathology occurred in less than 50% of the batches of rats that weighed more than 100 grams at the time of receipt by our laboratory; all batches of rats that weighed between 60 and 80 grams at the time of receipt by the laboratory developed systolic hypertension within approximately eight weeks when subjected to the experimental stressors.   Concomitant with the development of hypertension, the stressed rats demonstrated left ventricular hypertrophy and marked vacuolization of the three layers of the adrenal cortex with severe congestion of the sinusoids of the zona reticularis.

Reserpine, 0.1 mg/kg, i.p., daily not only failed to decrease the pressor effects induced by the stressors but also appeared to potentiate the lethal effects of experimental stress.   The reserpine treated animals also showed marked alterations in the adrenal cortex with zones of degeneration and possible necrosis in the outer layers of the zone fasciculata.

Two hundred and sixty male albino Sprague-Dawley rats were utilized to correlate alterations in serum corticosterone levels, serum FFA, brain NE, brain DA, and systolic blood pressure.   The mean systolic blood pressures of the stressed group rose progressively during the first eight weeks (Fig. 1), reached a peak of 150 mm Hg $\pm$ 1.01 by the eighth week and ranged between 150 and 160 mm Hg for the remaining 12 weeks of exposure to the experimental stressors.   The systolic blood pressures of the nonstressed control group ranged between 110 and 120 mm Hg throughout the same period.   The mean blood pressures of the stressed group were significantly higher

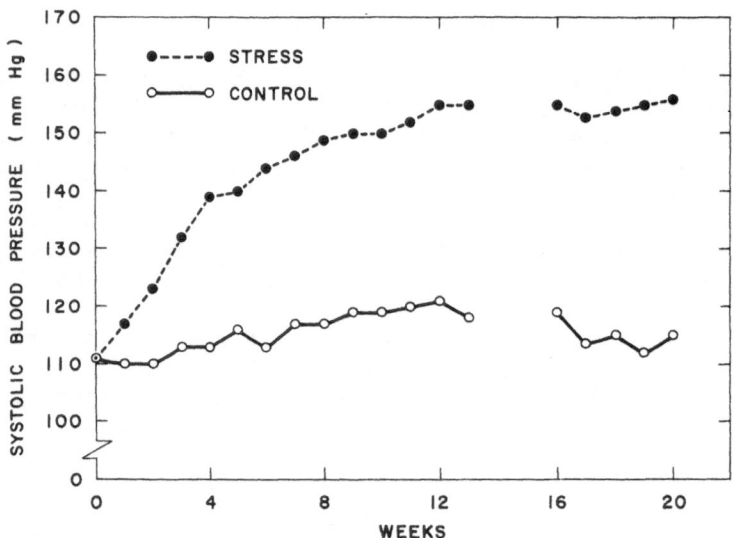

Figure 1. Effects of environmental stress on mean systolic blood pressure of rats (from Smookler, H.H. and Buckley, J.P., 1969, Int. J. Neuropharmacol. 8:33–41).

(p $<$ 0.01) than controls from the first week until the termination of the study.   There was approximately a three-fold increase in serum corticosterone (p $<$ 0.01) in the stressed animals during the first four weeks of exposure (Fig. 2).   By the end of week 5, the corticosterone levels of the stressed group markedly declined and were actually significantly lower than the values obtained in the

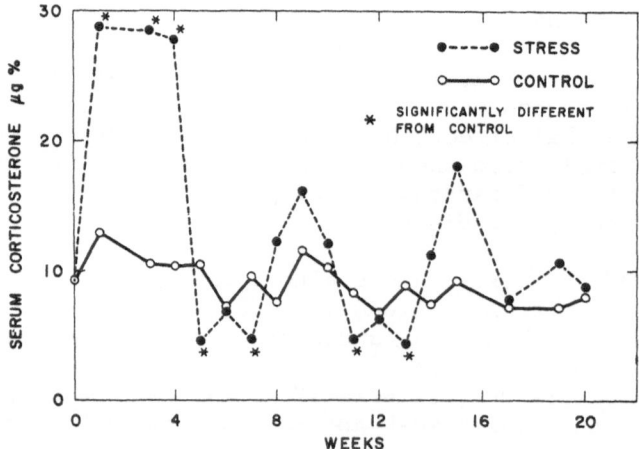

Figure 2.   Effects of chronic intermittent neurogenic stimulation on serum corticosterone levels in rats (from Smookler, H.H. and Buckley, J.P., 1969, Int. J. Neuropharmacol. 8:33–41).

control group (p < 0.05).   Thereafter, serum corticosterone levels
exhibited a cyclic pattern at six-week intervals with the trough of
the cycle at weeks 5, 6 and 7 and weeks 11, 12 and 13 with the peak
values at weeks 9 and 15.   There was no significant alterations in
brain NE and DA and serum FFA throughout the 20 weeks of exposure
to intermittent neurogenic stimulation.   Since the levels of catecholamines in brain represent a dynamic equilibrium between synthesis
and efflux, stress-induced alteration in brain amine utilization may
not have been observed due to rapid replenishment resulting from an
increase in synthesis.   The determination of brain NE and DA synthesis rate would provide a more significant indication in stress-
induced changes in central sympathetic activity.

     Stress-induced changes in turnover of brain catecholamines
were estimated by the difference in the degree of the depletion
of NE and DA after administration of alpha-methyltyrosine, an inhibitor of tyrosine hydroxylase, the rate limiting enzyme in
catecholamine biosynthesis.   The alpha-methyltyrosine was suspended
in 0.5% pectin solution and administered orally in a dose of 100
mg/kg.   The animals were treated on the day prior to stress and
again one hour prior to exposure to the experimental stressors.
Groups I and II, respectively, received the drug and the vehicle
prior to being stressed; Groups III and IV, respectively, received
drug and vehicle and served as controls.   Four animals from each
group were sacrificed on alternate weeks for a period of 14 weeks.
There was a 40 to 50% depletion in brain NE in the non-stressed
treated rats whereas more than 80% of brain NE was depleted in the
stressed treated group during the first four weeks of exposure.
(Figures 3 and 4).   By the end of week 6 to the termination

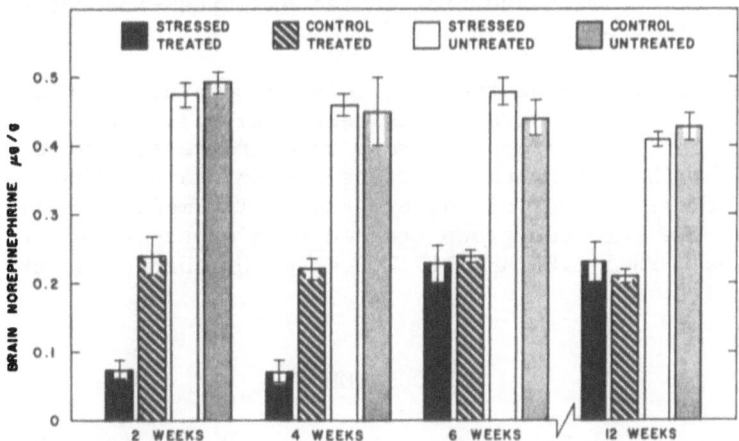

Figure 3.   Effects of stress on depletion of brain norepinephrine
by alpha-methyltyrosine, 100 mg/kg, p.o.   (from Smookler, H.H. and
Buckley, J.P., 1969, Int. J. Neuropharmacol. 8:33-41).

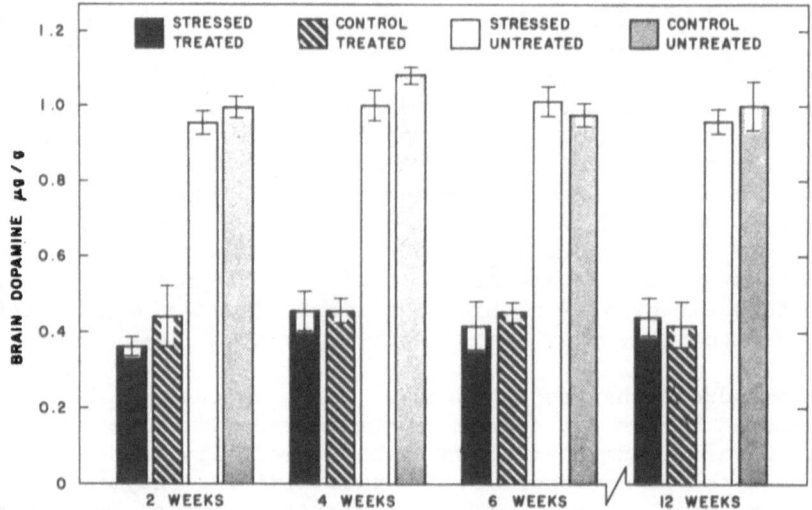

Figure 4.   Effects of stress on depletion of brain dopamine by
alpha-methyltyrosine, 100 mg/kg, p.o.   (from Smookler, H.H. and
Buckley, J.P., 1969, Int. J. Neuropharmacol. 8:33–41).

of the study at week 20, the per cent depletion of brain NE was
not significantly different between the alpha-methyltyrosine treated, stressed and unstressed groups.   In contrast, the per cent
depletion of brain DA by alpha-methyltyrosine was unaffected by
the stress procedure throughout the 20 weeks (Fig. 4).   These
data indicate a significant increase in turnover of brain NE during
the first four weeks of exposure.   The increase in NE turnover
closely correlated with the time course of stress-induced elevation
in serum corticosterone.

      Alpha-methyltyrosine treatment also prevented the production
of hypertension in the stressed animals (Fig. 5).   The blood pressures of the untreated stressed group rose progressively to 151 ±
2.01 mm Hg by the 10th week of stress and maintained that level
throughout the remainder of the program.   The mean systolic blood·
pressure of the stressed group treated with alpha-methyltyrosine
ranged between 98 ± 1.85 and 117 ± 2.41 mm Hg throughout the same
period of time.

                              DISCUSSION

      The exposure of male Wistar or Sprague-Dawley rats to chronic
intermittent neurogenic stimulation consistently produces hypertensive states accompanied by left ventricular hypertrophy and vacuolization of all three zones of the adrenal cortex.   Concomitant
with the rise in blood pressure, serum corticosterone levels increased

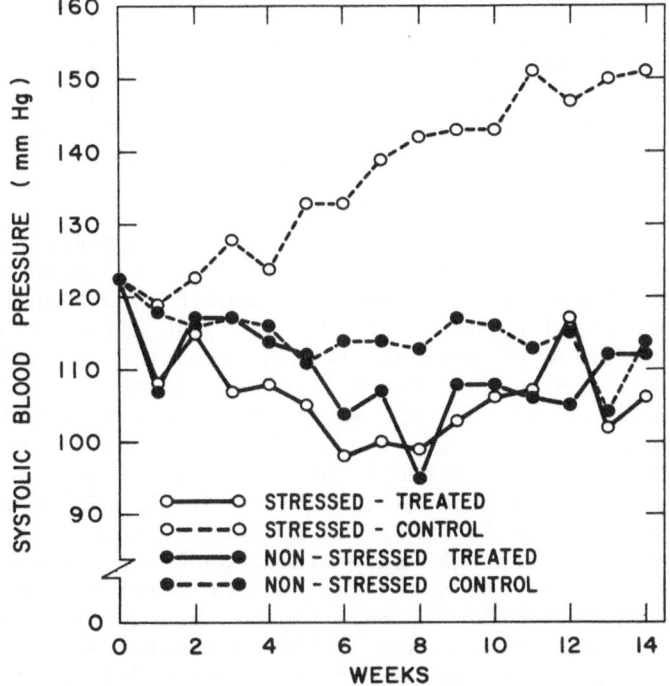

Figure 5. Effects of chronic intermittent neurogenic stimulation on the mean systolic blood pressure of rats treated with alpha-methyltyrosine, 100 mg/kg, p.o. (from Smookler, H.H. and Buckley, J.P., 1969, Int. J. Neuropharmacol. 8:33-41).

almost three-fold during the first four weeks of stress exposure. Since there was also a marked increase in brain norepinephrine turnover during the first four weeks of exposure to the stressors, these data suggest a close temporal relationship between the elevation in brain NE synthesis and stimulation of the anterior pituitary-adrenal cortical axis in response to environmental stressors. It appears that exposure to the experimental stressors activates central sympathetic centers releasing norepinephrine which is rapidly replenished by increased synthesis eliciting a stimulation of the anterior pituitary-adrenal cortical complex.

Alpha-methyltyrosine prevented the stress-induced elevation in systolic blood pressure. The compound reduces the level of catecholamines by inhibiting synthesis but leaves the integrity of the binding system intact. Since the reuptake of liberated amines is the major process of amine inactivation (21), the combination of a reduced amount of liberated transmitter plus rapid inactivation by an intact binding system may serve to dampen the stress-induced activation of central sympathetic centers. Reserpine, chlorpromazine, pentobarbital, morphine and acetylsalicylic

acid have failed to attenuate the development of the hypertensive
state induced by exposure to the environmental stressors and, in
many instances, potentiated the lethal effects of the stressors.
Alpha-methyltyrosine has been the only compound to date that we
have found to be capable of preventing the progressive development
of systolic hypertension induced by chronic exposure to intermittent neurogenic stimulation.

The response of the living organism to intermittent exposure
to environmental stressors can be divided into two distinct phases:
(1) the "acute" phase which involves a stimulation of central sympathetic centers and liberation of ACTH from the anterior pituitary
resulting in elevation of adrenal steroid secretion and (2) the
"chronic" phase which is associated with a reduction in the elevated
serum corticosterone and central sympathetic nervous activity to
pre-stress levels. The activation of central sympathetic centers
and the anterior pituitary adrenal cortical axis is important in
the initial phase of the stress reaction. Smelik (22) hypothesized
that the stress response is initiated by activation of hypothalamic
centers through afferents from the rhinencephalic or mesencephalic
limbic systems and that the posterior and the anterior regions of
the hypothalamus, respectively, mediate the subsequent phase of the
neuroendocrine emergency response. The data obtained in these
studies tend to support this hypothesis as related to the acute
phase of the physiological, biochemical, and endocrinological responses of the organism to environmental stressors. The factors
responsible for the progressive alterations in the cardiovascular
system involving the chronic phase are yet to be determined. These
data suggest that alpha-methyltyrosine may be a valuable therapeutic
agent if used in the very early stages of essential hypertension
which apparently involve a marked increase in a sympathetic outflow
as well as stimulation of the release of adrenal cortical steroids
without definite renal pathology. In these cases, the compound
may prevent the progression of the disease which eventually leads
to renal and cardiovascular pathology and sustained hypertension.

SUMMARY

Rats were subjected to a combination of audiogenic, visual and
motion stimuli in an attempt to better understand the physiological,
biochemical and endocrinological sequence of events which may contribute to the development of environmentally induced diseases.

Hypertension is consistently induced. The maximum increase
in blood pressure usually occurs within the first 8 to 12 weeks,
and serum corticosterone levels are increased almost threefold
during the first four weeks of stress exposure; however, the steroid
levels rapidly decline by the end of the fifth week. There is an

elevated turnover of brain norepinephrine which returns to control
levels by the end of the fourth to sixth week.   These data suggest
that the stress syndrome may be triggered by activation of central
sympathetic activity and stimulation of the anterior pituitary
adrenal cortical axis.   Pathology observed includes left ventri-
ular hypertrophy and hypertrophy of zona-fasciculata of the adrenal
cortex.

Such tranquilizers as reserpine and chlorpromazine not only
failed to prevent the effects of chronic exposure to these variable
stressors, but, to the contrary, markedly potentiated their lethal
effects.

Alpha-methyltyrosine, an inhibitor of tyrosine hydroxylase,
the rate-limiting enzyme for catecholamine biosynthesis, prevented
the development of hypertension in the animals subjected to the
variable stress program; it may be a valuable therapeutic agent
if used in the very early stages of essential hypertension.

REFERENCES

1. Weiss, E., 1942. Psychosomatic aspects of hypertension.
   Journal American Medical Association 120:1081-1086.

2. Alexander, F., 1939. Emotional factors in essential hypertension. Psychosomatic Medicine 1:173-179.

3. Saul, L.J., 1939. Hostility in cases of essential hypertension.
   Psychosomatic Medicine 1:153-161.

4. Selye, H., 1936. A syndrome produced by diverse nocuous agents.
   Nature, 138:32.

5. Selye, H., 1946. The general adaptation syndrome and the
   diseases of adaptation. Journal of Clinical Endocrinology
   6:117-230.

6. Selye, H., 1955. Stress and disease. Science 122:625-631

7. Barry, H., 111 and J.P. Buckley, 1966. Drug effects on animal
   performance and the stress syndrome. Journal of Pharmaceutical
   Science 55:1159-1183.

8. Selye, H., 1943. On the production of malignant hypertension
   by chronic exposure to various damaging agents. Revue Canadian
   Biologie 2:501-505.

9. Medoff, H.S. and A.M. Bongiovanni, 1945. Blood pressure in rats
   subjected to audiogenic stimulation. American Journal of
   Physiology 143:300-305.

10. Jurtshuk, P., A.S. Weltman, and A.M. Sackler, 1951. Biochemical responses of rats to auditory stress. Science 129:1424–1425.

11. Gupta, O.P., H. Blechman, and S.S. Strahl, 1960. Effects of stress on the periodontal tissues of young adult male rats and hamsters. Journal of Periodontology 31:413–417.

12. Herrington, L.P. and J.H. Nelbach, 1942. Relation of gland weights to growth and aging process in rats exposed to certain environmental conditions. Endocrinology 30:375–386.

13. Hudak, W.J. and J.P. Buckley, 1961. Production of hypertensive rats by experimental stress. Journal of Pharmaceutical Science 50:263–264.

14. Buckley, J.P., H. Kato, W.J. Kinnard, M.D.G. Aceto, and J.M. Estevez, 1964. Effects of reserpine and chlorpromazine on rats subjected to experimental stress. Psychopharmacologia 6:87–95.

15. Rosecrans, J.A., N. Watzman, and J.P. Buckley, 1966. The production of hypertension in male albino rats subjected to experimental stress. Biochemical Pharmacology 15:1707–1718.

16. Brodie, B.B., M.S. Comer, E. Costa, and A. Dlabac, 1966. The role of brain serotonin in the mechanism of the central action of reserpine. Journal of Pharmacology and Experimental Therapeutics 152:340–349.

17. Chang, C.C., 1964. A sensitive method for spectrophotofluorometric assay of catecholamines. International Journal of Neuropharmacology 3:643–649.

18. Novak, M., 1965. Colorimetric ultramicro method for the determination of free fatty acids. Journal of Lipid Research 6:431–433.

19. Guillemin, R., G.E. Clayton, H.S. Lipscomb, and J.D. Smith, 1959. Fluorometric measurement of rat plasma and adrenal corticosterone concentration. Journal of Laboratory and Clinical Medicine 53:830–832.

20. Zenker, N.E. and D.F. Bernstein, 1958. The estimation of small amounts of corticosterone in rat plasma. Journal of Biological Chemistry 231:695–701.

21. Hertting, G., J. Axelrod, I.J. Kopin, and L.G. Whitby, 1961. Lack of uptake of catecholamines after chronic denervation of sympathetic nerves. Nature, Lond. 189:66.

22. Smelik, P.G., 1969. The regulation of ACTH secretion. Acta physiologica et pharmacologica Neerlandica 15:123–135.

CARDIOVASCULAR AND TERATOGENIC EFFECTS OF CHRONIC INTERMITTENT

NOISE STRESS

WILLIAM F. GEBER
Department of Pharmacology
Medical College of Georgia
Augusta, Georgia 30902

Correlating the reactions of the various components of the
cardiovascular system to the stress of noise with the induction
of abnormal developmental pathways in the mammalian fetus by exposure of the pregnant animal to environmental noise is, at best,
a difficult and multidimensional problem.

This correlation will be discussed here by considering the
following factors separately: the effects of noise upon the cardiovascular system; the relation of maternal blood flow alteration in
the uterus to fetal development; and the role of maternal exposure
to noise stress to fetal abnormalities.

A rather extensive literature has developed over a period of
years which details the many effects that exposure to noise can
generate within the body of both man and animals.   Acute responses
to noise are extremely varied and complex and are beyond the scope
of this article.   While the responses to a single or limited number of noise events may be annoying, they probably do not constitute
a significant threat to the overall homeostasis of the body.   This
does not say these responses do not play a vital role in the bodily
response to a noise event if that event is intermittent or more or
less continuous.   Studies by a number of authors have indicated
the wide range of biochemical systems and organ tissues that are
affected by noise.   The work of Sackler et al.(1, 2), Biro et al.
(3-5), Anthony and co-workers (6, 7), Geber et al. (8-10), and
Anderson et al. (11) has demonstrated conclusively the response of
the hypophyseal-adrenocortical axis to noise stress in animals.

The studies of Gregorczyk et al.(12) and Jozkiewicz et al.(13)
on humans indicated both similarities and differences in response to

noise stress when their results were compared to data obtained from
their animal investigations.

The major emphasis in reports in the literature on the response of the cardiovascular system to noise has been related to
induction of an increased blood pressure, i.e. hypertension.
This effect of sustained noise stress has been reported by Rosecrans et al.(14), McCann et al.(15), and Medoff and Bongiovanni (16)
in animal studies.   Heart weights were not recorded in these
reports.   Using essentially the same type of audiogenic stimulation,
however, Geber et al.(17) demonstrated an increase, i.e. hypertrophy,
in the weight of the ventricles of the heart of rats and rabbits
following two weeks of intermittent noise stress.   Recently, Welch
and Welch (18) reported approximately the same percentage increase
in ventricular weights of mice exposed to another type of limited-
duration daily stress, consisting of 5-10 minutes per day of fighting for 5, 10 and 14 days.   It would appear that stresses of widely
divergent characteristics are capable of producing cardiac hypertrophy.

Studies in the author's laboratory (unpublished) have indicated
that uterine and placental blood flow are not consistently altered
in the same way either qualitatively or quantitatively by noise.
It was demonstrated in the sheep that regional blood flow in the
uterus and placenta were often altered in opposite directions
(increase vs. decrease), and that certain areas responded for
longer periods of time than others in the pregnant female exposed
to pure tones from an audio generator.   Exposure of the ewe to
frequencies ranging from 20 to 650 cycles per second at 95 decibels
produced no measurable uterine blood flow responses, but resulted
in a decrease in placental (cotelydon) flow.   Stimulation with
frequencies of 3000 - 10,000 cycles per second at 95 decibels
resulted in a decrease in blood flow had been reported in an
earlier study (19) utilizing various drugs and pain stress.

Another characteristic of the response of the uteroplacental
blood flow to maternal stress or drug injections noted in the
study by Geber (19) was the fact that a short stress, i.e. pain,
of 5-20 seconds duration was followed by a fetal blood flow response of approximately 15-20 minutes.   Together the facts of
variation in blood flow in different anatomic areas of the uterus,
and the prolonged fetal response to a given stress suggest that
alteration of blood flow (both maternal and fetal) could account
for abnormal fetal development, particularly if it occurred during
critical periods of fetal organogenesis.   The studies of Franklin
and Brent (20), Senger et al. (21), and Wigglesworth (22) clearly
indicated the role that decreased uterine blood flow played in
altering the normal course of events in fetal development.

Many of the previously cited studies utilized anesthetized

animals, or, what are considered by the author, unphysiologic forms of stress. Therefore, in an attempt to minimize these factors, unanesthetized pregnant rats were exposed to audiogenic stress throughout pregnancy for 6 minutes of each hour of each day. Complete loss of litters occurred in approximately 50 per cent of the stressed females, litter size was reduced by an average of two fetuses per litter, and a wide range of developmental abnormalities was observed in the surviving fetuses (23). It was postulated that the etiological factors involved included decreased uteroplacental and fetal blood flow, and the action of stress-released maternal neurohormones, primarily epinephrine, norepinephrine, and adrenal cortical hormones. It is further postulated that the mechanism of action of these two-parameter alterations involved the production of a relative hypoxemia which is known to be teratogenic (24), and disruption of normal fetal metabolic synthetic pathways involving RNA and DNA by the increased level of circulating catecholamines. The disruptive influence of catecholamines on RNA and DNA synthesis has been demonstrated by the studies of Barka (25, 26). Teratogenic activity of the catecholamines was recently presented in the data by Geber (27). Therefore, there is reasonable evidence in the literature for the participation of the two factors of hypoxemia and catecholamines in the etiology of congenital malformations.

The role of environmental noise on the total course of reproduction and fetal development has been evaluated by a number of investigators. Sontag (28) was one of the earliest workers to call attention to the fetal cardiovascular response to maternal audiogenic stress. Several studies by Zondek and Tamari (29-31) described the effect of audiogenic stress on reproduction and fetal development. These workers pointed out that noise stress reduced both the number of fetuses per litter and incidence of pregnancy.

Finally, in an attempt to develop an integrated overview of the effects of noise stress, we may consider the following scheme to describe the pathways or mechanisms involved in the production of adult or maternal stress response in correlation with the induction of abnormal fetal development. The noise stress is received by the auditory apparatus of the adult animal; the various higher brain centers integrate the signals; the hypothalamus and hypophyseal areas are activated; the adrenal cortex and medulla are stimulated to release their respective hormones; the combination of hormones and increased peripheral resistance raises blood pressure in the adult and decreases uterine blood flow, which in turn interferes with proper gaseous, nutrient, and waste product interchange between fetus and mother; the relative hypoxemia along with the catecholamines released by the maternal stress reflect their effects in the fetus by interfering with both normal fetal blood flow and certain biochemical pathways, thereby resulting in an

improper sequencing of developmental events in the fetus during
critical phases of organogenesis.

REFERENCES

1. Sackler, A.M., A.S. Weltman, M. Bradshaw and P. Jurtshuk, 1959.
   Endocrine changes due to auditory stress. Acta Endocrinologica
   31:405-418.

2. Sackler, A.M., A.S. Weltman and P. Jurtshuk, 1960. Endocrine
   aspects of auditory stress. Aerospace Medicine 31:749-759.

3. Biro, J., V. Szokolai and A.G.B. Kovach, 1959. Some effects
   of sound stimuli on the pituitary adrenocortical system.
   Acta Endocrinologica 31:542-552.

4. Biro, J., V. Szokolai and J. Fachet, 1962. Acta Physiologica
   Academiae Scientiarum Hungaricae 22:163-169.

5. Biro, J., 1963. The mechanism of acute eosinophilia. Acta
   Physiologica Academiae Scientiarum Hungaricae 23:105-114.

6. Anthony, A. and S. Babcock, 1958. Effects of intense noise on
   adrenal and plasma cholesterol of mice. Experientia 14:104-105.

7. Anthony, A. and E. Ackerman, 1955. Effects of noise on the
   blood eosinophil levels and adrenals of mice. Journal of the
   Acoustical Society of America 27:1144-1149.

8. Geber, W.F., T.A. Anderson, and V.VanDyne, 1966. Physiologic
   responses of the albino rat to chronic noise stress. Archives
   of Environmental Health 12:751-754.

9. Geber, W.F. and T.A. Anderson, 1967. Ethanol inhibition of
   audiogenic stress induced cardiac hypertrophy. Experientia
   23:734-736.

10. Geber, W.F., T.A. Anderson and B. VanDyne, 1966. Influence
    of ethanol on the response of the albino rat to audiovisual
    and swim stress. Experimental Medicine and Surgery 24:25-35.

11. Anderson, T.A., W.F. Geber, D. Duick, T. Pickard and B. VanDyne,
    1965. Andiovisual stress and blood serum proteins in the
    cholesterol-fed rat. Experimental Medicine and Surgery 23:
    375-382.

12. Gregorczyk, J., A. Lewandowska-Tokarz, J. Stanosek, and J. Hepa,
    1965. Effect of physical work and work under conditions of
    noise and vibration on the human organism. Acta Physiologica

Polonica 16:600–606.

13. Jozkiewicz, S., M. Puchalik, Z. Cygan, M. Drozdz, J. Gregorczyk, J. Grzesik, K. Krzoska, T. Zak, 1965. Studies on the effect of acoustic and ultra acoustic fields on biochemical processes. Acta Physiologica Polonica 16:620–628.

14. Rosecrans, J.A., N. Watzman and J.P. Buckley, 1966. The production of hypertension in male albino rats subjected to experimental stress. Biochemical Pharmacology 15:1707–1718.

15. McCann, S.M., A.B. Rothballer, E.H. Heakel and H.A. Shankin, 1948. Adrenalectomy and blood pressure of rats subjected to auditory stimulation. American Journal of Physiology 155: 128–131.

16. Medoff, H.S. and A.M. Bongiovanni, 1945. Blood pressure in rats subjected to audiogenic stimulation. American Journal of Physiology 143:300–305.

17. Geber, W.F. and T.A. Anderson, 1967. Cardiac hypertrophy due to chronic audiogenic stress in the rat, Rattus norvegicus albinus, and rabbit, Lepus cuniculus. Comparative Biochemistry and Physiology 21:573–578.

18. Welch, B.L. and A.S. Welch, 1970. Sustained effects of brief daily fights upon brain and adrenal catecholamines and adrenal, spleen and heart weights of mice. Proceedings of the National Academy of Sciences of the U.S.A. 64:100–107.

19. Geber, W.F., 1962. Maternal influences on fetal cardiovascular system in the sheep, dog and rabbit. American Journal of Physiology 202:653–660.

20. Franklin, J.B. and R.L. Brent, 1964. The effect of uterine vascular clamping on the development of rat embryos three to fourteen days old. Journal of Morphology 115:273–290.

21. Senger, P.L., E.D. Lose and L.C. Ulberg, 1967. Reduced blood supply to the uterus as a cause for early embryonic death in the mouse. Journal of Experimental Zoology 165:337–344.

22. Wigglesworth, J.S., 1966. Fetal growth retardation. British Medical Bulletin 22:13–19.

23. Geber, W.F., 1966. Developmental effects of chronic maternal audiovisual stress on the rat fetus. Journal of Embryology and Experimental Morphology 16:1–16.

24. Ingalls, T.H., F.J. Curley and R.A. Prindle, 1952.   Experiment-
    al production of congenital abnormalities: timing and degree
    of anoxia as factors causing fetal deaths and congenital ab-
    normalities in the mouse.   New England Journal of Medicine
    247:758-768.

25. Barka, T., 1965.   Induced cell proliferation: the effect of
    isoproterenol.   Experimental Cell Research 37:662-679.

26. Barka, T., 1968.   Stimulation of protein and ribonucleic acid
    synthesis in the rat submaxillary gland by isoproterenol.
    Laboratory Investigation 18:38-41.

27. Geber, W.F., 1969.   Comparative teratogenicity of isoproterenol
    and trypan blue in the fetal hamster.   Proceedings of the
    Society for Experimental Biology and Medicine 130:1168-1170.

28. Sontag, L.W., 1941.   The significance of fetal environmental
    differences.   American Journal of Obstetrics and Gynecology
    42:996-1003.

29. Zondek, B. and I. Tamari, 1960.   Effect of audiogenic stimu-
    lation on genital function and reproduction.   American Journal
    of Obstetrics and Gynecology 80:1041-1048.

30. Zondek, B. and I. Tamari, 1964.   Infertility induced by audit-
    ory stimuli prior to mating.   Acta Endocrinologica 90:227-234.

31. Zondek, B. and I. Tamari, 1964.   Effect of auditory stimulation
    on reproduction.   IV experiments on deaf rats.   Proceedings of
    the Society for Experimental Biology and Medicine 116:636-637.

EFFECT OF NOISE DURING PREGNANCY UPON FOETAL VIABILITY

AND DEVELOPMENT

A. ÁRVAY

Department of Obstetrics and Gynecology
University Debrecen, Hungary

Clinical experiences and observations have already indicated that the intensive stimuli manifested by external environment and affecting the exteroceptors and sensory organs result not only in specific organic reactions, but also bring about non-specific effect-manifestations in the organism. Bennholdt-Thomsen (1) was among the first to emphasize the significance of the so-called urbanization trauma which we may also call civilization damage. The accelerated and increased life-rhythm accompanying city-life, the intensive light effects, the traffic, the sound contrasts, the summing up of the various stimuli, the intensive and multi-directional psychic effects result, in sum total, in what Bennholdt-Thomsen call the phenomenon of acceleration.

This explains - with respect to the genital sphere - why the first menses presents itself sooner in girls living in the city and in those pursuing intellectual professions than in girls living in villages. But at the same time this also explains why the mean time of the first menses significantly advances in greater populations in consequence of the fact that the above-mentioned characteristic difference between city and country is becoming more and more equalized. The fact that a significant difference has been manifesting itself in the average height of adults as well as in the weight of new-borns has been ascribed to the steady and summarizing effect of urbanizational traumas - though the exclusive significance of this can rightly be argued. In the course of our therapeutic work we have been encountering more and more so-called psychosomatic diseases, the genesis of which may be ascribed to the numerous environmental neural effects (2).

Of the psychosomatic diseases it is perhaps those relating to

91

gynecology which dominate.   This is easy to understand, because
on the one hand the relation between the psyche and the sexual
functions is much closer in women than in men; on the other hand,
as a result of the changes occurring in the social status of women,
together with the accompanying milieu changes, the exteroceptive
effects exerted by the environment have become different in essence
as regards their character and intensity.

     The changed environmental effects and urbanization traumas,
besides causing various gynecological psychosomatic diseases, may
also cause multi-directional disorders in the reproductional
processes.   For many years we have been investigating at our clinic
the effects of intensive stimuli on the exteroceptors influencing
the reproductional processes, as well as its mechanism of action
(3).   In the course of these investigations we have examined the
effect of various cortical impulses as well as of unusual and intensive neural stimuli on the fertility capacity, on the pregnancy
of our experimental animals, on the development of the foetuses
and on the life conditions of the new-born (4).

     The investigation of the problem seemed to be interesting on
the basis of the results of our previous experiments.   In these
investigations we registered the effect of the complex neural
traumatization on the structure and function of the ovaries and
endeavoured to classify its mechanism of action (5-7).   In the
course of these experiments we found that far-reaching and characteristic structural and functional changes ensued in the ovaries under
the complex influence of neural stimuli: noise, light and electric
current.   The germinative and vegetative function of the ovaries
undergo changes.   The follicle-maturation processes become accelerated in the first phase of the complex traumatisation.   This is
followed by a massive stroma-luteinization in the second phase.
The follicle development slows down, the follicles do not reach
the state of maturation, they atretize and in the nucleolus of the
atretized follicles degenerative changes take place.   In the
vegetative neuro-network of the ovaries characteristic degenerative
changes are likewise observable.

     Under the influence of the employed stressogen agents - of
which the intensive sound-stimulus played a predominant role -
characteristic changes ensued in the hypothalamus-hypophysial
system in addition to the observed ovarial-structural and functional
changes.   Our investigations provide structural (histological,
cytological and cytochemical) and functional evidence of the fact
that the adenohypophysis displays increased gonadotropin activity
within the hypothalamo-hypophysial system as a consequence of the
intensive exteroceptive stimuli (8-10); but we have also put forward evidence of increased cotricotropin production (11, 12) and of
a transitional increase in thyreotropin production (13-15).

It may be mentioned that – on the basis of our experiments –
we attribute a role to the hypothalamic neuro-secretory material
which undoubtedly contains not only the so-called posterior lobe
hormones, but also the hypophysiotropic substance which takes
part in the cortico-hypothalamic regulation of gonadotropin function (16).

In our opinion the morphological and functional alterations in
female gonads influenced by the intensive exteroceptive stimuli we
employed may be causally related first and foremost to the increased
gonadotropin activity of the hypothalamo-hypophysial system.   But
presumably the adrenal cortex must also play a role in this reaction
in consequence of the close functional connection which exists between the ovary and the adrenal cortex.

In the first place – in connection with the theme to be discussed – we should like to give an account of those experiments
in which we investigated the incidental effect of the intensive
cortical impulses on the onset of sexual maturity.

METHODS

In our experiments we used a standard strain of white rats of
our own breeding.   During the experiments our animals were subjected to complex neural traumatization.   In the isolated rooms
serving as accommodation, bells of high pitch were sounded for 5
minutes every hour (95 phon. 1000 – 2000 hertz high frequency
sounds) and reflectors illuminated the animals (1900 – 2000 lux).
In addition, for 10 minutes every day the animals were exposed to
a combination of intense light (reflector providing 1,250 lux)
sound (105–115 phon. with an average frequency of 500–1,000 hertz)
and electrical stimuli (70–80 v. and 35–40 mA) in an apparatus
constructed especially for this purpose.

In our latest, hitherto not yet reported, experiments we investigated one by one the effect of all the factors of the complex
traumatization employed by us.   A group of our animals was subjected to a combination of the above-described stimuli for 42 days.
The animals of the other experimental group were exposed daily for
10 minutes to the described sound stimuli, the next group only to
light stimuli, the last group exclusively to electric stimuli.
Conclusions were drawn from the cytological examination – performed
twice daily – of the vaginal secretion of the female rats subjected
to the various stressors, from the histological picture of the
endocrine organs of the animals sacrificed on the 3rd, 10th, 20th,
30th and 42nd day of the stimulation period, and from the histo-
chemical examinations (elaborated organs) adrenohypophysis, thyroid,
adrenal cortex, ovary.   The changes of the hypothalamic neuro-

secretory material were likewise examined. In the course of the
experiments the incidental behaviour-changes of the animals were
also observed.

Our experimental series verified that the separately applied
sound stimulation, as well as the isolated light, or electric
stimulus, elicited essentially the same effect as the complex
traumatization, but differed from each other, and also from the
effect of the three combined stimuli only in the degree of intensity. The behavioural changes were the most pronounced after
the electric stimulation. The most intensive effect on the
structure and function of the endocrine glands was elicited by the
isolated sound stimulus. This was followed by the effect of the
light stimuli, and the least intensive effect was observed when
isolated electric stimulation was applied. The effect of complex
neural traumatization is no less than a summarizing of the influence
of the three stressing agents, resulting in effects of fundamental-
ly identical character.

Though it is far removed from the theme under discussion, I
should like to refer to the fact that in the case of identical
species of animals too, different stressing agents may produce non-
specific effects of dissimilar character. Characteristically
different, non-specific effects were observed e.g. when employing
intensive olfactory stimuli than when employing noise, light, or
electric stimuli separately or combined. Allow me to refer to
the classical experiments of Selye (17, 18) in which the employed
formalin-stress resulted in reactions not quite identical with
our complex traumatization. On the other hand, Zondek and Tamari
(19) registered effects corresponding entirely to those observed
by us, using exclusively sound stimuli in the course of his investigations on reproductive processes.

EXPERIMENTS AND RESULTS

After these methodical observations, I should like to give an
account of those introductory experiments in which we investigated
the effect of complex neural stimulation from the point of view of
how far it influences the onset of sexual maturity. In these experiments (4) juvenile female rats were exposed to complex neural
trauma, and the beginning of sexual maturation was estimated from
the time of the opening of the vaginal membrane, from the cytological
examination of the vaginal secretion and from the histological
appearance of the ovaries. The results showed that in animals
under the influence of intensive exteroceptive stimuli the onset of
puberty differs significantly from that of control animals. Puberty begins earlier if the stimuli are applied shortly before physiological puberty is due to begin. It is significantly delayed,
however, if the neural stimuli are applied well in advance of this time

I could refer to those of our investigations in which we examined the significance of complex neural traumatization and of exclusively employed sound stimuli in the genesis of certain functional uterine bleeding. In our model experiments we were able to demonstrate an intensive follicle-maturation process verifying increased FSH influence in the ovaries in the first phase of the effect caused by lasting sound stimuli, while the histological examination showed that a great number of follicles in the most diverse stage of maturation dominated the histological picture. The histological picture of the endometrium indicated a hyperestrogen state, showing a very marked hyperproliferational structure, which corresponded to the histological picture of the glandular cystic hyperplasia in women. By means of our model experiments we also gave an explanation of those clinical observations – among others of our own experiences – according to which the frequency of those bleeding-disorders increase under the influence of intensive psychic effects where the endometrium shows the picture of glandular cystic hyperplasia.

We investigated the effect of intensive sound, light and electric stimulation on the fertility conditions of animals, on the course of the gestational period, on the development and life conditions of the foetuses and new-born in two series of experiments.

In the first experimental series, we used 206 nulliparous 5-6 month old white rats. Of these 169 became pregnant during a 49 day experimental period and 1020 cubs were born of these 160 pregnant animals.

Our animals were divided into 3 experimental groups. Besides the 31 control animals (A group), 54 animals (B group) were exposed to intensive sound, light and electric stimuli for 35 days in the manner already described. After the traumatization period of 5 weeks the females were put with males. Thus the intensive neural stimulation was only applied prior to conception for the females of this group. Our inferences concerning fertilization and the life-conditions of the foetuses and new-born pups may be brought into connection in endocrinium with the structural and functional changes ensuing before pregnancy under the influence of complex neural stimulation.

Complex neural traumatization was begun with the 121 animals of group "C" when the females were brought together with the males. In this case therefore neural stimuli exerted direct influence during pregnancy in addition to exerting an acute effect on the ovaries.

Our experimental results are shown in Table 1. It can be seen that of the 31 control animals belonging to group "A" 30, i.e. 96.7% of the animals littered within 49 days after being mated.

Table 1. Effect of complex neural trauma prior to and subsequent to conception upon fertilization and reproduction.

	Number of animals	Within 49 days after mating		Of the non-littered animals at the end of the experiment			Ratio of sterility	Fertilization after the mating in the week				Number of offsprings	Number of offsprings per pregnancy	Of the offsprings was		
		littered	did not litter	was not pregnant	resorptional pregnancy	normal pregnancy		1st	2nd	3rd	4th			living	dead	stillborn ratio
A. Control animals	31	30=96.7%	1=3.3%	-	-	1	0%	20=66.8%	5=16.6%	4=13.3%	1=3.3%	208	7.2	204	4	1.9%
B. Traumatised before mating	54	49=90.6%	5=9.4%	3	2	-	5.5%	22=44.9%	15=30.6%	7=14.3%	5=10.2	302	6.1	271	31	10.3%
C. Traumatised during mating and pregnancy	121	90=74.3%	31=25.7%	18	12	1	14.9%	28=22.2%	25=27.7%	26=28.8%	19=21.3	510	4.6	415	95	18.6%

In contradistinction to this, of the 54 animals of group "B" pre-
conceptionally traumatized, 49 (=90.6%), of the 121 animals of
group "C" only 90 (=74.3%) littered during the experimental period
of 49 days.   On the basis of these numbers it is unquestionable
that the <u>fertility and birth ratios of our white rats decreased</u>
<u>when intensive sound, light and electric stress was applied pre-</u>
<u>conceptionally, but even more markedly when the neural stress was</u>
<u>applied subsequently to mating.</u>

    Even more striking is the difference between the three groups
if we also take into consideration the obductional findings of the
females not having delivered during the 49 day experimental period.
According to the data of the table the only animal of the control
group that had not littered proved to be pregnant on dissection.
Thus the fertility index of the control "A" group was 100, while
the sterility ratio was 0% during the experimental period.   On the
other hand, of the 54 animals of group "B", 3 (=5.5%), of the 121
females of group "C" 18 (=14.9%) did not become pregnant within 49
days after having been put with males - the dissection of the
animals not having delivered bears evidence of this fact.   It may
be remarked that deceased foetuses and resorptional pregnancy was
found in 2 animals of group B and in 12 animals of group C.   Thus,
taking into consideration the dissection findings, <u>the sterility</u>
<u>ratio of the control group was 0%, whereas in the groups of animals</u>
<u>exposed to complex traumatization the sterility ratio was 5.5 and</u>
<u>14.9% respectively.</u>

    A significant difference manifests itself under the influence
of intensive sound, light and electric stimuli in the animals of
each group at the time of fertilization also.   In this respect I
can refer to the data of Table 1. which shows quite clearly that
66.8% of the control animals became pregnant one week after being
put with males, whereas in 44.9% of the females of group B, and

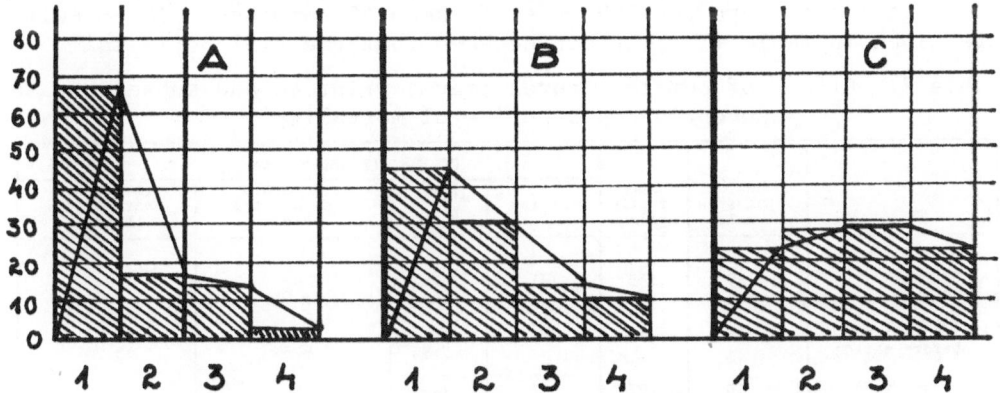

Figure 1. The chronological distribution of fertilization of female
rats.   Abscissa: the length of the experiment in weeks.   Ordinate:
the percentage of fertilized animals.

only in 22.2% of the animals of group C did pregnancy ensue.
Beyond the characteristic difference manifested at the time of
fertilization in the first week after being put with males, a
further obvious difference is that the fertilization ratio suffers
a delay in time as a consequence of the traumatization.   Besides
the relevant data of the Table 1, also refer to Figure 1.   It can
be seen that in the 4th week after mating only 3.3% of the control
animals became pregnant.   In sharp contrast to this, the ratio in
group B was 10.2%, in group C it was 21.3%.   The maximum of the
fertility ratio in the control group was prominently high in the
first week after having been put with males, this ensued in group
C only in the 3rd week.   These results prove therefore that under
the influence of pre- and post-conceptionally applied complex neural
stimulation the conception period of the females suffers a delay.

    Further establishments of our series of experiments are as
follows: First, the average number of cubs born per delivery was
smaller in the case of pre- and post-conceptionally traumatized
females: in contrast to the 7.2 average of the control group, the
average of the cubs from one litter was 6.1 in group B, in group C
it was 4.6.   Second, the ratio of still-born was higher; of the
208 cubs born in the control group only 4 were still-born (=1.9%),
whereas 31 (=10.3%) of the 302 cubs of group B, and 95 (=18.6%) of
the 510 cubs of group C were still-born.   Third, the perinatal
mortality of the born cubs rose characteristically also, i.e. the
ratio of the still-born and those which died in the first three
days following birth.

    The latter statement was made on the basis of a second group
of experimental animals selected according to similar viewpoints,
since the cubs born in the previous groups were immediately killed
and elaborated after delivery in the interests of our investigation.
In this series 49 females were divided into 3 groups (A-B-C); our
observations concern the 299 cubs of these 49 animals.   The results
are given in Table II.   It can be seen that the 4.5% perinatal

Table 2. Effect of complex neural trauma prior to and subsequent to
            conception upon perinatal mortality.

	Number of animals	Number of offsprings	Of the offsprings was			
			living	still-born	died within 3 days after birth	stillborn + death after birth
A. Control animals	19	133	131	2=1.4%	4=3.0%	6=4.5%
B. Traumatized before mating	15	89	82	7=7.9%	7=8.5%	14=15.7%
C. Traumatized during mating and pregnancy	15	77	66	11=16.9%	10=15.6%	21=29.9%

mortality recorded in the control group jumps to more than three-fold (15.7%) in group B, and more than six-fold (29.9%) in group C. It is evident therefore that the pre- and post-conceptionally applied intensive sound, light and electric stimulation raises perinatal mortality more than six-fold, and within this the still-born ratio from 1.4% to 16.9%, the deaths occurring in the first three days of life from 3.0% to 15.6%. Hence, complex neural traumatization exerts a marked effect on the life conditions of the foetuses and new-born respectively.

As supplement to the experimental series, I should like to mention that we carried out a separate investigation as to whether the neural stimuli employed by us influence the fertilizing ability of the male rats. For this purpose we set up two experimental groups: in one of them (group B_1) the pre-conceptionally traumatized females were put with males traumatized in the same manner. In the other group (C_1) the traumatization was performed with both sexes after the females had been put with males - in the self-same manner as the females of group C had been treated. So - in this respect group B became the control of group B_1, while group C became the control of group C_1. In the course of these experiments no significant difference was found between groups B and B_1, or between C and C_1. This seems to prove that intensive sound, light and electric stimulation did not influence the fertilizing ability of male rats to a measurable extent.

A subsequent effect of the intensive stimuli applied by us was verified by the series of experiments in which such females were mated which - similarly to the members of group C - had been exposed to complex neural traumatization. After a rest period of 1-2 months they were put with males and we began to apply the type of stimuli corresponding to that of group C. In this group 24.2% of the cubs were still-born, as opposed to the 18.6% still-born ratio of group C which may be considered to be control. This series of experiments appears to prove that in female rats sound, light and electric stress applied before allowing them to begin sexual life affects the reproductional processes and the life conditions of foetuses even after a prolonged period of rest.

Summarizing the results of our experiments described above, we may ascertain that intensive sound, light and electric stress significantly influence the fertility conditions of experimental white rats, the development and life-expectancy of foetuses and new-born.

In the course of these experiments we found it conspicuous that in the cubs of traumatized mothers (groups B and C) various developmental disorders could be observed even macroscopically. On the basis of our previous experiments we could rightly suspect that there may be some connection between the frequency of the developmental disorders and the applied intensive neural effects. This question could reasonably be raised on the basis of these ex- periments in which we were able to demonstrate degenerative changes in the ovary – partly in the follicle system, partly in the vegeta- tive nerve fibers in consequence of the applied neural stress. In order to study this problem more thoroughly, a new series of experi- ments was started (20,21).

These experiments comprised a group of 55 control and a group of 105 experimental animals. The white rats were nulliparous and were 4-5 months old at the outset of the experiment. Until the beginning of the experiment, all the females lived under identical environmental conditions and on identical standard diet. After establishing cycle stability, complex traumatization was begun with the animals of the experimental group and on the 21st day the traumatized females were mated. The intensive sound, light and electric stress were continued until the pregnant animals were placed in separate dropping-boxes four to five days before litter- ing. The complex traumatization was stopped and only an hourly ringing of the bell and the strong light of a reflector served as stimulation.

After littering had taken place, the new-born were at once killed and divided into two groups on the basis of a macroscopic examination. Group A animals were dissected and any macroscopic- ally registrable abnormality was noted. Group B animals were fixed in formalin and embedded whole in paraffin. Serial sections 10u thick were prepared, stained with haematoxylin-eosin, and evaluated microscopically.

Table 3 gives the results of macroscopic examination of 396 offsprings of control animals and 525 offsprings of experimental animals. The table shows that of the 206 control animals belonging to group A - examined only macroscopically - 3 (=1.4%) were found to have congenital malformations: one had syndactylia, two had cleft palate. The members of the B control group, after being examined macroscopically, were elaborated for the purpose of histological ex- aminations also. The 180 new-born belonging to this group were found to have no developmental disorders when examined either macroscopic- ally or microscopically. As regards the 396 members of the two con- trol groups, therefore, developmental disorders were found in 3 (=0.7%) In contrast to this, of the two groups of 525 cubs born of trauma- tized mothers, 60 (=11.4%) were found to have 84 kinds of develop- mental disorders. Group B offspring of experimental animals showed the following anomalies when examined microscopically: malformation of the eye (lack of crystalline lens) in five animals, cleft palate in 16, developmental abnormity of the heart (defect of atrial wall) in three, double ureter in four, hydronephrosis alone or with hydroureter in 18. It is to be remarked that when

Table 3. Number of malformations observed macroscopically in 921
new-born rats.

	Number of malformations in offspring of			
	Control animals		Experimental animals	
	Group B[x] 216 animals	Group B 180 animals	Group A 283 animals	Group B[+] 242 animals
Syndactylia one or more extremity	1		9	10
Rudimentary development of extremities			4	3
Vascular lesion of the extremities and tail, and spontaneous amputation			3	6
Cleft palate	2		12	8
Fissure of diaphragm			6	8
Diaphragmatic hernia			2	
Hydronephrosis alone or with hydroureter			6	5
Cortical haemorrhage beneath kidney capsule			1	
Anencephalia			1	
Total malformations	3	0	44	40
No. /and percentage/ of animals with above abnormalities	3 = 1.4%	0 = 0%	29 = 10.2 %	31 = 12.8%

x No abnormalities seen in this group either macroscopically or microscopically.
+ I.e. before microscopic examination.

the elaboration of the new-born was performed in section series,
the examination of the developmental disorders did not include the
genitalia.

The most frequent developmental disorder, cleft palate, was
found in 3.8% of the new-born. In a part of our cases (8 new-born)
bi-lateral partial (Figure 2), in 3 animals unilateral, and in 9
whole cleft palate was found (Figure 3). Of the 525 cubs born of
traumatized mothers 19 (=3.7%) had syndactylia. This developmental
disorder was confined to one extremity in 6 new-born, in 13 however
it was detected in several extremities (Figure 4). In a few cases
in addition to syndactylia the developmental disorder involved the
whole extremity. In 9 animals, besides the rudimentary development of the tail and hind extremities, vascular lesion of the extremities and tail, as well as spontaneous amputation was seen after
birth. Two of the new-born with spontaneous amputation of extremity were allowed to survive and were brought up. In a picture made
when they were half a year old and in an X-ray, the spontaneous

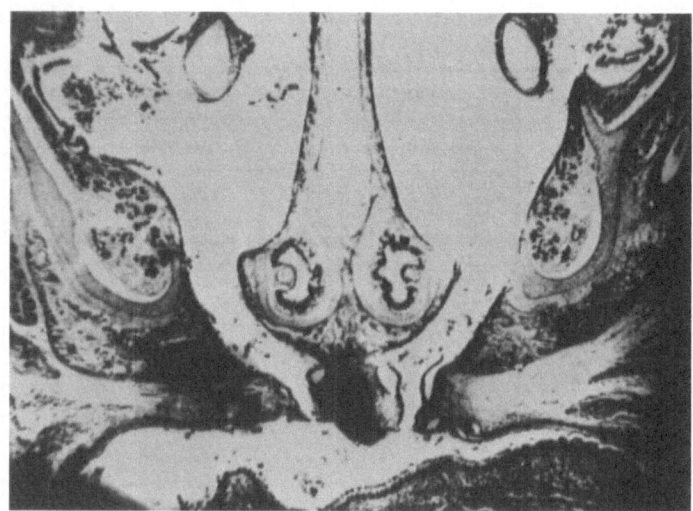

Figure 2. Bilateral partial cleft palate on a new-born rat, the mother of which has been traumatized.

amputation of the anterior extremity can well be seen (Figure 5 a,b). Developmental disorder of the heart was observed in 3 of the 525 new-born.   Each of these cases was an atrial wall defect (Figure 6). Developmental disorder of the eye was registered only by means of microscopic examination in 5 animals.   In the picture of Figure 7 the crystalline lense could not even be demonstrated in series sections in the left eye-socket.   The picture shows that the whole cavity of the bulb is filled by the infolding retina and choriod like a cyst.   In the other eye-socket, besides the otherwise intact crystalline lense, the anterior chamber was found to be missing.

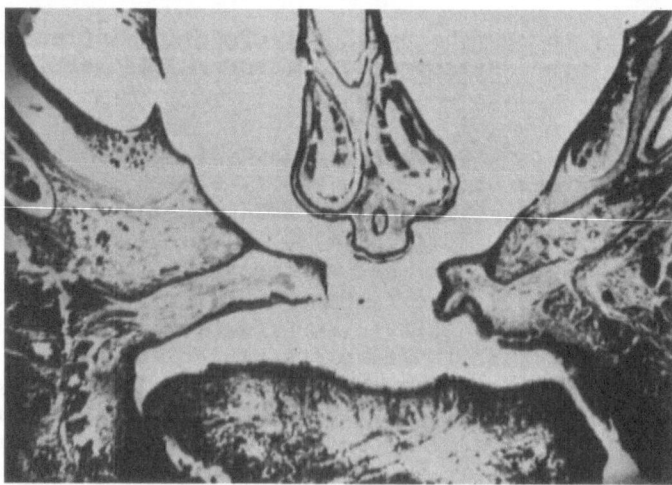

Figure 3.   Total cleft palate.

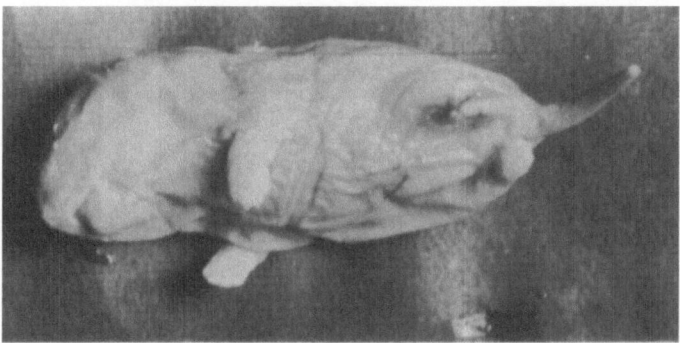

Figure 4.   Syndactyly of the posterior right limb.   The posterior
left limb is incompletely developed.

In 2% of the new-born pronounced hydronephrosis was to be seen together with enormously dilated hydroureters.   The alterations in
most cases – in 8 animals out of 11 – involved both kidneys (Figure 8).   In 2 new-born the kidneys were dilated balloon-like, the
kidney papillas were flattened and compressed and only the thin

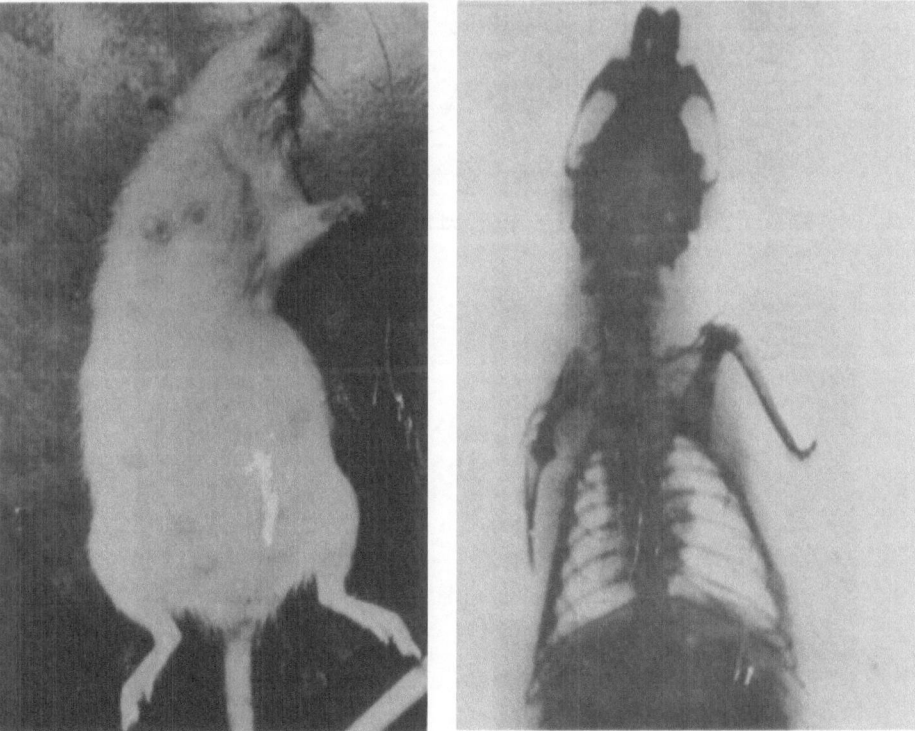

Figure 5 a and b.   Animal with spontaneous amputation of a limb,
aged 5 months.   The corresponding shoulder has a rudimentary
development.

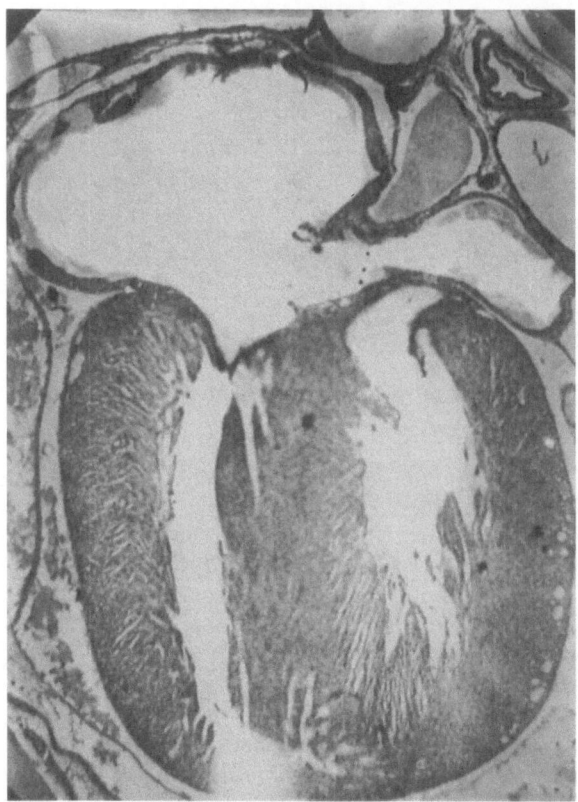

Figure 6. Heart of a new-born rat with interruption of the atrial wall.

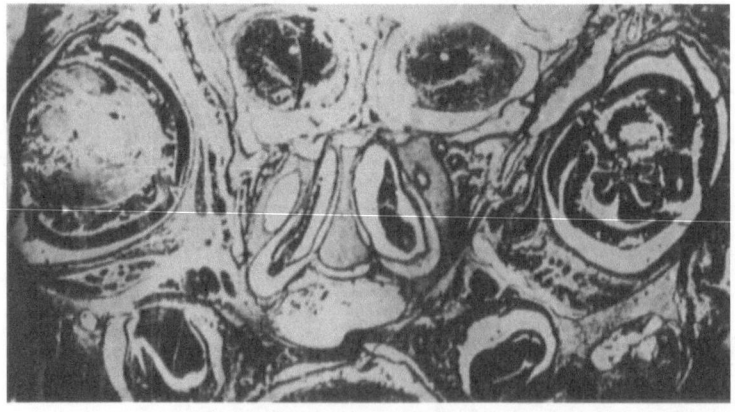

Figure 7. Bi-lateral ocular malformation. Total absence of the left lens. The whole cavity of the bulbe is filled with choroid and retina as a cyst. The right anterior chamber is absent, where-as the lens is normally developed.

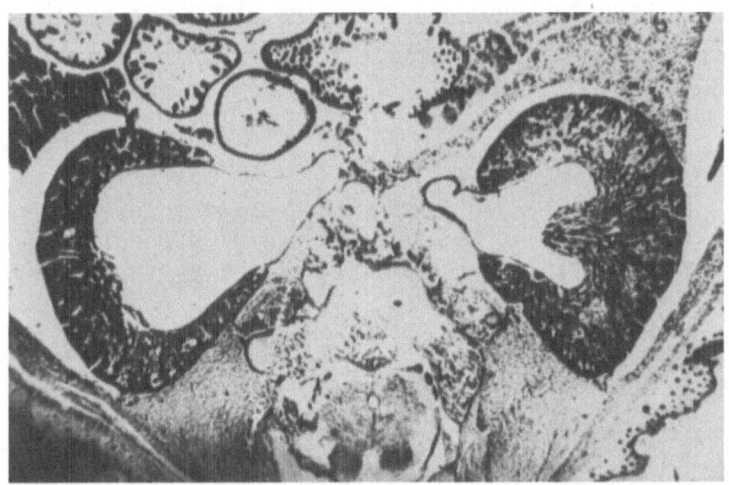

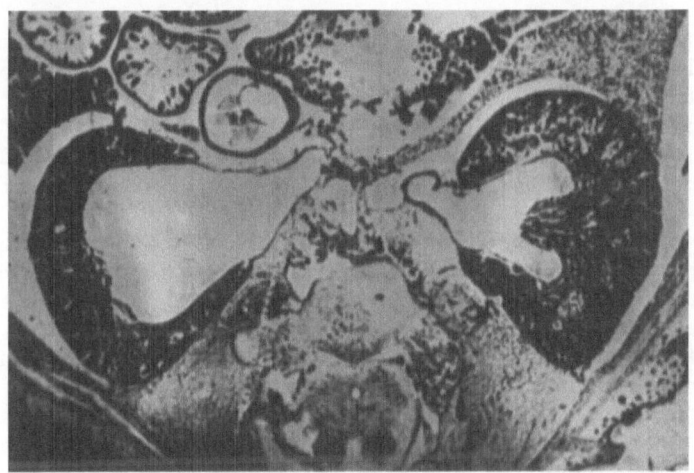

Figure 8. Unilateral hydronephrosis. Compression of the renal cortex. Enormous dilatation of the renal pelvis. The other one is only slightly distended.

kidney parenchyma seemed to constitute the capsule of the enormous-ly dilated pyelon. Besides such developmental disorders we also found a doubling of the ureters in two of our animals.

In our experimental series we investigated the 921 (396 + 525) new-born of 160 (55 + 105) female rats macroscopically, and in series sections, in histological preparations, respectively, and ascertained that <u>intensive sound, light and electric stimulation</u>

of the females preconceptionally and during pregnancy results in
various developmental disorders in 11.4% of the new-born as opposed
to 0.7% of the control.

DISCUSSION

We have already indicated from a methodical point of view that
the effect of the complex application of neural stress: intensive
sound, light and electric stimulation is - as regards its character -
identical with the effect of isolated sound stimuli, there being only
a difference in intensity.   Of the factors of the complex neural
stimulation it was the separately applied sound stimulus which
exerted the most intensive effect on our white rats, as could be
judged by the parameters taken into account.   Thus, in our experiments, from the effect collectively elicited by intense sound,
light and electric stimulation we can - on the basis of our establish-
ments - rightly conclude as to the isolated effect of sound stimuli.

In our first experimental series we proved that the intensive
cortical impulses employed by us influence the commencement of
sexual maturity: the time of the onset of puberty is brought forward if the application of the stimuli is begun a short time before
normal physiologically maturation; on the other hand, it is delayed if the neural traumatization is begun very early in life.

On the basis of our previous experiments the explanation of
these experimental results can be given as follows: we have demonstrated in our multi-directional investigations that neural stimuli
increase the activity of the hypothalamus-hypophysis system, including also the gonadotropic functions.   During the first phase
of stimulation the production of FSH increases together with the
other trophic hormones.   Influenced by this, the follicle-maturation
processes are accelerated in the ovary.   If the intensive follicle
stimulating hormone (FSH) effect reached "adequately prepared"
receptive organs, puberty will begin earlier.   If, however, the
ovaries are not yet sufficiently receptive to respond adequately
to increase FSH effects - that is, if the intensive exteroceptive
stimuli are applied in the very first days of life - then the degenerative changes which we demonstrated in our earlier investigations will lead to delay in the onset of puberty.

In connection with the human relations of our experiments, I
should like to mention that clinical observations and reports provide evidence for the two-directional effect of intensive psychic
traumas on the onset of sexual maturity.   According to one study,
the psychic stimuli, acting or urbanization traumas, together with
the increased rate of physical development, significantly bring
forward the time of the first menses.   According to other observations, the grave psychotraumatic situations of the war years

resulted in a retardation of the age of puberty – at least according
to the reports coming from Germany (22-24), while in Finland it was
precisely the advancement of the menarche that was observed (25).

Though we always insist that the results of the animal experiments – especially as concerns the nervous system – can only be
applied with the strictest criticism in human physiology or pathology, yet we believe that on the basis of our animal experiments
we can accept the view that the ever increasing urbanization traumas,
the accumulating psychic stimuli, cause an increase of activity of
various trophichormones in the hypothalamus-hypophysis system –
the clinical manifestation of which is the observed general physical
acceleration and within this the ever increasing advancement of the
age of puberty.

The contradictory effect of the war years as ascertained by
various authors – taking into consideration the incidentally different reactions of the various stressor agents – can be explained
by the fact that the authors who give an account of the retardation
of the age of puberty made their observations under such circumstances
when the qualitatively and quantitatively inadequate nutrition
accompanying the protracted war conditions caused a decrease of
the receptive ability of the ovaries, thereby postponing the onset
of the menarche.   This establishment is generally accepted in the
literature (26-27).

In the second experimental series reported in this paper, the
significance of intensive neural stimuli was verified in the genesis
of certain bleeding disorders of functional origin.   In the first
phase of the overloading neural stimulation applied by us in this
series of experiments, we demonstrated the increased trophichormone
activity of the hypothalamus-hypophysis system by means of the
histological examination of the uterus and of the endometrium,
respectively, in addition to the morphological and functional examination reported in the foregoing.   During the first phase of
the complex neural stimulation the endometrium presented the same
histological picture which, in human pathology, we designate as
glandular cystic hyperplasia and which we see at the histological
examination of endometrium obtained from curettages performed on
account of disorderly menses characteristic of the hyperoestrogen
state.

According to several clinical reports – among them our own
observations – in the case of continuously exerted psychotraumatic
situations there is an increase of those bleeding disorders when
the endometrium shows a picture of glandular cystic hyperplasia.
In the material of our clinic also during those months of World War
II when our territory became the scene of war events and grave
psychotraumatic situations affected the health of the population,
the ratio of patients admitted with bleeding disorders of such

nature significantly increased (28-32).

On the basis of the results of our animal experiments, I believe that in the genesis of the bleeding disorders caused by grave
psychic stimuli, psychotraumatic situations, we may rightly ascribe
a role to the increased gonadotropin activity of the hypothalamus-
hypophysis system and within this the increased FSH production
brought about by the effect of intensive neural stimuli and to the
hyperfolliculinaemia causing this structural condition of the endometrium.

In our third experimental series we investigated the significance of the complex neural effects on the fertility conditions of
animals, and on the development and life conditions of the foetus
and the new-born, respectively. We deem that our observations
concerning this experimental series can be brought into causal
connection with the results of our previous investigations. Under
the influence of continuously applied sound, light and electric
stimulation the histological picture of the ovaries during the
second phase of effect is dominated by corpora lutea filling in a
significant part of the histological picture of the ovaries. The
follicle maturation processes are strongly thrust into the background. Significantly fewer follicles can be seen than in the
ovaries of control animals and even in a part of these more or
less grave degenerative alterations can be detected. On the basis
of the findings concerning this experimental series, it is under-
standable that the fertilization of our animals suffers a delay
on account of the protraction of the follicle maturation processes.

One can also understand that fewer eggs available for fertilization become liberated and that fewer become fertilized on account
of the fact that fewer follicles reach the stage of complete maturity, the rupture of the follicle. That is why the average number
of offspring per litter is decreased. It is assumed that the less
damaged nucleoli can reach the stage of fertilization and become
fertilized, but such disorders can ensue during intrauterine development which may cause the foetus to die, to be resorbed or to become
stillborn, or even if it is born at all it will die in the first
three days of postnatal life. Hence the rise in the ratio of still-
born and the increase in perinatal mortality.

After our experiments had been reported in 1956, the paper of
Zondek and Tamari (19) appeared in 1960 on the effect of intensive
audiogenic stimuli in which their observations fully confirmed the
results of our experiments.

We were not able to find any report of relevant clinical observations, but it would be very difficult to collect conclusive
observations in this respect which would bear all criticism. And
yet if we take into consideration the observations of Stieve (33),

according to whom intensive psychic effects bring about demonstrable
degenerative alterations in the ovary of women, we cannot exclude
the assumption that in analogy to our animal experiments, especially
in neuro-labile persons, the continuously exerted stressing psychic
effects also influence the germinative functions of the ovaries in
human pathology, hence in women too.   Apart from the fact that the
so-called urbanization trauma accompanying the ever increasing urbanization are making their effect felt more and more, the number of
women participating in the different industries, very often in the
heavy industry, is growing all over the world.   Within these oc-
cupations the women are often exposed to intensive sound, light
and other stimuli, so that we must reckon with an accumulation of
psychic, neural stimuli under such circumstances.   Of course, these
findings have no conclusive force clinically, but they may assign
us tasks from the viewpoint of prevention.

In the last experimental series we ascertained that 11.4% of
the new-born of female rats exposed preconceptionally and during
pregnancy to intensive sound, light and electric stimulation had
various congenital malformations.

In the field of teratogenesis, the very intensive investigations
of the last 1-2 decades brought to light numerous teratogenic factors.
It would surpass our possibilities if on the basis of the investigations carried out so far we attempted to make known - if only
briefly - the very important and practical problems of teratogenesis
and the results attained hitherto.   It must be pointed out, however, that in the course of the extensive investigations, despite
the discovery of diverse teratogenic noxae, only clinical observations indicate that the teratogenic effect of intensive psychic
stresses may also be reckoned with.   We were the first to succeed
in demonstrating by means of animal experiments that under the influence of complex neural traumatization the new-born of traumatized
female animals were found to have a remarkable number of congenital
malformations.

From a clinical viewpoint Klebanow's (34-36) observations de-
serve attention.   He had an opportunity to control the reproductive conditions of numerous women who had been kept in concentration
camps during World War II.   After the life conditions of these
women had returned to normal, even when their ovarian cycles had
become regular, endocrine disorders and sterility were quite frequent.   When they did become pregnant a significant number of
foetuses died in utero, or else they gave birth to new-borns with
congenital malformation in a much higher ratio than the average.

I can also refer to the numerous observations of authors -
mostly German - according to whom the ratio of infants born with
congenital malformations, particularly neuroectodermal, significantly increased in the last years of World War II, and even more during

the years following the war (37-41).   We have drawn the same conclusion regarding the material of our clinic (42).   At any rate,
it is to be remarked that when we evaluate the effect of the war
years and the grave psychotraumatic situations accompanying them,
we must by all means also take into account the question of the
multiplicity of the agents exerting influence.   We must reflect
that besides the grave psychic effects, the quantitatively and
qualitatively inadequate nutritional conditions must also have
played a very significant role as regards teratogenesis.

     In our latest experiments we may emphasize the role of two
factors to which we ascribe significance in the mechanism of action.
The first: the demonstrated degenerative alterations in a part of
the follicle system, but perhaps in the vegetative plexus of the
ovary too, brought about under our experimental conditions by the
effect of the complex neural stimuli.   The other factor is the
functional change of the adrenal gland (both the cortex and the
medullary substances) caused by stress.

     The increase of the adrenal cortex activity - a consequence of
the effect of stressing agents - was verified by means of our multi-
directional experiments too (11-12).   The functional increase of
the medullary substance brought about by stress effect, hyperadrenal-
inaemia, has been verified in classical experiments.   In consequence
of the applied complex stimulation the level of the ACTH, Cortisone,
17-hydrocorticosterone of the organism increases, i.e. the level of
those hormones whose significance in experimental teratogenesis has
been demonstrated by several authors (43-46).

     To conclude with, in the genesis of the malformations caused
by the prolonged application of intensive sound, light and electric
stimuli, we attribute a role on the one hand to the degenerative
changes demonstrated in the ovaries - germ-damage - on the other
hand we may raise the question of the possible teratogenic significance of the elevated levels of endogenous ACTH, cortisone and
adrenaline elicited by the neurotropic stimuli.   Thus, the interaction of genetic and exogenic factors may also play a role, en-
hancing each other's effectiveness and thereby creating a greater
likelihood of malformations.

SUMMARY

     The three agents of complex neural stimuli, when applied sep-
arately, resulted in effects identical in character to those of the
complex neural stimuli and differing merely in their degree of intensity.   Of the three stressors employed, it was the sound stimuli
which dominated.

We have already made known the results of our four series of
experiments.   In the first we described the influencing effect of
the complex neural stimulation on the onset of sexual maturation.
In our second model experiment we verified the role of continuous
stimulation in the genesis of certain functional uterinal bleedings.
In the third series we demonstrated the significance of complex
neural stimulation on the fertility of animals, the development and
life conditions of the foetus and new-born, and finally, in the
last series of experiments we verified the significance of neurotropic stimuli in teratogenesis.

In connection with each series of experiments we made known -
on the basis of our previous experiments - our concept concerning
the mechanism of action of the asserted effects.

As regards our findings concerning the mechanism of action,
we should like to emphasize that the applied stimuli increase the
function of the hypothalamus-hypophysial system.   This has been
verified by our multi-directional experiments.   Correspondingly,
mentioning only the influences affecting the reproductional organs,
the follicle maturation processes become more and more intensive
in the ovaries during the first phase of the effect.   The second
phase is characterized by a massive stroma-luetinization such as
the manifestation of the increased FSH, LH and LTH effect respectively.   Simultaneously, characteristic degenerative changes run
their course in the follicle system and in the vegetative plexus
of the ovaries.

In the investigations concerning the genesis of malformations
with respect to their mechanism of action, we referred to the
change in the endogenous hormone-milieu, to the elevation of the
ACTH, cortisone and adrenaline level elicited by the effect of the
stimulation.

In connection with each series of experiments we indicated
the clinical, human pathological aspects as well.

REFERENCES

1. Bennholdt-Thomsen, C., 1938. Über die Acceleration der Entwicklung der heutigen Jugend. Klinische Wochenschrift (Berlin) 17:
   865-870.

2. Árvay, A., 1967. Neuroendocrine aspects of gynecologic and
   obstetric diseases. In: E. Bajusz (Ed.), An Introduction to
   Clinical Neuroendocrinology, S. Karger, Basel-New York, pp. 380-
   402.

3. Árvay, A., 1967. The role of environmental factors in reproductional processes. In: K. Lissák (Ed.), Symposium on Reproduction,
Akadémiai Kiadó, Budapest, pp. 65-80.

4. Árvay, A., T. Nagy, 1956. Der Effekt intensiever Nervenreize
auf die Fertilität und auf die Lebenserwartung der Frucht.
Schweizerische Medizinische Wochenschrift (Basel) 86, Beiheft
zu Nr. 37:1070-1074.

5. Árvay, A., T. Nagy, S. Kovács-Nagy, 1956. Die Wirkung der auf
die Sinnesorgane ausgeübten extrem starken Reize auf die Funktion und Morphologie des Ovars. Zeistschrift für Geburtshilfe
und Gynaekologie (Stuttgart) 147:371-388.

6. Árvay, A., 1960. Die Wirkung der belastenden neurotropen
Reize auf die weiblichen Sexual funktionen. Zentralblatt fur
Gynaekologie (Leipzig) 82:886.

7. Árvay, A., I. Nyiri, 1961. Die Rolle von nervösen Unweltein-
flüssen auf die weiblichen Sexualfunktionen. Zeitschrift fur
Aerztliche Fortbildung (Jena) 55:722-728, 878-884, 901-913.

8. Árvay, A., L. Balázsy, 1960. Changes in the gonadotrophic
function in response to nervous stress. Acta Physiologica
Academiae Scientiarum Hungaricae (Budapest) 14:317-325.

9. Árvay, A., L. Balázsy, 1960. Experimenttel und klinische
Beweise der hypergonadotropen Funktion der Adenohypophyse
unter der Wirkung belastenden neurotropen Reize. In: I. Törő
(Ed.), Hypothalamus-Hypophysensystem und Neurosekretion,
Akadémiai Kiadó, Budapest, pp. 93-103.

10. Árvay, A., F. Bölönyi, L. Balázsy, S. Jakubecz, 1960. Changes
of hypothalamic neurosecretion under the influence of overloading nervous impulses. Acta Anatomica (Basel) 40:256-272.

11. Árvay, A., L. Balázsy, S. Jakubecz, I. Takács, 1959. Effect
of severe nervous stimulation on the morphology and function
of the adrenal cortex. Acta Physiologica Academiae Scientiarum
Hungaricae (Budapest) 16:267-284.

12. Árvay, A., L. Kertész, L. Lampé, 1959. The effect of severe
nervous stimulation on pituitary, adrenal and ovarian function:
Studies with P^{32} . Acta endocrinological (Kobenhavn) 30:
585-592.

13. Árvay, A., L. Lampé, L. Kertész, L. Medveczky, 1960. Changes
of thyroid function in response to severe nervous stimulation:
Studies with I^{131}. Acta endocrinological (Kobenhavn) 35:
469-480.

14. Árvay, A., I. Nyiri, M. Rákosi, L. Buris, 1960. Die Wirkung
    der belastenden neurotropen Reize unter der chemischen Blockade
    der Adenohypophyse. Endokrinologie 40:39-53.

15. Árvay, A., 1964. Cortico-Hypothalamic control of gonadotrophic
    functions. In: E. Bajusz, G. Jasmin (Eds.), Major Problems in
    Neuroendocrinology, S. Karger, Basel New York, pp. 307-321.

16. Árvay, A., I. Nyiri, M. Rákosi, L. Buris, 1962. Contribution
    a l'étude des modifications du fonctionnement de l'adénohypo-
    physe par des substances pharmacologiques, Annales d'Endocrinol-
    ogie (Paris) 23:186-290.

17. Selye, H. 1950. The Physiology and Pathology of Exposure to
    Stress, Acta Inc., Medical Publishers, Montreal, Canada.

18. Selye, H. and E. Bajusz, 1960. Recent progress in stress
    research and the role of stress theory in modern pathophysio-
    logical work. (Hungarian). Orvosi Hetilap (Budapest) 101:1-6.

19. Zondek, B., I. Tamari, 1960. Effect of audiogenic stimulation
    on genital function and reproduction. American Journal of
    Obstetrics and Gynecology 80:1041-1044.

20. Árvay, A., T. Nagy, J. Bazsó, 1961. L'importance des excita-
    tions cumulatives neurotropes dans la genese des malformations
    congénitalles. Biologia Neonatorum (Basel) 3:1-23.

21. Árvay, A., 1967. Effects of exteroceptive stimuli on fertility
    and their role in the genesis of malformations. In: G.E.W.
    Wolstenholme and Maeve O'Connor (Eds.), The Effects of External
    Stimuli on Reproduction, J. and A. Churchill, Ltd., London,
    pp. 20-28.

22. Nevinny-Stickel, H., 1950. Acceleration und Menarrche.
    Archiv für Gynaekologie (munchen) 178:300-308.

23. Häberlin, C., 1918. Über die körperliche Entwicklung von
    Kindern im Frieden und Krieg. Archiv für Kinderheilkunde
    (Stuttgart) 66:370-379.

24. Grimm, H., 1948. Der gegenwärtige Verlauf der Pubertät bei
    der weiblichen Berufsschuljugend in Mitteldeutschland.
    Zentralblatt für Gynaekologie (Leipzig) 70:8-17.

25. Simmel, G., 1952. Über das Menarchealter in Finnland. Acta
    Pediatrica (Turku), Suppl. 84:1-67.

26. Gaethgens, G., 1943. Mangelernährung und Generationsvorgänge

im weiblichen Organismus, G. Thieme, Leipzig.

27. Guggisberg, H., 1935. Die Bedeutung der Vitamine für das Weib.
    Urban und Schwarzenberg, Wien.

28. Benedetti, G., 1956. Beziehungen zwischen endokrinen und psycho-
    pathologischen Geschehen bei der glandulär-zystischen Hyperplasie
    des Endometriums. Schweizer Archiv für Neurologie, Neuro-
    chirurgie und Psychiatrie (Zurich) 78:1-14.

29. Froewies, I., E. Islitzer, 1949. Eklampsie und zweiter Welt-
    krieg. Geburtshilfe und Frauenheilkunde (Stuttgart) 9:572-581.

30. Dech, H., 1949. Das vermehrte Auftreten der glandulär-
    zystischen Hyperplasie der Uterusschleimhaut in der Nachkriegs-
    jahren. Geburtshilfe und Frauenheilkunde (Stuttgart) 9:208-219.

31. Gitsch, E., 1951. Uber die Bedeutung der Umwelt für die
    Entstehung der glandulären Hyperplasieblutung. Wiener Klinische
    Wochenschrift 63:601-605.

32. Árvay, A., I. Nyiri, 1959. Die Bedeutung nervaler Einflüsse
    in der Genese einzelner funktioneller Blutungsstörungen.   Acta
    Neurovegetativa (Wien) 20:76-93.

34. Klebanow, D., 1948. Hunger und psychische Erregungen als Over-
    und Keimschädigungen. Geburtshilfe und Frauenheilkunde
    (Stuttgart) 8:812-820.

35. Klebanow, D., 1949. Fertilitätsstörungen als Spätfolge chro-
    nischen Hungers und schwerer seelischer Traumen. Geburtshilfe
    und Frauenheilkunde (Stuttgart) 9:420-429.

36. Klebanow, D., H. Hegnauer, 1950. Zur Frage kausalen Genese
    von angeborenen Missbildungen. Medizinische Klinik (Munchen)
    45:1198-1203, 1233-1246.

37. Aresin, N., K.M. Sommer, 1950. Missbildungen und Umweltfaktoren.
    Zentralblatt für Gynaekologie (Leipzig) 72:1392-1398.

38. Giroud, A., 1955. Les malformations congénitales et leurs
    cause. Biologie Medicale (Paris) 44:524-609.

39. Kühnelt, H.J., P. Rotter-Poll, 1955. Die Missbildungen an der
    Universitäts-Frauenklinik Berlin. Wiener Klinische Wochenschrift
    77:893-900.

40. Flegenheimer, F.A., 1956. Zur Frage der Häufigkeitszunahme der
    Missbildungen in den Nachkriegsjahren. Wiener Klinische Wochen-
    schrift 68:468-470.

41. Gesenius, H. 1951. Missbildungen im Wechsel der Jahrhunderte. Zeitschrift für Klinische Medizin (Berlin) 26:359-362.

42. Nagy, T., J. Bazsó, L. Lampé, 1961. Häufigkeit der Missbildungen im Krankengut unserer Klinik von 27 Jahren. Zentralblatt für Gynaekologie (Leipzig) 83:866-874.

43. Choukroun, J., A. Minkowski, 1952. Enquete sur l'etiologie des malformations congénitales a la clinique obstetricale Baudeloque en 1949-1950. Semaine des Hopitaux de Paris 28:3030-3033.

44. Fraser, F.C., T.D. Feinstat, 1951. The experimental production of congenital defects with particular reference to cleft palate. Etudes Neo-natales 2:43-48.

45. Jost, A., 1953. Dégénérescence des extrémités du foetus de rat provoquée par l'adrénaline. Comptes Rendus Hebdomadaires des Seances de l'Academie des Sciences; D:Sciences Naturelles (Paris) 236:1510-1512.

46. Giroud, A., M. Martinet, 1954. Altérations del'épithélium et des fibres du cristallin apres thyroxine. Archives d'Ophtalmologie (Paris) 14:247-258.

AUDIOGENIC STIMULATION AND REPRODUCTIVE FUNCTION*

I. TAMARI

Hebrew University
Hadassah Medical Centre
Jerusalem, Israel

F.H.A. Marshall in 1936 was the first to emphasize the dependence of the breeding season, in many species, on environmental or "exteroceptive" factors. He suggested that such factors exert their effects by nervous and reflex stimulation of the secretion of gonadotrophic hormones, primarily FSH from the anterior pituitary. The term "exteroceptive" was first used by Hartridge in the same year to describe "external local factors" affecting the pituitary.

The aim of our studies during the past nine years was to determine whether and how far auditory stimuli influence the reproductive function, with special emphasis on their effect on fertility and maintenance of pregnancy.

Some studies reported so far indicate that auditory stimuli induce regressive changes resembling those produced by other stressful stimuli. The weight of the adrenal cortex increases under auditory stimulation (D'Amour and Shaklee, 1955) combined with an associated increase of corticosteroids (Duncan, 1957). The studies of Sackler et al. (1950, 1959, 1960) also provide considerable evidence for increased production of adrenocorticosteroids, and inhibition of gonadotrophic and ovarian hormones, and possible inhibition of the thyrotrophin and thyroid hormones. Fortier (1951) induced rapid discharge of ACTH as evidenced by marked depletion of adrenal ascorbic acid in male rats by intense sound stimulation. Miline et al. (1951, 1954) dealt with the effect of noise on testicular function.

---

*This work was carried out in collaboration with the late B. Zondek in the Department of Obstetrics and Gynecology, Hebrew University Hadassah Medical Center, Jerusalem, Israel.

Árvay et al. (1956, 1959) applied prolonged "nervous traumatisation" by intensive noise, light and electric stimuli to rats,
whereby catatonic conditions were observed.   These workers described increased luteinization of the ovaries and degenerative
changes of the follicular apparatus.

PART I.   EFFECTS OF AN ALARM BELL

Our experiments (Zondek and Tamari 1958, 1960, 1964; Zondek
1959) were carried out on 48 mature rabbits and over 4000 infantile
and mature rats.   The animals were kept close to an electric alarm
bell which rang for periods of 1 minute, at 10 minute intervals,
during day and night.   The audio-frequency spectrogram of the
bell revealed that the acoustic energy is concentrated in a relatively small frequency band, between 3000 to 12,000 cps.   The bell
shows a pronounced peak of nearly 100 db at a frequency of 4,000
c/s and a second one of 95 db at 10,000 c/s* (Figure 1).

Some of the rats reacted quite soon, showing restlessness,
increased motor activity, and running around in circles, whereas
most of the animals did not react at all behaviorally.   After
some time all rats appeared to become accustomed to the sound
stimuli and remained quiet.   Rabbits were not at all influenced
in their behaviour.

Effect of Auditory Stimuli on the Genital Organs

Experiments in rats.   Mature female rats reacted to the continual sound stimuli after 1-2 weeks by changes in the oestrous rhythm,
as manifested by prolonged or sometimes persistent oestrus.   After
two months of stimulation their uterus and ovaries, respectively,
were increased in weight by 33 to 62 per cent.   The ovaries -
and this is important - were composed essentially of corpora lutea.
The number of corpora lutea varied in different experimental animals
from 8 to over 20 as compared to 6 to 10 in the control animals
(Figure 2).

The enlargement of the ovaries (62.7 per cent weight increase)
with a considerable increase in the number of corpora lutea indicates
a stimulating effect of sound on the anterior pituitary gland and
the consequent release of gonadotrophin (especially LH).

In the next experiments we tried to stimulate gonadotrophic
function in immature animals.   Infantile 3 to 4 week old rats,
weighing 30 to 35 grams were exposed for 3 to 24 hours to auditory
stimuli and killed 24 hours later.   Slight hyperaemia of the
ovaries was found in about a third of the experiments, but was

*L = sound pressure level db; F = frequency cps.

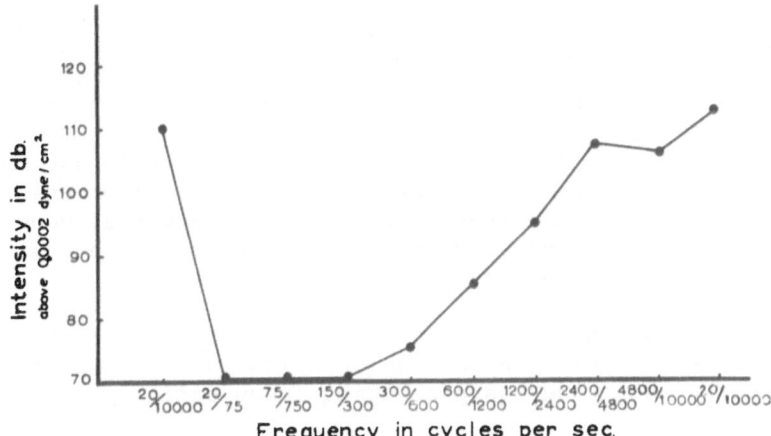

Figure 1. Audio-frequency spectrogram of electric alarm bell to which rats and rabbits were exposed for periods of 1 minute at 10 minute intervals for various lengths of time.

certainly not as distinct as that following injection of human pregnancy urine.

Premature sexual development could not be achieved. Infantile rats were exposed to stimulation by the alarm bell until the first oestrous reaction occurred, whereby no difference was observed as compared to the control animals.

The adrenals in rats exposed to auditory stimuli during 60 days were enlarged, their average weight being 61.5 mg as compared to 39.5 mg in controls.

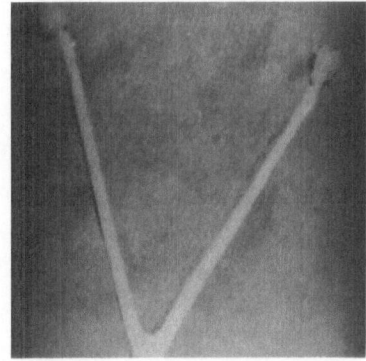
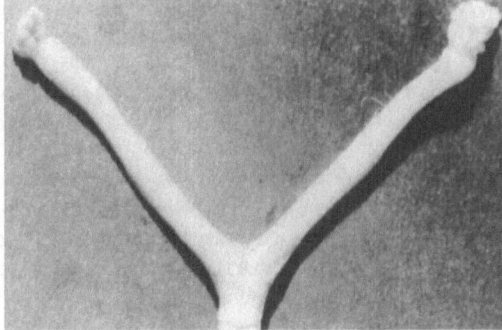

Figure 2. Uterus of mature rats not stimulated by electric alarm bell (left) and stimulated for two months (right).

These experiments show that sound stimuli induce an increase
of the anterior pituitary gonadotrophic function in adult animals.
In immature animals, however, in whom gonadotrophic function has
not yet commenced, auditory stimuli have little or no effect.

There is a remarkable sex difference.   In infantile and mature
male rats no distinct stimulating effect on the sexual organs could
be achieved by continual auditory stimuli.   There was also no distinct inhibitory action on the weight of the testes and accessory
sex organs, or on spermatogenesis.

The late Dr. Warren O. Nelson kindly agreed to re-examine the
testes of our experimental animals and confirmed: "The spermato-
genic process in each case was proceeding in an entirely normal
fashion."

Experiments in rabbits.   The stimulating effect of auditory
stimuli on anterior pituitary gonadotrophic function was demonstrated
even more convincingly in mature rabbits.   In contrast to rats,
ovulation and corpus luteum formation do not occur spontaneously
in rabbits, but only following fertile or sterile mating or simu-
lation of the coital stimulus by mechanical irritation of the
cervix (Long & Evans, 1922).   Sexual excitement affects the secretion of LH by nervous reflex activation of the anterior pituitary
via the hypothalamus (Marshall and Verney, 1936; Fee and Parkes,
1929; Deanesley, Fee and Parkes, 1930).   The same effect can be
achieved in rabbits by administration of various gonadotrophins
(Zondek, 1929).

Forty-eight mature rabbits with an average weight of 2,200 g
were exposed to auditory stimuli by the alarm bell for 6 to 60 days.
In 16 rabbits, the weight of the ovaries was doubled to about 400
mg and in four animals even exceeded.   The ovaries contained small
and large follicles, several of them with follicle haematomata
blood (blood points) and one or several (up to ten) corpora lutea,
changes described as indicative of gonadotrophic reactions (Zondek
and Aschheim).   In six of the sixteen positive cases, enlargement
of the breast with abundant secretion of milk was observed.   By
sound stimuli the same anatomical changes could thus be achieved
in the ovaries and breasts as in pregnancy (Figure 3).

The importance of these experiments seems to lie in the fact
that by auditory stimuli we were able to induce the same changes
in the female genital system as those induced by stimulation of the
hypothalamus or as by administration of pituitary or chorionic
gonadotrophin.

These findings speak against the assumption of a systemic stress
reaction, which would be manifested by an inhibitory action on the
anterior pituitary gonadotrophic functions with resultant atrophy of

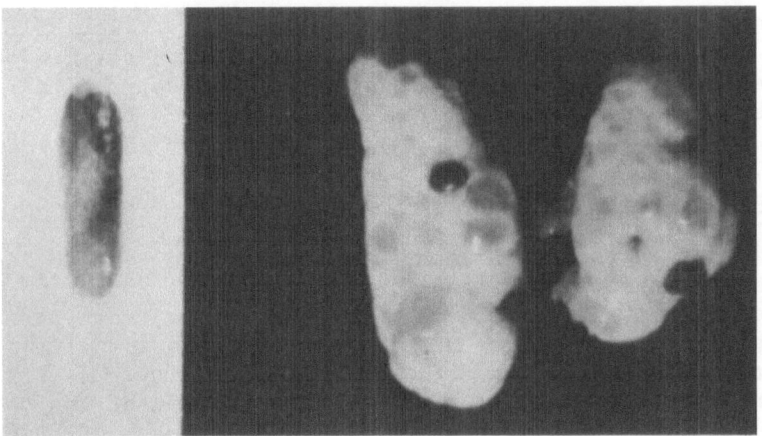

Figure 3.   Ovaries of mature rabbits not stimulated by electric alarm bell (left) and stimulated (right).

the sex organs.

Effect of Auditory Stimulation on Fertility

Stimulation in the pre-mating period.   In apparent contrast to the stimulating effect on the genital organs is the inhibitory effect on the reproductive function.   80 mature male rats weighing 160 to 180 g, were exposed to auditory stimulation for 9 days and each male was then housed together with an unexposed female for 4 days.   The fertility of the males was greatly reduced, only 9 of the 80 males (11.2 per cent) were able to fertilize females, as compared to 64 of 80 control animals.*   Ejaculation took place but the sperm apparently lost its fertilizing capacity.   Following the 9 days of sound stimulation, a rest period of 12 days was necessary to restore the fertility of the males.

In further experiments females were found to react in the same way as males.   Fifty females were exposed to auditory stimuli by the alarm bell for 9 days and then housed together with unexposed males for 4 days, each couple being kept in a separate cage.   Only 9 of the 50 females proved to be fertile as compared to 40 of 50 control animals.   Copulation took place as evidenced by the presence of sperm ejaculate, and the oestrous reaction persisted.

In contrast to the male some histological changes occurred in the female reproductive tract.

---

*Similar results were achieved by reduction of the stimulation period to 4 days.

By hormonal treatment during the 9 day auditory stimulation period the damaging effect on fertility – in studies carried out so far – could not be prevented by the use of gonadotrophin from the serum of pregnant mares and from human pregnancy urine, by hydrocortisone and Vitamin E.

Stimulation in the mating period. In control experiments the fertility rate of our strain of rats has been determined over several years and no seasonal influences were recorded. We define fertility as the number of pregnancies in the total number of females mated. Productivity is defined as the number of foetuses independent of their size or state of development over the total number of females mated. In several hundred rats the fertility rate – for a 4 day period of mating without prior determination of the oestrous cycle – ranged from 70 to 100 per cent with an average of 80 per cent, with an average productivity of 6.5 per rat mated. In the first experiments, carried out monthly during seven months, each of ten pair were housed together in a separate cage – and exposed to auditory stimuli for 2 to 3 weeks. A striking effect was obtained, the fertility was diminished to 5.1 per cent and the productivity to 7 per cent of that in control animals.

In further experiments mating and sound stimulation time were reduced from 2 to 3 weeks to 4 days. Sound stimulation during the 4 days of the mating period has been our standard experiment in the past years. For instance, in one group of 20 animals only 1 was found to be pregnant, and the four foetuses were considerably diminished in size and were in a state of resorption. Among 20 control animals 14 were pregnant with 120 foetuses. This experiment was repeated on hundreds of rats during different seasons of the year and the results were always similar, the fertility rate decreased to an average of 20 per cent as compared to 80 per cent in controls. Figure 4 shows dramatic differences in their uteri.

It might be possible that the sound stimulus induced an irritability of the male partner and disturbed his mating behavior. This, however, was not the case. During the 4 days mating and stimulation period, copulation took place, as evidenced by the presence of the sperm ejaculate in the vagina of the female partner.

Stimulation in the post-mating period. Audiogenic stimulation during pregnancy results in interruption of pregnancy. After mating for four days groups of female rats were exposed to auditory stimuli for 48 hours, starting at varying intervals from the beginning of the mating period. It should be recalled that the rats were mated without prior determination of the oestrous phase, so that conception could have occurred on any of the four days of the mating period (Figure 5). Stimulation during the 4th to 6th day resulted in a

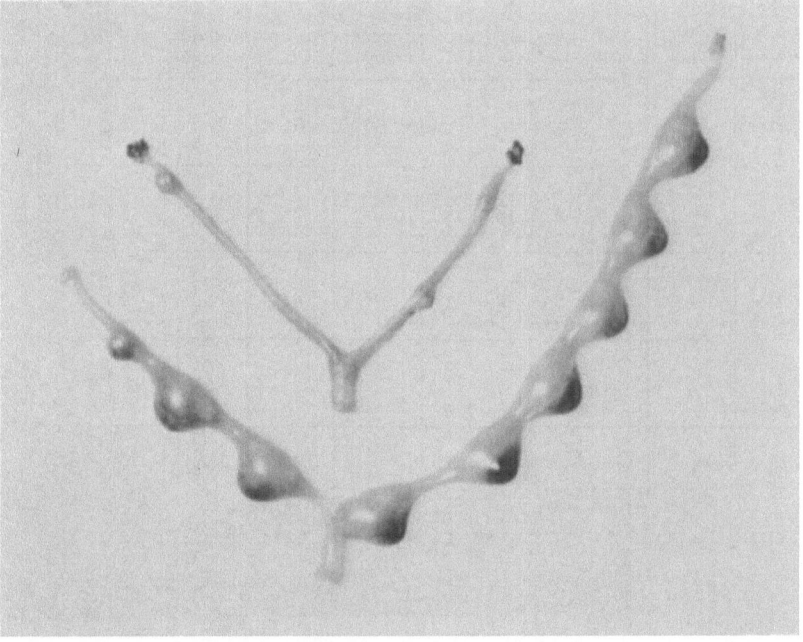

Figure 4. Uteri of rats stimulated (above) and not stimulated by
sound during the mating period.

decrease in the number of pregnancies from 80 to 25 per cent, and
in the productivity index from 6.2 to 1.5. The effect was more
pronounced following stimulation during the 7th to 9th and 8th to
10th day periods, following which fertility was decreased to 15 and
5 per cent, and the productivity index to 0.75 and 0.5 respectively.

We were unable to prevent the damaging effect of auditory stimuli on reproduction by administration of gonadotrophins derived
from pregnant mares' serum (Gestyl), from postmenopausal urine
(Pergonal), and from human pregnancy urine (Pregnyl).

Our observations upon the hormonal factors involved indicate:
a) that auditory stimuli increase uterine and ovarian weight with
increased number of corpora lutea in rats and induce follicle
haematomata and corpora lutea in rabbits, effects which are indicative of gonadotrophic action, and
b) that oestrogens and gonadotrophins are able to block and to in-
terrupt pregnancy, favouring the assumption that secretion of anterior pituitary gonadotrophin may be a cause for the inhibitory
effects on reproduction in our experiments.

A possibility to be considered is a stimulating effect on the
hypothalamo-pituitary posterior axis, inducing secretion of pitocin
and leading to hypermotility of the uterine musculature with ex-
pulsion of the blastocysts from the uterus. By electric stimulation

RATS	MATING days	Auditory stimuli days	Pregnancy	Per cent	Fetus number	Per rat
20	4	control	16	80	125	6.2
20	4	4 - 6 (96 -144 hr)	5	25	31	1.5
20	4	6 - 8 (144-192 hr)	8	40	63	3.1
20	4	7 - 9 (168 -216 hr)	3	15	15	0.75
20	4	8 - 10	1	5	10	0.5

Figure 5.   Effect upon reproductive performance in rats of 48 hours auditory stimulation at various times during pregnancy.

of the supraoptico-hypophyseal tract it is possible to evoke secretion of the oxytocic hormone (Harris, 1954).   Bargmann and Scharrer (1951) assumed that hormones of the posterior pituitary lobe are produced in neurosecretory cells of the hypothalamus and are transported along the axons of the supraoptico-hypophyseal tract.

   The mechanism of the effect of audiogenic stimulation on the reproductive tracts seems to be of complex nature and needs further studies.

Experiments on Deaf Rats

   Experiments were carried out by Zondek and Tamari in 1964 on rats rendered deaf by several weeks' treatment with kanamycin as described by Hawkins in 1959.   The deaf rats were exposed to auditory stimuli by the alarm bell, but their fertility was not diminished, in contrast to the effect in intact rats.   Stimulation in the post-mating period did not induce interruption of pregnancy.

PART II.   EFFECTS OF PURE TONES IN THE AUDIO AND SUPERSONIC RANGE

   Rats may react to complex sounds such as the noise of the alarm bell with unrest, nervous excitement and increased motor activity. We, therefore, studied the effect of pure tones in the audio and supersonic range (Figure 6).*   These sounds do not influence the general behaviour, or eating and sleeping habits of the rats. Their effects on the reproductive functions cannot, therefore, be influenced by other factors.

---

*In cooperation with L.H. Schaudinischky, Head of the Department of Applied Acoustics, Israel Institute of Technology, Haifa.

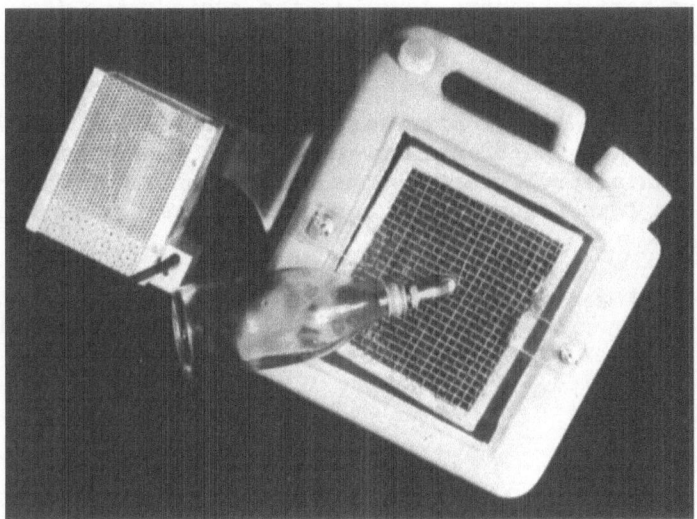

Figure 6.   Equipment for subjecting rats to supersonic stimulation.
Sound waves were transmitted into the cage through an open window.

The following equipment was used:

(a) The sound source was a sinoisoidal pure tone voltage generator
    (Krohn & Hite, Model 430-AB Cambridge, U.S.A.) having a range
    of 5 to 500,000 cps.

(b) In order to produce the necessary acoustic power, a broad band
    amplifier (Model UA 10, Elron, Haifa) of 10 watt undistorted
    power output (0.5 per cent average distortion at maximum power
    of any chosen frequency) was added to the generator, capable
    of delivering only 50 milliwatts.

(c) The supersonic frequency-voltage was transformed into super-
    sonic sound by Ionovac loudspeakers of Electrovoice and DU-
    Kane Corporation.

(d) The control procedure of sound distribution in the supersonic
    range was carried out with a 1/4" Bruel and Kjoer (Copenhagen)
    condensor microphone connected to their wide range amplifier
    (2 to 200,000 cps, mod.).

    The range of hearing in rats may extend up to 40,000 cps (accord-
ing to the Handbook of Biological Data (W. B. Saunders Co., Phila-
delphia, Pa. and London, 1956).

    The results of exposing rats to pure sounds in the audio-
and supersonic range were as follows: In 160 females exposed in

the 4 day mating period to pure tones in the audio range (L = 80 - 90 db, F = 10,000 to 40,000 cps) the decrease in fertility and productivity was 73.7 and 84 per cent respectively.

In 70 rats exposed in the post-mating period to pure sounds in the audio range and in 182 animals exposed to supersonic sounds (as above) the inhibitory effect on the maintenance of pregnancy was 75 and 72 per cent respectively and on productivity 81.8 per cent.

It is interesting to note that the inhibitory effect of auditory stimuli on the reproductive functions does not considerably differ whether complex sounds or pure tones in the audio- and supersonic range were applied.

SUMMARY

Exposure of adult female rats to auditory stimulation by an electric alarm bell induces changes in the oestrous rhythm, manifested by prolonged or persistent oestrus. It, furthermore, increases the weight of the uterus and the ovaries, and the latter contain mature follicles and increased number of corpora lutea. In immature female rats audiogenic stimulation apparently has little or no effect.   There is a distinct sex difference since auditory stimuli do not affect male rats as indicated by normal spermatogenesis.

In mature rabbits changes can be produced which do not occur spontaneously, namely: increase in size of the ovaries up to three times their normal weight, and formation of follicle haematomata and corpora lutea, changes which are indicative of gonadotrophic reactions.   In some rabbits enlargement of the breasts with copious milk secretion occurs.

The fact that auditory stimuli induce the same changes in the genital system as those induced by stimulation of the hypothalamus or as by administration of gonadotrophins is against the concept of a systemic stress reaction, which would be manifested by an inhibitory action on the anterior pituitary gonadotrophic function with resultant atrophy of the female sex organs.

In apparent contrast to the stimulating effect on the sex organs is the inhibitory action of auditory stimuli on reproductive functions.

The reproductive functions of rats rendered deaf by kanamycin are on the other hand not inhibited by auditory stimulation.

Pure tones in the audio- and supersonic range were administered in the pre-mating, mating, and post-mating periods.   No changes are observed in the behaviour of these animals, but disturbance in fertility, productivity and maintenance of pregnancy nevertheless occur.

It seems that auditory stimuli stimulate the hypothalamo-pituitary-ovarian axis leading to secretion of gonadotrophins (FSH and LH) which increase oestrogen production and thus interfere with the development of the fertilized ova and the maintenance of pregnancy.   If the implantation of the fertilized ovum is well established in the endometrium, auditory stimuli have no effect.   Failure of the production or the release of anterior pituitary luteotrophin, which normally follows mating, also seems to play a role in the inhibitory effect of sound stimuli on the reproductive functions.

A further possibility to be considered is stimulation of the supraoptico-hypophyseal tract with possible secretion of oxytocin from the posterior lobe of the pituitary.

ACKNOWLEDGEMENT

Supported by grants from The Population Council, Bio-Medical Division, Rockefeller Institute, New York, and from the Israel Academy of Science and Humanities, Jerusalem.

REFERENCES

1. Amoroso, E.C. and F.H.A. Marshall, 1960.   External factors in sexual periodicity.   Marshall's Physiology of Reproduction, edited by H.S. Parkes 1960, Vol. 1, part 2, 707.   Longmans Green & Co., Ltd., London.

2. Árvay, A. and T. Nagy, 1956.   Der effekt intensiver Nervenreize auf die Fertilität und auf die Lebenserwartung der Frucht.   Schweizer Med. Wschr. 86, 1070.   Ibid. 1959. Der Einfluss belastender Nervenreize auf den Zeitpunkt des Eintrittes der Geschlechtsreife.   Acta neuroveg. 20, 57-75.

3. Árvay, A., T. Nagy and Kovacs-Nagy, 1956.   Die Wirkung der auf die Sinnesorgane ausgeübten extrem starken Reize auf die Funktion und Morphologie, des Ovars.   Zeitschrift für Geburtshilfe und Gynaekologie (Stuttgart) 147,371.

4. Aschheim, S. and B. Zondek, 1927.   Ei und Hormon.   Klinische Wochenschrift 28, 1-3.

5. Bainbridge, F.A. and J.A. Menzies, 1936.   Essentials in Physiol-
    ogy, 8 ed., rev. by H. Hartrige, London, Longmans, Green.

6. Bargman, W. and E. Scharrer, 1951.   The site of origin of the
    hormones of the posterior pituitary.   American Scientist 39, 255.

7. Bruce, H.M., 1959.   An exteroceptive block to pregnancy in the
    mouse.   Nature (London) 184; 105.

8. Bruce, H.M., 1960.   A block to pregnancy in the mouse caused
    by proximity of strange males.   Journal of Reproduction and
    Fertility 1:96.

9. Bruce, H.M. and A.S. Parkes, 1960.   Hormonal factors in extero-
    ceptive pregnancy block in mice.   Journal of Endocrinology
    20:19-20.

10. Parkes, A.S. and Bruce, H.M., 1961.   Olfactory stimuli in
    mammalian reproduction.   Odor excites neurohumoral responses
    affecting oestrus, pseudopregnancy and pregnancy in the mouse.
    Science 134:1049-1054.

11. D'Amour, F.E. and A.B. Shaklee, 1955.   Effect of audiogenic
    seizures on adrenal weight.   American Journal of Physiology
    183:269.

12. Deanesly, R., A.R. Fee and A.S. Parkes, 1930.   Studies on
    ovulation II.   The effect of hypophysectomy on the formation
    of the corpus luteum.   Journal of Physiology 70:38.

13. Duncan, J.W., 1957.   The effect of audiogenic seizures in
    rats on the adrenal weight, ascorbic acid, cholesterol and
    corticosteroids.   Journal of Biological Chemistry 229:563.

14. Everett, J.W., 1947.   Hormonal factors responsible for de-
    position of cholesterol in the corpus luteum of the rat.
    Endocrinology 41:364.

15. Fee, A.R. and A.S. Parkes, 1929.   Studies on ovulation.
    I. The relation of the anterior pituitary body to ovulation
    in the rabbit.   Journal of Physiology 67:383.

16. Fortier, C., 1951.   Dual control of adrenocorticotrophin
    release.   Endocrinology 49:782.

17. Harris, G., 1954.   Hormone produced by neurosecretory cells.
    Recent Progress in Hormone Research 10:234.

18. Hawkins, J.E., Jr., 1959.   The otoxicity of kanamycin.
    Annals of Otology, Rhinology and Laryngology 68:698.

19. Jurtshuk, P., Jr., A.S. Weltman and A.M. Sackler, 1959. Biochemical responses of rats to auditory stress. Science 129:1424.

20. Long, J.A. and H.M. Evans, 1922. The oestrous cycle in the rat and its associated phenomena, by J.A. Long and Herbert McLean Evans, Berkeley, Univ. of California Press 1922. Memoirs of the Univ. of California v.6.

21. Marshall, F.H.A. and E.B. Verney, 1936. The occurrence of ovulation and pseudopregnancy in the rabbit as a result of central nervous stimulation. Journal of Physiology 86:327.

22. Marshall, F.H.A., 1936. I. Sexual periodicity and the causes which determine it. The Coronian Lecture. Philos. Trans. B.226, 423. II. Exteroceptive Factors in sexual periodicity. Biological Reviews 1942, vol. 17, 68.

23. Miline, R. and O. Kochak, 1951. L'Influence du bruit et des vibrations sur les glandes surrenales. Comptes Rendus de l'Association des Anatomistes 38:692-703.

24. Miline, R. and Chtchepovitch, 1953. Influence du bruit et des vibrations sur les testicules de rats. Comptes Rendus de l'Association des Anatomistes 40:173-178.

25. Nelson, W.O., 1929. Oestrus during pregnancy. Science 70:453.

26. Parkes, A.S. and C.W. Bellerby, 1926. I. Studies on the internal secretions of the ovary. II. The effects of injection of the oestrus producing hormone during pregnancy. Journal of Physiology 62:145.

27. Parkes, A.S., E.C. Dodds and R.L. Noble, 1938. Interruption of early pregnancy by means of orally active oestrogens. British Medical Journal 2:557.

28. Parkes, A.S., 1961. Animal Reproduction. Proc. IVth Intern. Congr. The Hague, 163.

29. Sackler, A.M., A. St. Weltman and P. Jurtschuk, 1960. Endocrine aspects of auditory stress. Aerospace Medicine 31:749.

30. Sackler, A.M., R.R. Sackler and J.H.W. Van Ophuijsen, 1950. The research background of a system of neuroendocrinologic formulations, part 1 of physiodynamics and some major metabolic disorders. Journal of Clinical Psychopathology 11:1.

31. Sackler, A.M., A.S. Weltman, M. Bradshaw and P. Jurtshuk, Jr., 1959. Endocrine changes due to auditory stress. Acta Endocrinologia 31:405.

32. Scharrer, E. and B. Scharrer, 1954.   Hormones produced by
    neurosecretory cells.   The relationship between the hypo-
    thalamus and the posterior lobe of the pituitary.   Recent
    Progress in Hormone Research 10:188.

33. Swezy, O. and H.M. Evans, 1930.   Ovarian changes during preg-
    nancy in the rat.   Science 71:46.

34. Zondek, B., 1926.   Über die Funktion des Ovariums, Ztschrift
    für Geburtshilfe und Gynaekologie (Stuttgart) 90:372-380.

35. Zondek, B. and S. Aschheim, 1927.   Hypophysenvorderlappen
    und ovarium.   Beziehungen der Endocrinen Drüsen zur ovarial-
    funktion.   Archive für Gynakologie 130:1.

36. Zondek, B. and S. Aschheim, 1928.   Ovulation der Graviditat
    ausgelöst durch Hypophysenvorderlappen hormone.   Endokrino-
    logie 1:10.

37. Zondek, B., 1929.   Hypophysenvorderlappen und Schwangerschaft.
    Endokrinologie 5:425.

38. Zondek, B., 1929.   Weitere Untersuchungen zur Darstellung,
    hypophysenvorderlappen Hormons (Prolan).   Klinische Wochen-
    schrift (Berlin) 4:157.

39. Zondek, B. and I. Tamari, 1958.   Stimulation of the anterior
    pituitary function with pronounced decreased infertility by
    stimulation of the auditory organs.   Bulletin Research Council
    Israel, v. 7E, 155.

40. Zondek, B., 1959.   Studies on the mechanism of the female
    genital function.   Fertility and Sterility 10:1-14.

41. Zondek, B. and I. Tamari, 1960.   Effect of audiogenic stimu-
    lation on genital function and reproduction.   American Journal
    of Obstetrics and Gynecology 80:1041-1048.

42. Zondek, B. and I. Tamari, 1964.   Effect of auditory stimulation
    on reproduction. IV. Experiments on deaf rats.   Proceedings of
    the Society for Experimental Biology and Medicine 116:636-637.

43. Zondek, B. and I. Tamari, 1964.   Effect of audiogenic stimu-
    lation on genital function and reproduction.   Acta Endocrinol-
    ogica, Supplement 90:227-234.

EFFECT OF NOISE DURING PREGNANCY UPON FOETAL AND SUBSEQUENT ADULT BEHAVIOR

LESTER W. SONTAG

The Fells Research Institute
Yellow Springs, Ohio 45387

It is certainly most appropriate that the sonic boom of supersonic jets be considered as an environmental intrusion. Had we
been concerned with this problem 40 years ago, I am sure that the
human fetus would not have been given much consideration as a human
being. At that time, to an unbelievable degree, the fetus was
considered to be immune from any but the most severe of stresses.
It was known that syphilis in the host mother could be transmitted
to the fetus with resulting congenital syphilis. A report of fetal
rickets (1) observed by a missionary during a severe famine in China
was in the literature. Anophthalmia (2) and cleft palate have been
reported in newborn pigs whose mothers were sustained on a vitamin
free diet. There were various other fragments of evidence or suggestions in the literature of environmental factors which could have
deleterious effects on the fetus. By and large, however, the unborn
child was considered to be immune from the variety of stresses which
were known to have damaging effects later in life upon the health
and development of a young infant. Investigations in recent years
have revealed a variety of relationships between the behavior, health
and development of the offspring of both human mothers and laboratory
animals and a variety of stresses during pregnancy.

In this paper I will first establish the vulnerability of the
fetus and its behavior pattern to a myriad of environmental changes,
and then I will review the relatively small amount of information
that is available concerning the effect of sound stimuli, as for
example the sonic boom, upon the behavior of the fetus and its
possible developmental and behavioral pattern after birth. I shall
try to show that such concepts as expressed some years ago by
Nandor Fodor (3) in his treatise The Search for the Beloved, are
very, very far from reality. Dr. Fodor wrote, "For nine months the

131

child lives and grows in the quiet contentment of its mother's womb.
Without struggle or effort, all its wants are gratified.   The sense
of perpetuity with which its requirements are taken care of makes
the child feel like a god in a universe of its own.   Nothing comes
to it in overwhelming abundance or in frustrating meagerness.   Conditions for its development are ideal.   The eternity of the fetal
night knows of no breaks.   The idea of change is beyond conception.
Growth is a sensation giving strength and power.   It not only
leaves the child's environment unaltered, but enhances the feeling
of omnipotence which the perfection of its environment inspires.
The existence of an overshadowing, all-enveloping supreme being
called mother, the awe-inspiring reality of a vast outer universe
peopled with incomprehensible giants are infinitely removed from
the cosmos of the tiny god of the womb."   I shall try to marshall
for you some of the evidence which may have perhaps motivated Rene
Dubos (4) to devote himself to this subject of fetal vulnerability.
"Until quite recently," stated Dr. Dubos in an interview, "it was
the custom to regard the fetal and neonatal states as relatively
uninvolved in the external environment except for such instances
as teratogenesis, overt infection or other obvious threat.   But
we now know that subtle variations in the biological environment,
factors germane to the environment of the mother, can profoundly
affect growth, development, and even the personality of the child."

     Among the earlier findings of the ability of the fetus to re-
act to noxious stimuli, the work of Sontag and Wallace (5) demonstrated an increase in fetal heart rate that resulted from the
mother's smoking a cigarette.   This experiment, which proved conclusively that there was an increase in heart rate and, therefore,
presumably that toxic products were entering the fetal circulation,
was later repeated by a number of investigators, including Hellman
et al (6).   The latter workers, however, used a somewhat different
technique in that they used pregnant women who were heavy smokers
and deprived them of cigarettes for 24 hours prior to the experiment.
They found the same increase in fetal heart rate which had been
previously described, but it occurred at the time the cigarette
was introduced to the mother and before it was lighted.   Their
interpretation was that the psychosomatic effects of anticipating
smoking the cigarette had created changes in the hormones of the
mother's blood which passed into the placental blood stream and
caused the accelerated heart rate.   Such an interpretation is
tenable also in view of the fact that certain drugs, including
epinephrine, produce cardiac acceleration when injected into the
mother's circulation.   However, it does not negate the possibility
and the probability that the toxic products of cigarette smoke do
enter the fetal circulation and exert a direct effect upon it.

     In a later study (7), it was shown that the fetal death rate
was 2½% higher among smoking mothers in a population of 3000 than
in non-smokers.   So also was the birth weight less in infants born

of smoking mothers, facts which emphasize the importance of placental transfer of these toxins.

That maternal emotional peaks could affect the activity of the
fetus was reported by Whitehead (8), an obstetrician, as early as
1867.   A more comprehensive set of observations have been published
by Sontag since 1935.   In this later study (9), in which the levels
of kicking and writhing movements of fetuses had been under observation daily or weekly for some weeks before such tragedies as the
accidental death of a husband occurred, it was demonstrated conclusively that fetal activity was increased during these periods
of emotional stress by a factor of 5 or 6.   It must be assumed
that hormone levels, blood sugar level or other changes in maternal
blood passing through the placenta, produced fetal blood changes
and thereby stimulated fetal motor activity (and heart rate).   It
was also observed that infants born of such emotionally disturbed
pregnancies exhibited greater postnatal activity, cried more and
suffered more frequently from disturbed gastro-intestinal function.
Unfortunately, the human is not a very good laboratory animal in
which to observe long time effects, so the results of such prenatal trauma are impossible to assess accurately at the adult level.
There are just too many postnatal differences which express themselves throughout the long human childhood to permit identification
of those differences in behavior or function resulting from such
prenatal environmental situations.

Ader and Conklin (10) have, however, influenced the behavior
of young in utero by producing emotional differences in pregnant
experimental rats.   In a well-controlled series of observations,
they have demonstrated significant differences at rat adolescence
in the behavior of the offspring of pregnant mothers handled for
10 minutes three times a day, as compared with those of unhandled
rats.   The experimental group had offspring which showed less
"emotionality" on the open field test than did the controls.   W.R.
Thompson (11), in a series of experiments created "neurotic" pregnant rats by subjecting them to a conflictual situation in which
their learned electric shock avoidance procedure was frustrated
in an unpredictable manner.   He, like Ader and Conklin, found
that the offspring of the neurotic rats at adolescence, performed
differently from the control group.   They were slower in leaving
their cages, were less free in exploration and had more frequent
urination and defecation, patterns of behavior commonly ascribed
to "anxiety".

Hockman (12), in a similar experiment, found that the offspring of the experimental animals showed reduced patterns of
activity and increased defecation at 30 and 45 days, but not at
180 and 210 days.   The mortality in the experimental group was
much higher during stress.

Ferreira (13), using a questionnaire inventory, assessed the attitudes of 103 pregnant women and categorized their infants after birth as deviant or non-deviant on the basis of criteria that included amount of crying, amount of sleep, degree of irritability, intestinal function and feeding behavior. He found that infants born of mothers who expressed fear of harming the baby and showed extreme rejection of the pregnancy showed more deviant behavior than did those of the mothers who served as controls.

The teratogenic effects of some drugs and infections have been well established as, for example, those of rubella and thalidomide (14). Other drugs used during the period of gestation have been shown to modify the fetal EEG tracings.

The above evidence of the vulnerability of the fetus, particularly during the early weeks of pregnancy, but also up to the time of birth, establish without question, I believe, the need for concern and further exploration of fetal-maternal environmental experiences which may leave a disadvantaged newborn. Actually, one might expand the statement to include a search for experiences during the prenatal period which could prove advantageous to the optimum development of the fetus and its later health and performance.

This session today has been called to discuss the impact of the sonic boom as produced by supersonic planes on the emotional state, health and performance of human beings. I have been asked to present to you what little information is available which would indicate the importance of assessing the effects of this violent sound stress on the individual who is not yet born. Since we have as yet no adequate evaluation of the effects of this sound stress on children and adults, it is not surprising that we know little or nothing about its effect on the fetus.

There are, however, certain observations and experiments on the effect of sound on the fetus which are, at this point, worthy of a review and which may have implication for further studies. Peiper (15), in 1925, reported increased kicking and body movement of the fetus during late months of pregnancy in response to loud sound applied to the mother's abdomen. Since similar sounds produce convulsive response of the young infant, a response known as the Moro reflex, it is not surprising that the fetus so responds. Peiper reported that immediately after the sound stimulus and the movement response, there was a refractory period during which the fetus did not again respond to the sound. He believed that the evidence of this refractory period indicated that the fetus is capable of receiving "sensory impressions". Peiper also called attention to the severity and amount of fetal movement felt by some women while listening to concerts. Forbes and Forbes (16) mentioned the case of a woman who, while attending a concert, experienced a sensation of great fetal activity at the time the

audience applauded.   Preyer (17), however, rather arbitrarily concluded that because the middle ear contains no air during the fetal
period, that sound was imperceptible to the fetus.

Sontag and Wallace (18), using an ordinary electric doorbell
clapper which struck a wooden disk placed on the mother's abdomen
for five seconds at one-minute intervals, elicited startle responses
from a none-month fetus in 28 of 29 trials.   In a series of subjects
to whom such stimuli were presented at two-minute intervals, there
was a 92% movement response.   There was also a substantial increase
in heart rate.

Bernard and Sontag (19) worked with eight-month and nine-month
fetuses using tones relatively pure in character that were generated
by a Jackson audio frequency oscillator amplified by a Setchell-
Carlson amplified Model 13-B and brought to the maternal abdomen
through a Utah G8P speaker resting on a rubber baffle centered over
the head region of the fetus.   The rubber baffle seemed to insure
that the sound energy was conducted through the air rather than
through a solid medium and, therefore, clearly "sound" was being
introduced as a fetal stimulus.   The audio frequency range was
from 20 to 4000+ vibrations per second.   At each frequency, there
was a significant increase in fetal heart rate.   That this "direct"
audio stimulation of the fetus produces both a cardiac and a skeletal-
muscular contractual response is now fully established.

Spelt (20) reports having achieved fetal conditioning.   His
conditioning procedure consisted of presenting a tactile stimulus
(US) to the maternal abdomen over a fetal part for five seconds
followed by a loud sound.   The sound (CS) was produced by a wooden
clapper striking a wooden box.   After repeated trials, the fetus
learned to anticipate the sound after presentation of the US.

Habituation, or fetal learning, was reported by Sontag and
Newbery (21).   In this instance a loud 120 double vibrations per
second sound was created by a doorbell knocker hitting a wooden
plate resting on the maternal abdomen for three seconds at one-
minute intervals for five minutes.   625 such stimuli were presented
to the fetus in the course of 125 such fourteen-minute sessions.
All records which could have been influenced by fetal hiccough or
other extraneous factors were excluded.   The records of heart rate
for each session were divided into prestimulation, stimulation and
poststimulation periods.   The heart rate response of the experimental fetus was compared to the composite response of 15 fetuses only
occasionally stimulated (a total of 51 trials distributed over the
last nine weeks of pregnancy).   In the non-experimental (control)
group, the heart rate acceleration in response to stimulation is
least in younger fetuses and becomes progressively greater and persists longer as term approaches.   In the experimental fetus, the
opposite was true.   The response decreased as the experimental

procedure continued and term approached.   The experimental fetus
apparently became habituated to the stimulus.

Thompson and Sontag (22), using female rats of a strain known
to be subject to audiogenic seizures were mated and then divided
into two groups, experimental and a control.   In the experimental
group, on the fifth day of pregnancy and continuing through the
eighteenth day, each female received two auditory stresses a day
placed eight hours apart.   The procedure consisted of placing the
animal in a metal enclosure containing an electric bell, which
produced a sound level of high intensity and composite frequency.
Females in the control group received no special handling.   Both
groups were weighed daily to determine any difference in weight
gain.   Both groups were permitted to develop normally and, by
random selection, each litter group was reduced to two males and
two females.   Litter switches were made as soon as possible to
foster mothers.   Both groups were handled and weighed every other
day.   Litter size in the two groups was not significantly different.   Maze learning testing was begun at eighty days of age.   The
experimental group was inferior in maze learning, a fact which suggests greater disorganization of behavior among the rats of the
experimental group.   Whether the slower learning was a product of
increased anxiety level or of an actual different metabolic learning capability of the brain cells is, of course, not clear.

This experiment introduces another very interesting question.
Was it the sound stimulation of the fetus which produced the differences in behavior and performance, or was it the changed maternal
environment created by the audiogenic stress which was responsible?
In instances of human maternal anxiety, and the different emotional
environments under which experimental and control groups of pregnant
rats have been maintained, we are obviously not dealing with a
direct external stimulation of the fetus.   The differing postnatal
behavior cannot be attributed to direct application of noxious
auditory stimuli to the fetal sensory system.

We must then, when considering the possible effects of the
supersonic boom on the later behavior and development of a human
fetus, be concerned both with the ways in which the boom modifies
the host-mother's environment around the fetus and how it stresses
fetal sensory systems directly.   Recently, sonar has been introduced as a method for studying the prenatal development.   So has
habituation of the fetus to sound stimuli (23) been used in an
attempt to identify defective fetuses, i.e., those fetuses which
fail to habituate.   Ian Donald (24) suggests that such procedures
do not present known hazards to the fetus.   However, in view of
our increased knowledge of the vulnerability of the fetus, the external and maternally introduced stresses, it might be well to go
slow in introducing such possible stresses as standard procedure
both in experimentation and evaluation.

Some of the work on the effects of emotionally induced environmental differences in pregnant rats and of naturally occurring
emotional stress in humans, on the behavior of the newborn and in
some instances, of the adolescent, is highly important.   It should
stimulate further exploration of the role of fetal environment on
differences in prenatal and postnatal behavior and function.   The
radio and television have become prominent aspects of environment
as stimulators of emotional status or as a quieting influence on
the psyche and psychophysiological mechanisms of human beings.
The radio, record player or television blares its sound for a considerable part of the day in a vast majority of homes in this country.
We decided, therefore, to attempt to determine whether music itself
is capable of producing physiological changes in the host mother to
a degree that there are demonstrable changes in some aspect of fetal
function.   If so, we could demonstrate that any such physiological
changes were being relayed to the fetus, without implying that
postnatal behavior was affected either favorably or unfavorably.

Seventeen women in the last six weeks of pregnancy served as
volunteer subjects for a study (25).   Each was seen for five successive days as a subject for our experiment.   The experiments
were conducted from two to forty-one days before birth.   Heart
rate was measured by recording fetal ECG's on equipment previously
described.   The beat-by-beat rate of the fetal heart was calculated
by measuring the distance between each R-R interval on a Benson-
Lehner Oscar E record reader.   The instrument converts the distance
between each two beats into time, which is punched onto IBM cards
for computer processing.   The maternal R-R interval data were
recorded directly onto IBM punch tape by means of a SETAR recorder.

Each subject was informed of the nature of the experiment and
its objective before serving as a subject.   She was asked to select
her own favorite piece of music which was then used as a music
stimulus in her experiment.   After electrodes were attached to her
abdomen, a ten-minute period of fetal and maternal heart rate were
recorded after which the musical selection was played for a period
of ten minutes, being presented through two AR2 floor speakers.
The music level was held to a level of 65 db to 100 db, with an
average of 75 db measured at the subject's head.   Following the
ten minutes of music, both maternal and foetal heart rates were
recorded for a rest period of fifteen minutes.

Both maternal heart rate and fetal heart rate rose apparently
in response to the music.   A comparison of heart rate during the
control period with that of the heart rate after two minutes of
music stimulation showed a significantly elevated fetal heart rate
without elevated fetal movement.   While the maternal heart rate
was accelerated, the increase was not significant.   Because the
increase in fetal heart rate was appreciably slower in appearance
in this music stimulation experiment than it was in the doorbell-

wooden-block-on-the-maternal-abdomen experiment previously described,
and since there was no increase in fetal activity as occurred in the
doorbell experiment, we are inclined to believe that the stimulation
was mediated through a maternal psychophysiological reaction.   It
was a part of the emotion engendered in the mother by the music.
This is in contrast to the response to the directly perceived sound
as occurred in the doorbell=wooden block experiment.   In other
words, it was not a true Moro reflex.

As I have stated before, there are many reports of changed
adult behavior as a result of modified prenatal environment.   Our
work presents no such evidence, possibly because we have been dealing with human beings, animals whose developmental span is so long
as to permit the intrusion of innumerable and highly varied stresses
and impacts along the years from birth to maturity.   In addition,
we are dealing with a minimum stress stimulus which, if there were
postnatal sequelae, we would expect to be within the range of normal patterns of behavior.

We have, however, been able to demonstrate a statistical relationship between quick movement or activity during the fetal
period and social apprehension defined as hesitation to join groups,
anxiety in the face of threatened peer aggression, reluctance to
enter nursery school car, etc., at two-and-one-half years.   The
relationship is stronger in male children than in females.   We
have also been able to establish tentatively that there is a relationship between the beat-by-beat irregularity in the rate of
the fetal heart at the end of pregnancy and the beat-by-beat
lability-stability in the same individuals at post adolescence (26).
There is also demonstrated relationship between lability-stability
of heart rate and the degree of affect in the perception of young
adults (27).   The fact that there appear to be similar patterns
of lability-stability in young adults to those they exhibited
during a fetal period cannot, of course, be interpreted with the
knowledge we have now as products of prenatal stress.

Nevertheless, affect, a part of personality which is the
degree of response of the individual to perceived stimuli and,
therefore, also a part of the way he perceives stimuli, has a
physiological correlate, cardiac lability-stability.   And this
lability-stability is relatively fixed over at least the first
thirty years of life including the fetal period.   It seems logical
to explore further what fetal and neonatal factors influence this
lability or whether it is a gene determined factor.   It seems
important to learn whether cardiac lability-stability is merely
statistically correlated with some aspects of personality or
whether it is an expression of autonomic function which is really
a determinant of affective response.

I believe I have mustered enough evidence of the vulnerability

of the mammalian fetus to environmental stress and the child-adult
consequences of such stress to justify our concern about the possibility of fetal damage from such violent sounds as sonic booms.
It seems not unlikely that adults are not alone in their objection
to such noxious stresses.   The fetus, while he cannot speak for
himself, may have equal or greater reason to object.

SUMMARY

There is ample evidence that environment has a role in shaping
the physique, behavior and function of animals, including man, from
conception and not merely from birth.   The fetus is capable of
perceiving sounds and responding to them by motor activity and
cardiac rate change.   It is capable of undergoing conditioning in
the late months of pregnancy, anticipating from one stimulus a
second stimulus, and responding to the first stimulus as if the
second had been applied.   Its physiological system is capable of
responding to physiological changes in the mother, changes induced
by a variety of sounds or emotions and mediated through the placental
interchange.

Extreme maternal anguish produces greatly increased motor
activity of the fetus.   Emotion-inducing music causes a change in
the fetal heart rate.   Emotional or mood differences created in
laboratory animals by frequent handling and petting produce offspring which, at maturity, function better and are "less anxious"
than those of mothers isolated from human beings during pregnancy.

Human beings in severe emotional states in the later months of
pregnancy produce hyperkinetic, uncomfortable and poorly functioning infants.   Hyperkinetic fetuses later exhibit a pattern of
apprehension in peer situations at the age of two or three years.
Such evidence, and more, suggests that we must be concerned about
the advent of the sonic boom as a daily occurrence in the lives of
pregnant women because of its possible damaging effect upon the
fetus through alteration of the maternal environment and its
possible consequent alteration of the total behavior and adjustment
pattern in later life.

REFERENCES

1. Sontag, L.W. and R.F. Wallace, 1934.   Preliminary report of the
   Fels Fund.   American Journal of Diseases of Children 48:1050-57.

2. Hale, Fred, 1935.   The relation of vitamin A to anophthalmos
   in pigs.   American Journal of Ophthalmology 18:1087-1091.

3. Fodor, Nandor, 1949.   The Search for the Beloved, Hermitage

Press, Inc., New York.

4. Dubos, Rene, 1968.   Neonatal deprivations can be permanent
   ones.   Journal of the American Medical Association 205:34-35.

5. Sontag, L.W. and R.F. Wallace, 1935.   The effect of cigarette
   smoking during pregnancy upon fetal heart rate.   American
   Journal of Obstetrics and Gynecology 29:77-82.

6. Hellman, L.M., H.L. Johnson, W.E. Tolles and E.H. Jones, 1961.
   Some factors affecting the fetal heart rate.   American Journal
   of Obstetrics and Gynecology 82:1055-1063.

7. Frazier, T.M., G.H. David, H. Goldstein and I. Goldberg, 1961.
   Cigarette smoking and prematurity: a prospective study.
   American Journal of Obstetrics and Gynecology 81:988-996.

8. Whitehead, J., 1867.   Convulsions "in utero".   British Medical
   Journal 2:59.

9. Sontag, L.W., 1941.   The significance of fetal environmental
   differences.   American Journal of Obstetrics and Gynecology
   42:906-1003.

10. Ader, R. and P.M. Conklin, 1963.   Handling of pregnant rats:
    effects of emotionality on their offspring.   Science 142:411-412.

11. Thompson, W.R., 1957.   Influence of prenatal maternal anxiety
    on emotionality in young rats.   Science 125:698-699.

12. Hockman, C.H., 1961.   Prenatal maternal stress in the rat: its
    effect on emotional behavior in the offspring.   Journal Compara-
    tive and Physiological Psychology 54:679-684.

13. Ferreira, A.J., 1965.   Emotional factors in prenatal environ-
    ment.   Journal of Nervous and Mental Disease 141:108-114.

14. Lieberman, N.W., 1963.   Early developmental stress and later
    behavior.   Science 141:824-825.

15. Peiper, A., 1925.   Sinnesempfindungen des Kindes vor seiner
    Geburt.   Monatschrift für Kinderheilkunde 29:236-241.

16. Forbes, H.S. and H.D. Forbes, 1927.   Fetal sense reactions:
    Hearing.   Journal of Comparative Psychology 7:353.

17. Preyer, W., 1885.   Spezielle Physiologie des Embryo. Leipzig:
    T. Grieben. pg. 481.

18. Sontag, L.W. and R.F. Wallace, 1935.   The movement response of

the human fetus to sound stimuli. Child Development 6:253-258.

19. Bernard, J. and L.W. Sontag, 1947. Fetal reactivity to fetal stimulation: a preliminary report. Journal of Genetic Psychology 70:205-210.

20. Spelt, D.K., 1948. The conditioning of the human fetus "in utero". Journal of Experimental Psychology 38:338-346.

21. Sontag, L.W. and H. Newbery, 1940. Normal variations of the heart rate during pregnancy. American Journal Obstetrics and Gynecology 40:449-452.

22. Thompson, W.D. and L.W. Sontag, 1956. Behavioral effects in the offspring of rats subjected to audiogenic seizure during the gestational period. Journal of Comparative and Physiological Psychology 49:454-456.

23. Electronics in Perinatal Studies, 1969. Bio-Medical Engineering 4:120-121.

24. Donald, Ian, 1969. Sonar as a method of studying prenatal development. Journal of Pediatrics 75:326-333.

25. Sontag, L.W., W.G. Steele and M. Lewis, 1969. The fetal and maternal cardiac response to environmental stress. Human Development 12:1-9.

26. Sontag, L.W., 1963. Somatopsychics of personality and body function. Vita Humana 6:1-10.

27. Lacey, J.I., J. Kagan, B.C. Lacey and H. Moss, 1963. The visceral level: situational determinants and behavioral correlates of autonomic response patterns. In: P.J. Knapp (Ed.), Expression of Emotions in Man, International University Press, New York.

EXTRA-AUDITORY EFFECTS OF SOUND ON THE SPECIAL SENSES

JOSEPH R. ANTICAGLIA*

Bureau of Occupational Safety and Health
1014 Broadway
Cincinnati, Ohio   45202

The auditory pathways of the nervous system consist of specific projections to the auditory cortex and non-specific projections to the reticular formation. Impulses from the reticular formation are transmitted to different areas of the brain and also influence the peripheral autonomic nervous system (Figure 1) (1,2). In this way, acoustic stimulation may influence both auditory and non-auditory processes. This paper reviews the effects of sound (and noise) upon the sensory processes and its use as an audio-analgesic.

SPECIAL SENSES

Non-visual sensory stimulation, such as labyrinthic excitation (3) or exposure to sound (4) can affect visual functions. The not infrequent association of retinitis pigmentosa (5) with deafness and chronic simple glaucoma with hypoacusia (6) has prompted some researchers to investigate the effects of sound on the field of vision (8,9).

Grognot and Perdriel (10) briefly exposed healthy young subjects to noise of different spectra in the 98-105 dB range.≠ Such exposure caused color perception to be modified with a tendency to protanomalia, and night vision to be diminished. Others report that sound adversely affects depth perception (11), provokes mydriasis (12,13), and influences intraocular pressure (14). These effects were transitory in nature and conditions returned to normal within a few hours after cessation of the sound.

*Director, National Noise Study

≠Unless otherwise noted, all references to decibels will be in sound pressure level re 0.0002 microbar measured on the C scale.

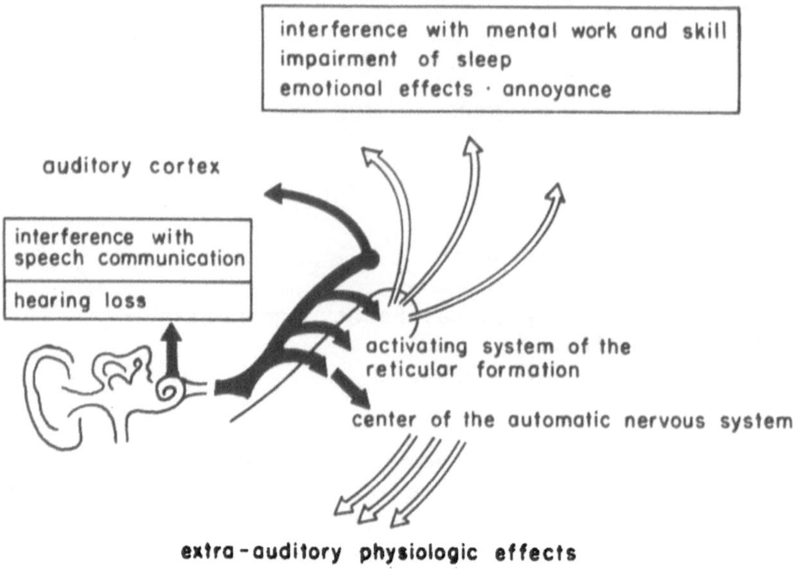

interference with mental work and skill
impairment of sleep
emotional effects · annoyance

auditory cortex

interference with
speech communication

hearing loss

activating system of the
reticular formation

center of the automatic nervous system

extra-auditory physiologic effects

Figure 1. Physiologic organization of the auditory pathways in
the brain and their relation to the effects of noise on man (modified from Grandjean (1,2).

However, Benko (15) examined workers who had been employed for
one to four years where they were exposed to noise levels in the
110-124 dB range for eight hours each workday and found a narrowing of the field of vision that appeared to be permanent. The
controls (workers who had been employed for only six months) also
demonstrated a progressive narrowing of the field of vision.

These findings were supported by Ogielska (4) who concluded
that a noise intensity of about 100 dB narrowed the field of vision
for red in 10.5 percent of the boilermakers examined. In 18 percent
of these cases, the narrowing of the field of vision was constant.
Also, in a series of one hundred wire mill workers exposed to
noise levels of 100-105 dB for an average of sixteen years,
Vynckier (8) observed alterations in color perception which he
felt confirmed a pathologic relationship between excessive noise
exposure and certain visual functions.

An interesting correlation exists between auditory stimulation
and the condition of the iris. The concentration of melanin in
the stria vascularis of the cochlear is directly proportional to
that of the iris (16) and inversely proportional to that of dehydro-
genase in the cochlear (17). The stria vascularis is almost de-
pigmented when the iris is blue, but has higher concentrations of
pigment when the pigmentation of the iris is dark brown. This pigment, according to Bonaccorsi (18), acts in the respiratory chain

and seems to protect the neuroepithelium from intense stimulation.

Tota and Bocci (19) studied the relationship between the color of the iris and the temporary threshold shifts for hearing that are caused by exposure to sound. One hundred healthy young subjects were subdivided according to the color pigmentation of their iris and were exposed to a sound pressure level of 100 dB centered at 1000 Hz for three minutes. The authors reported the greatest threshold shifts and the longest recovery time in subjects with a lightly pigmented iris while the opposite was noted in those individuals with a darkly pigmented iris. They interpreted these results as support for the hypothesis that there is a relation between iris melanin and resistance to acoustic trauma. They concluded that the resistance to auditory fatigue is directly proportional to the pigment concentration in the iris.

It has been suggested that noise might influence the critical flicker frequency, the frequency of flicker above which a flickering light may be perceived as steady (20,21). Noise appears to lower the critical flicker frequency, but the effects of different noise levels, of subject age and of test conditions are too variable to permit generalization.

The vestibular organs are in close proximity to the cochlea of the inner ear. Powerful or moderate auditory stimulation can elicit nystagmus, vertigo, and disruption of equilibrium. Sounds of modest intensity elicit lateral eye movements in normal subjects which Hennebert termed "audiokinetic nystagmus (22,23). Intense noise levels of 130 dB or higher induced nystagmus and vertigo even in deaf subjects (24). Bekesy reported vertigo in normal persons exposed for brief periods to intermittent sound stimulation of 100 Hz at 120 dB (25). Complaints of disturbances to equilibrium have been noted after jet aircraft exposure (26), and difficulty in maintaining balance has been reported in a laboratory situation during exposure to noise intensities of 120 dB (27).

The transitory vertigo experienced by some people when exposed to loud sounds (the Tullio phenomenon) is considered an uncommon reaction to sudden, intense noise exposure. It is occasionally observed after a fenestration operation if the footplate is mobile (28). No reports have associated excessive noise exposure with either ageusia or anosomia.

Exposure to high intensity sound can evoke changes in the electrical potential and resistance between two points of the skin (Tarchanow and Fer Effects, respectively) (29). Plutchik exposed eighteen subjects to brief periods of intermittent sound in the 100-120 dB range. There was no effect on skin temperature, but galvanic skin response showed a linear decrease in voltage with an increase in the intensity of the sound (30).

AUDIO-ANALGESIA

Gardener and Licklider (31) described an audio-analgesic device which suppressed pain experienced by dental patients. These patients listened to music or white noise through earphones but operated a switch button to control the sound intensity, increasing the levels as needed to suppress pain. In a series of one thousand patients, Gardener et al. (32) reported that audio-analgesia was complete or sufficiently effective in nine out of ten patients so that no ordinary anesthetic was required. Others corroborated the clinical efficacy of audio-analgesia in pediatric patients (33), during minor and major dental operations (34) and in obstetrical cases (35). However, no controlled clinical studies were performed while evaluating the effectiveness of audio-analgesia in the latter situations.

Most controlled experiments that have been performed under laboratory conditions have gauged the ability of sound to influence dental pain. In one study, nineteen dental patients were intermittently exposed to a noise level of 100 dB with a commercial apparatus (36). An electrode was placed on the patients' teeth, half of which were electrically stimulated in the presence of noise and half of which were stimulated without noise. If a higher voltage was needed to evoke a response when the patient was exposed to sound, this was interpreted as evidence that noise suppressed pain. In order to study the effects of suggestion, eight of these patients were informed that noise could reduce pain sensations. The results obtained did not support the contention that pain suppression can be attributed to the influence of noise. Negative results also have been reported in attempts to use audio-analgesia to decrease the pain response to intense radiant energy (37) and also to other types of superficial pain (38). Yet Morosko and Simmons (39) contend that the subjects in the latter experiments were not allowed to control the volume of white noise and that this influenced the results obtained. They presented white noise in the 100-120 dB range to their test group. A stimulating electrode was placed on the right central incisor, and each of forty dental students was instructed to turn up the noise when the pain was noticeable. Here, audio-analgesia significantly elevated threshold and the tolerance for pain. The authors also noted that eliminating apprehension and anxiety by suggestion reduced the overall painfulness of the experience.

DISCUSSION

Various explanations have been offered for the effects of sound upon the special senses. Neural impulses generated by sound are known to spread throughout the gray and white matter of the brain, though it is recognized that the responses of certain areas of the brain are highly specialized while those of others are relatively non-specific (40).

The specialized area for vision is located in the occipital cortex, but analytical elements of vision are dispersed over the surface of the hemispheres. Ogielska (4) suggested that certain visual responses to color are found in these areas and are adversely affected by excessive noise stimulation. Others postulate that noise impairs visual functions by influencing the reticular formation (15) or a specific tempro-occipital tract (3).

Nausea and vertigo have been described secondary to intense noise exposure on aircraft carriers (41). Under laboratory conditions, 135 dB at 1000 Hz was reported by Ades to be the threshold of vestibular stimulation in humans (41). In histological studies, McCabe and Lawrence (42) found complete collapse of the saccular epithelial tube in guinea pigs exposed to intense sound, and they suggested that "the symptoms reported by individuals placed in intense sound areas are mediated through the saccule". Hence, the saccule is likely the site of acoustic damage in the vestibular labyrinth.

Obviously, sound has important effects upon the special senses. But further work is required to determine the prevalence of its effects and to evaluate their pathologic significance.

SUMMARY

The literature on the extra-auditory physiological effects of sound on the special senses is reviewed. Evidence is presented that noise can adversely affect certain visual functions, induce nystagmus and vertigo, disrupt equilibrium, influence the galvanic skin response, and act as an audio-analgesic. Possible mechanisms of such effects are discussed.

REFERENCES

1. Anticaglia, J. and A. Cohen, 1970. Extra-auditory effects of noise as a health hazard, Journal of the American Industrial Hygiene Association, in press.

2. Grandjean, E. Biological effects of noise. Department of Hygiene and Work Physiology, Swiss Federal Institute of Technology, Zurich (undated).

3. Pressburger, I. and I. Sommer, 1937. Uber die beeinflussung des gesichtsfeld durch labyrinthare reize. Zentralblatt für die Gesamte Ophthalmologica 37:667.

4. Ogielska, E. and K. Brodziak, 1965. L'influence due bruit sur le champ visuel. Annales d'Occulistiane (Paris) 198:115-122.

5. Guerrier, Y. et al 1960. Retinite pigmentaire et surdite. Journal Francais d'Oto-Rhino-Laryngologie, Audio-Phorologie et Chirurgie Maxillo-Faciale (Lyon) 8:1025-1035.

6. Ceroni, T. et al 1958. A study in the vestibular function in patients with chronic simple glaucoma. Rivista Oto-Neuro-Oftalmologica (Bologna) 33:172-194.

7. Beretta, L. and E. Cerabollini, 1957. The problem of hypoausia in glaucoma. Rivista Oto-Neuro-Oftalmologica (Bologna) 32:662-677.

8. Vynckier, H., 1967. Surdite professionelle et champ visuel. Acta Oto-Rhino-Laryngologica, Belgica (Bruxelles) 21:213-222.

9. Kravkov, S., 1936. The influence of sound upon the light and color sensibility of the eye. Acta Ophthalmologica (Kobenhavn) 14:348-360.

10. Grognot, P. and G. Perdriel, 1959. Influence du bruit sur certaines fonctions visuels. Medicine Aeronautique 14:25-30.

11. Grandpierre, R., Grognot, P., and G. Perdriel, 1965. Le bruit est il nocif pour la vision? Presse Thermale et Climatique (Paris) 120:164-165.

12. Jansen, G., 1969. Effects of noise on the physiologic state. In: Noise as a public health hazard, American Speech and Hearing Association, Washington, D.C., Report 4:89-98.

13. Roth, N., 1966. Startling noise and resting retractive state. British Journal of Physiology 23:223-231.

14. Kishida, H., 1960. Effect of low frequency current on the eyes. Acta Societatis Ophthalmologicae Japonicae (Tokyo) 64:695-700, 1150-1156.

15: Benko, E., 1959. Objeckt- und farbengesichtsfeldeinengung bei chromischen larmschaden. Ophthalmologica 138:449-456.

16. Bonaccorsi, P., 1965. Il colore dell' iride come "test" di valutazione quantitative, nell' uomo, della concentrazione di melanina nella stria vasculare. Annali di Laringologia, Oto-logia, Rinologia, Faringologia (Torino) 64:725.

17. Bonaccorsi, P., 1965. Studio comparativo sulla distribuizione della melanina della stria vasculare nella varie spire della coclea umana. Bollettino delle Malattie dell'Orecchio, della Gola, del Naso (Firenze) 83:829.

18. Bonaccorsi, P., 1963. Comportamento delle barriere emolabirin-tica, liquorale ed oftalmica nell' albinismo. Annali di Larin-gologia, Otologia, Rinologia, Faringologia (Torino) 62:432.

19. Tota, G. and G. Bocci, 1967. L'importanza del colore dell' iride nella valutazione della resistanza dell' udito all' affaticamento. Rivista Oto-Neuro-Oftalmologica (Bologna) 42:183-192.

20. Levinson, J., 1968. Flicker fusion phenomena. Science 160:21-28.

21. Besser, G.M., 1967. Some physiological characteristics of auditory flutter fusion in man. Nature 214:17-19.

22. Hennebert, P., 1960. Nystagmus audiocinetique. Journal of Auditory Research 1:84-89.

23. Weber, B. and W. Milburn, 1967. The effects of a rhythmically moving auditory stimulus on eye movements in normal young adults. Journal of Auditory Research 7:259-266.

24. Ades, H.W. et al, 1957. Nystagmus elicited by high intensity sound. Research Project NM 130199, Subtask 2, Report 6, U.S. Navy Research Laboratory, Pensacola, Florida.

25. Bekesy, G., 1935. Acoustic stimulation of the vestibular apparatus. Archiv für die Gesamte Physiologie 236:59-76.

26. Dickson, E.D. and D. Chadwick, 1951. Disturbance of equilibrium and other symptoms induced by jet engine noise. Journal of Laryngology and Otology 65:154-165.

27. Von Gierke, H.E., 1965. On noise and vibration criteria. Archives of Environmental Health 11:327-339.

28. Lange, G., 1966. The tullio phenomena and possibility for its treatment. Archiv für Klinische und Experimentelle Ohren-, Nasen-und Kelhkopfheilkunde (Berlin) 187:643-649.

29. Hirsch, I., 1952. The measurement of hearing, McGraw-Hill Book Cp., Inc., New York.

30. Plutchik, R., 1962. Physiological responses to high intensity

intermittent sound. Contract Number NONR-2252 (01), Office
of Naval Research, Hofstra College, Hempstead, New York.

31. Gardner, W. and J. Licklider, 1959. Auditory analgesia in den-
    tal operations. Journal of the American Dental Association
    59:1144-1149.

32. Gardner, W., Licklider, J. and A. Weisz, 1960. Suppression of
    pain by sound. Science 132:32.

33. Sidney, B., 1962. Audio-analgesia in pediatric practice: A
    preliminary study. Journal of the American Pediatric Association
    7:503-504.

34. Monsey, H., 1960. Preliminary report of the clinical efficacy
    of audio-analgesia. Journal of the California Dental Associa-
    tion 36:432-437.

35. Burt, R. and G. Korn, 1964. Audio-analgesia in obstetrics.
    American Journal of Obstetrics and Gynecology 88:361-365.

36. Carlin, S. et al, 1962. Sound stimulation and its effect on
    dental sensation threshold. Science 138:1258-1259.

37. Camp, W., Martin, R., and L. Chapman, 1962. Pain threshold
    discrimination of pain intensity during brief exposure to intense
    sound. Science 135:788-799.

38. Robson, J. and H. Davenport, 1962. The effects of white sound
    and music upon superficial pain threshold. Canadian Anaesthesi-
    ology Society Journal 9:105-108.

39. Morosko, T. and F. Simmons, 1966. The effect of audio-analgesia
    on pain threshold and pain tolerance. Journal of Dental Research
    45:1608-1617.

40. Rosenblith, W., 1968. In: Noise as a public health hazard.
    American Speech and Hearing Association, Washington, D.C.,
    Report 4:12-17.

41. Ades, H., 1953. Benox Report, Orientation in space. Office of
    Naval Research Report 144079, University of Chicago.

42. McCabe, B. and M. Lawrence, 1958. The effects of intense sound
    on the non-auditory labyrinth. Acta Oto-Laryngologica (Stockholm)
    49:147-157.

HUMAN STUDIES OF EPILEPTIC SEIZURES INDUCED BY SOUND AND THEIR CONDITIONED EXTINCTION

FRANCIS M. FORSTER

Department of Neurology
School of Medicine
University of Wisconsin
Madison, Wisconsin

In considering the untoward effects of auditory stimulation the most important and obvious instances are those where the auditory stimulus is unpleasant for the vast majority of people (for example, sonic booms or the screaming of sirens) or where it is reasonably unpleasant for a percentage of the population yet may be harmful even to those who find it pleasant (rock and roll music). The untoward and unpleasant effects of these auditory stimuli affect large numbers of people. This paper deals with a small select group of patients who by their own predisposition are harmed by auditory stimulation which for others is entirely neutral (for example, telephone bells or even pleasurable music). In these patients the auditory stimulus evokes a seizure because they are in that group of patients afflicted with what is referred to as sensory-evoked or reflex epilepsy.

Reflex epilepsy is not a large problem in the sense of numbers of patients. While epilepsy affects close to one percent of the population it is estimated by Symonds (1) that only six and one-half percent of the epileptic patients have their seizure evoked by sensory stimulation. The vast majority of these patients have their seizures evoked by stimulation in the visual modalities but auditory stimulation as a cause of seizures is by no means rare. The auditory stimulation may be relatively simple and non-specific as unexpected noise, or highly sophisticated and indeed difficult to determine as occurs in musicogenic epilepsy.

The most common of the auditory induced seizures is the so-called startle or acoustico-motor epilepsy. In this form the patient when presented with a sudden loud noise which is unexpected will have a seizure, usually a short minor seizure, usually akinetic

151

and with loss of consciousness.   There may be only movement artifact in the EEG (Figure 1) indicating a subcortical type of seizure
or a dysrhythmia as well defined as a three-per-second wave and
spike dysrhythmia (Figure 2).   The seizures may be elicited by the
presentation of any sudden unexpected noise, for example, the ringing of a door bell, telephone, gunshot, dropping a pan, or ringing
a buzzer.

There is a second group of startle patients, usually with
brain damage and with hemiplegia and in whom the presentation of
noise evokes clonic movements on the paralyzed side usually with
little or no EEG changes.   These clinical responses seem to be
release phenomena rather than true seizures.

Rarely this type of sensory stimulus evokes a major seizure
or convulsion. A brain wave tracing of a patient was studied some
twenty years ago at Jefferson Memorial College.   This child had small-
ness of body parts, hemiatrophy of the left hemisphere, and marked
mental retardation. When an emesis basin was dropped behind him, such
that he could not observe it, he jumped and thus produced artifact in
the record followed by some muscle artifact, and then there was the
beginning of spiking dysrhythmia in the left temporal region with a
gradual build-up and spread to other cortical areas.   The patient had
a focal seizure with head and arm deviation to the opposite side and
proceeded into a generalized tonic clonic seizure.   However, convulsions due to startling auditory stimuli are rare in humans.

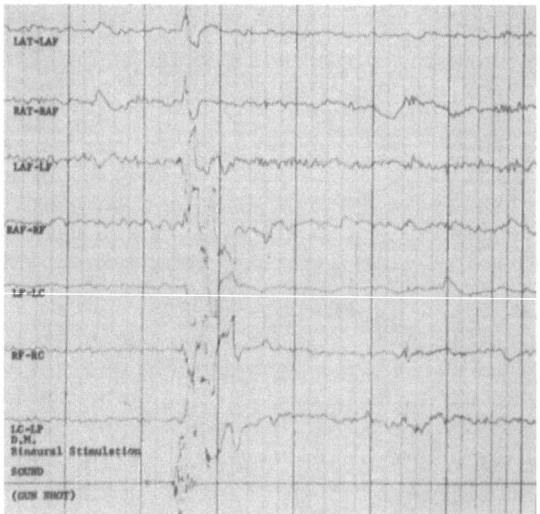

Figure 1. EEG of patient with startle epilepsy.   Lowest tracing is
stimulus marker indicating the presence of sound (gunshot).   Other
seven tracings are scalp EEG recordings.   The disturbance in the
EEG is artifactual and was not present during anectin administration.
This is a subcortical type of seizure.

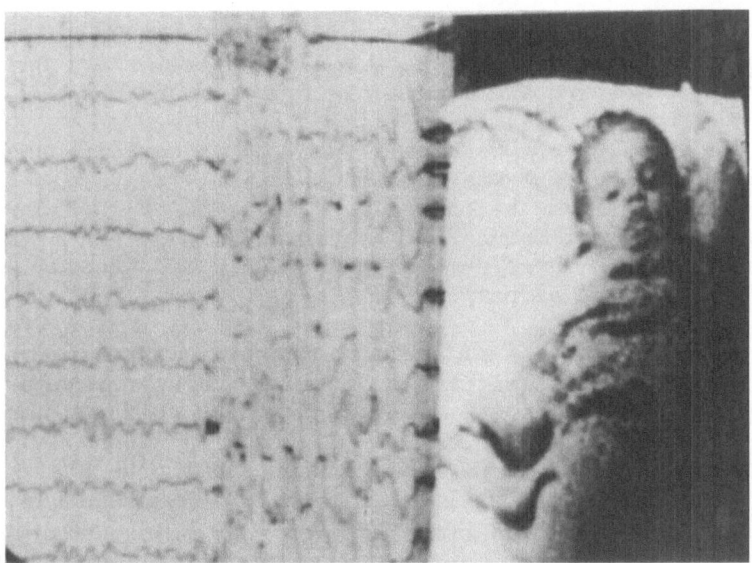

Figure 2. Filmed from AV tape, hence loss of resolution.   Infant
with startle epilepsy. Top channel of EEG is incidence marker and
indicates presentation of sound of bell. Record is running from
left to right. Three per second spike wave activity appears immediately after presentation of sound. Infant had rhythmic myoclonic movements.

      In some patients it is difficult if not impossible to reproduce
the situation in the laboratory.   One young lady who has seizures
with the ringing of the telephone bell in her home or her doctor's
office was brought to the University Hospital and did not have
seizures during the presentation of similar stimuli at the University of Wisconsin. The Telephone Company kindly taped the phones
in the local community which were different instruments from those
in the Madison area.   When these were played to her she identified
the ring of her own home phone but no seizure or dysrhythmia occurred.
In another instance a patient has had ten convulsions.   Eight of
these occurred in the exact situation of his business telephone
ringing, one of his employees answering it, calling his name, and
handing him the phone, and he announcing his name over the phone.
He immediately had an aura of a dreamy state type and had time to
get to the backroom of his shop where he proceeded into a generalized convulsion.   Calls from the particular telephone in his store
to the laboratory telephone did not evoke a seizure or EEG abnormality.   Upon taking the telemetry apparatus to his place of business
and telephoning him there with calls from the same people over the
same instruments also failed to evoke an EEG change or dysrhythmia.
There is no doubt in our minds that the patient's seizures are precipitated by something in the telephone connection but we have not
been able to discover what.   Playing telephone company training
tapes containing the artifacts of transmission also failed to evoke

a seizure.   Despite the expenditure of time and effort to recreate
the situation in which seizures occurred we sometimes are unable
to reproduce the seizure and induce brain wave changes.

     That the stimulus can be highly specific in certain cases is
indicated by a patient who was studied in an Army Hospital.   He
had seizures only when he heard billiard balls click.   From the
neurologic study it was found that he had an astrocytoma of the
temporal lobe, and this was removed.   Our detailed studies were
not possible after the surgery.

     The most complex types of auditory evoked seizures are those
elicited by voice characteristics or musical characteristics.
In the literature there is one case (2) in which seizures were
induced in a woman by certain voices (Figure 4).   These were all
male voices and included three radio announcers in the State of
Wisconsin, one of them in her local community.   She had had a
serious head injury followed after two years by major seizures and
focal seizures.   The major convulsions were controlled by anticon-
vulsive therapy.   The minor seizures were not and only occurred
when listening to a radio during programs including these radio
announcers.   They occurred four to ten times a week.   The minor
seizures consisted of short periods of confusion, loss of consciousness for a few seconds followed by a prolonged aphasic period.
It was determined by repeatedly playing the tape recordings of
these announcers that the seizures and electrical discharges were
not evoked by the pronunciation of particular sound or sequence of
sounds.   Analysis of voice characteristics of these three announcers
as well as of the voices of those around the laboratory which had
not evoked a seizure showed no significant difference.   Indeed the
voice prints could not be put into two groups, those noxious and
those innocuous.

     Somewhat similar to this in a way is the language induced
epilepsy described by Geschwind and Sherwin (3).   Similar situations
can be seen occasionally in reading epilepsy.   The three patients
having reading epilepsy that we have studied at the University of
Wisconsin all show no difficulty in reading memorized material or
repeating this material out loud (4).   A fourth case, however,
was reviewed by Bennett and myself (5) at the University of Utah
after he had read repeatedly a noxious paragraph and memorized it;
he still had seizures if he repeated it aloud from memory without
reading.

     The most esoteric of the auditory induced seizures is the
musicogenic.   Since 1937 there have been some 48 reported in the
literature, probably becoming more apparent in this day of tran-
sistorized techniques which make music available almost everywhere,
thus exposing musicogenic patients to liberal doses of the noxious

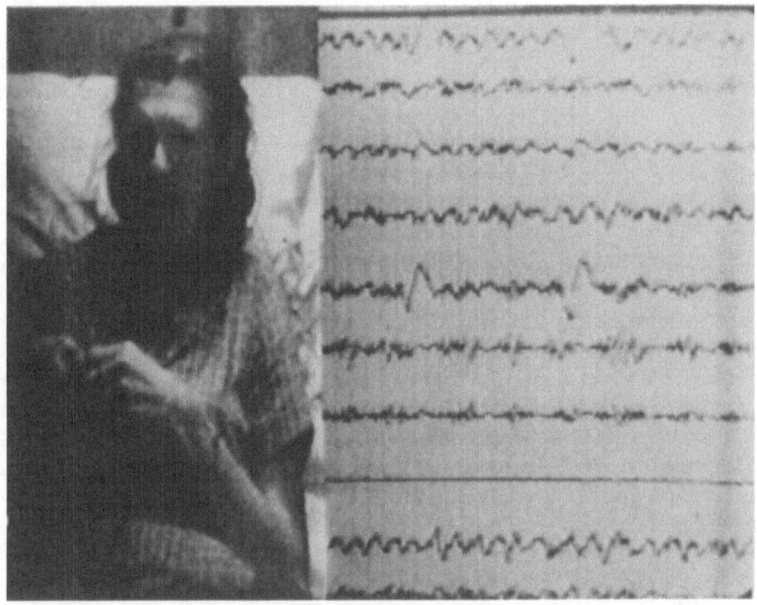

Figure 3. Filmed from AV tape, hence loss of resolution.   Voice
induced epilepsy.   Spiking dysrhythmia best developed in second to
last channel (left temporal).   Patient is aphasic.

stimuli.   These patients have seizures when listening to a particular type of music (6, 7).   Again, it is not the sound of a
particular instrument or the sequence of notes played.   This was
established by reviewing all the tapes and films of our three musicogenic epilepsy patients.   From these tapes and films we determined
the exact instrumentation, note, chord, and key, that was being
played at the occurrence of the dysrhythmia, and there was no
common denominator in a given patient in these musical factors.
On replaying the audio-visual tapes for the patients so they could
hear the music on the AV tape while we compared their taped EEG
with the second EEG tracing we could establish no synchrony between
the dysrhythmia in the two sets of tracings.   In other words, the
spiking dysrhythmia occurred at different parts in the music.   One
of the patients was a musician and said that we shouldn't expect
to find the noxious factor in the mechanics of the music and thought
it was more likely to be within the theme.   This was confirmed by
the fact that after the cessation of the music at the close of a
noxious musical number the spiking dysrhythmia still occurs for
the few seconds when the "music rings through your head."

These types of seizures like other forms of reflex epilepsy
are amenable to clinical therapeutic conditioning.   This type of
therapy has been described in the various visually induced sensory

seizures as well as in those evoked by auditory stimulation.   In
essence, the treatment consists of either (a) the repeated presentation of a modified stimulus so that it is innocuous, or (b) the
presentation of the stimulus in order to evoke a seizure followed
by repeated presentations during the seizure and the postictal
period and again after return to the normal clinical and electroencephalographic state (8).

These methods are much more similar than they seem to be on
the face of the matter.   The second method, repeated presentation
through the seizures and postictal period, can be used successfully
only in the instances where psychomotor seizures have been evoked,
and where there is a period of postictal confusion and postictal
slowing in the electroencephalogram in the area of the original
focus.   In other words while the stimulus is being continually
presented at a noxious level, it is being presented at a time when
the brain itself has an elevated seizure threshold; therefore although the stimulus is of usual strength the receptivity of the brain
is decreased.   In the startle types the unilateral presentation
of the stimulus is usually innocuous provided the level of stimulation is held below 60 decibels.   Above this level even though
the stimulus has been administered monaurally there is binaural
input due to bone conduction.   The stimulus can be presented repeatedly monaurally and then eventually binaurally either directly
in free fields or gradually introduced into the second ear.   Stereo-
phonic earphones and stereo tape recorders are used with identical
material on the two bands but the volume as presented to each ear
is altered.   The voice induced and the musicogenic epilepsy are
treated in the second method that is with the presentation of the
noxious stimulus evoking the seizure and then the continuous repetition of the noxious sounds.

There is a certain degree of specificity in the results of the
conditioning.   A startle patient who is sensitive to the sound of
bells, buzzers, and gunshots, if conditioned to the sound of the
bell, will still have seizures on the presentation of rifle shots
or buzzers.   There is also some generalization, for bells that
are somewhat similar to the one to which the patient was conditioned
will not evoke a seizure.   The generalization is seen more clearly
in the musicogenic type of evoked seizure.   One patient was sensitive to the music of the Strauss waltzes, but after the Emperor
Waltz was made innocuous none of the Strauss waltzes evoked a
seizure.   The specificity was also evident for parts of Handel's
"Messiah", and other music quite dissimilar to the Strauss waltzes
still evoked a seizure.

While the clinical therapeutic conditioning is effective in
the laboratory it is not permanent and it is therefore necessary
to reinforce it by daily presentation of the formerly noxious

stimuli.   This is accomplished for the patients by preparing
tapes in the laboratory of those auditory stimuli which evoked
seizures and for which they have been conditioned in the laboratory.
These tapes are listened to at least twice a day by the patients.
The results of this reinforcement are more satisfactory in the complex types, the musicogenic and voice induced, than they are in the
startle epilepsy.

SUMMARY

This paper presents a review of studies with a relatively small
number of patients who were affected adversely by sound stimuli that
was not of sufficient volume, or indeed of type, to be unpleasant
or disturbing to the vast majority of citizens of this country.   In
these relatively few patients, however, these particular sounds have
seriously interfered with their way of life.   These effects are beyond a doubt pathophysiological and are not on an emotional or
functional basis; they are ameliorated by a behavioral type of
therapy, clinical therapeutic conditioning.

ACKNOWLEDGEMENT

Supported by U.S.P.H.S. Grant BO3360

REFERENCES

1. Symonds, C., 1959. Excitation and inhibition in epilepsy.
   Brain 82: 133-146.

2. Forster, F.M., P. Hansotia, C. Cleeland, and A. Ludwig, 1969.
   A case of voice-induced epilepsy treated by conditioning.
   Neurology 19: 325-331.

3. Geschwind, N. and J. Sherwin, 1967. Language-induced epilepsy.
   Archives of Neurology 16: 25-31.

4. Forster, F.M., W. Paulsen and F. Baughman, 1969. Clinical
   therapeutic conditioning in reading epilepsy. Neurology 19:
   717-723.

5. Bennett, D.R. and Forster, F.M. Unpublished data.

6. Forster, F.M., H. Klove, W.G. Peterson and A.R.A. Bengzon, 1965.
   Modification of musicogenic epilepsy by extinction technique.
   Transactions of the American Neurological Association, 1965:
   179-182.

7. Forster, F.M., H.E. Booker, and G. Gascon, 1967. Conditioning
   in musicogenic epilepsy. Transactions of the American Neurological Association, 92: 236-237.

8. Forster, Francis M., 1969. Conditional reflexes and sensory-
   evoked epilepsy: the nature of the therapeutic process.
   Conditional Reflex: 103-114.

THE FUNCTIONAL STATE OF THE BRAIN DURING SONIC STIMULATION

L.V. KRUSHINSKY, L.N. MOLODKINA, D.A. FLESS, L.P.
DOBROKHOTOVA, A.P. STESHENKO, A.F. SEMIOKHINA,
Z.A. ZORINA, L.G. ROMANOVA

Laboratory of the Physiology and Genetics of Behavior
Moscow Lomonosov State University, Moscow, U.S.S.R.

Audiogenic epilepsy of rodents is a striking example of the pathogenic action of strong sounds on a living organism. The ascertainment of the physiological mechanisms of this reaction, and above all of the protective brain mechanisms which prevent or weaken it, is of great significance for the elaboration of measures aimed at struggling against the noxious influence of noises on the human organism.

In particular, it proved possible to study different forms of inhibition, which hinder the development of convulsive seizures, on a model of audiogenic epilepsy (6,7,12,31,35,37-39,63,64,66). A. Anthony and E. Ackermann (2) used this model for studying stress factors. Audiogenic epilepsy made it possible to investigate some physiological mechanisms in the development of a shocklike state which leads to death as a result of an acute disturbance of blood circulation (13,14,35,40,42,45,47).

Investigations carried out by G.N. Krivitskaya (30) revealed some morphological changes which arise in the neurons of the rat's brain as a result of the action of an acoustic stimulus.

Bures (8,9), V.M. Vasilieva (81), A.F. Semiokhina (67), K.G. Guselnikova (24,25), M. Niaussat and P. Laget (57), M. Niaussat (56) established the specific features of the bioelectrical brain activity which are peculiar to different phases of a convulsive seizure in reflex epilepsy of rats and mice.

The investigations of F.A. Beach and T.H. Weaver (4), H. M. Weiner and C.T. Morgan (85), B.I. Kotlyar (27,28) as well as Van Bin (79), who extirpated different parts of the cerebral cortex, disclosed

the role of the latter in the development of various types of convulsive seizures in audiogenic epilepsy.

The use of an experimental model of reflex epilepsy in rats
proved to be very useful for evaluating and studying a number of
pharmacological substances (1,3,10,11,19,20,23,41,50,52,53,61,77,
78).

The results obtained from investigations of reflex epilepsy
in rodents proved also essential in the study of general questions
of pathological nervous activity connected with the physiological
mechanisms of epilepsy, as well as of other pathological states
(5,16,26,34-37,62,71,72).

The purposes of the present investigation are: (a) to describe
various types of convulsive seizures developing under the action of
an acoustic stimulus; (b) to examine the physiological mechanisms
of these seizures; (c) to establish the pathological consequences
of reflex epileptical seizures in rats; (d) to investigate the
protective brain mechanisms which prevent or weaken the convulsive
reactions to sonic stimulations and their consequences.

The whole material set forth below is predominantly based on
experimental investigations conducted in our laboratory since 1947.

AUDIOGENIC EPILEPTIC SEIZURES IN RATS AND THEIR CONSEQUENCES

Our experimental investigations showed that several types of
audiogenic seizures with a different aetiology can be established
in rats.

We distinguish the following three essentially different types:

(a) Seizures which arise in rats as a result of sonic stimulation during a period of one or one and a half minutes. This type
was disclosed in mice in 1924 by Studentsov in the Pavlov Laboratory
(see Y.A. Vasiliev, ref. 80). It was described in rats by H.H.
Donaldson (1924) in the same year. At present it is widely studied
in different countries under the name of "audiogenic seizures".

(b) Seizures (sometimes convulsive, but more often only in the
form of marked motor excitation) which arise under an intermittent
action of an acoustic stimulus applied after the end of a seizure
of the first type. This second type was first discovered in our
laboratory in 1951 and since that time has been thoroughly investigated. These seizures essentially differ from seizures of audiogenic epilepsy both in their genotypic foundation and in their
entire pathophysiological complex.

(c) Myoelonic convulsions which fully differ from the two above-mentioned types of motor activity in their external manifestation, as well as in their aetiology. They were detected in our laboratory in 1949-1950, and their study continues up to the present time. Such convulsions arise only during repeated reproductions of seizures of the first type.

I. Audiogenic convulsive seizures.

Audiogenic seizures manifest themselves in the shape of three basic successive reactions.

(a) Intensive motor excitation which is expressed in disorderly running and jumping; (b) a convulsive tonic-clonic seizure during which the animal falls on its belly and exhibits a tonic tension of the musculature with a subsequent clonic twitch of the extremities; (c) a convulsive tonic seizure during which the animal falls on its side and manifests a pronounced tonic tension of its whole musculature.

The stage of motor excitation develops in two forms; in some animals it is interrupted by inhibition which lasts from 5 to 20 seconds.

Numerous experiments performed in our laboratory (31,35,36,38, 43) showed that the development of a seizure in a rat during the action of an acoustic stimulus is determined by a quantitative correlation between the processes of excitation and inhibition.

The emergence of the afore-mentioned inhibitory pause between two waves of excitation reflects the participation of a protective inhibitory mechanism which temporarily stops the arising pathological excitation.

As regards the inhibitory phase, our investigations showed that it may appear, disappear or persist during an audiogenic seizure under the action of neurotropic substances which intensify or weaken the motor excitation and convulsions (18-20). This led to the assumption that the development of the inhibitory pause is due to the activation of those brain structures which do not directly participate in the production of the motor phases of the seizure. A study of the hippocampus as one of such structures showed that it may exert inhibitory - regulating influence on the course of the seizure and that its convulsive readiness is different in rats in which the inhibitory phase is present or absent, or which are insensitive to sonic stimulation (21,22,86).

Investigations of the inhibitory phase showed that it is determined by the development of inhibition which is similar to the active inhibitory process which was elaborately studied in I.P. Pavlov's

laboratories.   This inhibition stops the arising pathological excitation and becomes exhausted after the action of the sound stimulus during 5-20 seconds; after that, the excitation irradiates from
its initial focus and brings about a convulsive seizure.

Electrophysiological investigations carried out in our laboratory (24,68,84) demonstrated that during an audiogenic convulsive
seizure the focus of pathological excitation arises in the subcortical structures from where it irradiates along the whole brain.
(Figure 1).

According to the data obtained by K.G. Guselnikova, it is the
region of the auditory and vestibular nuclei of the medulla oblongata
where the initial focus of excitation is formed.

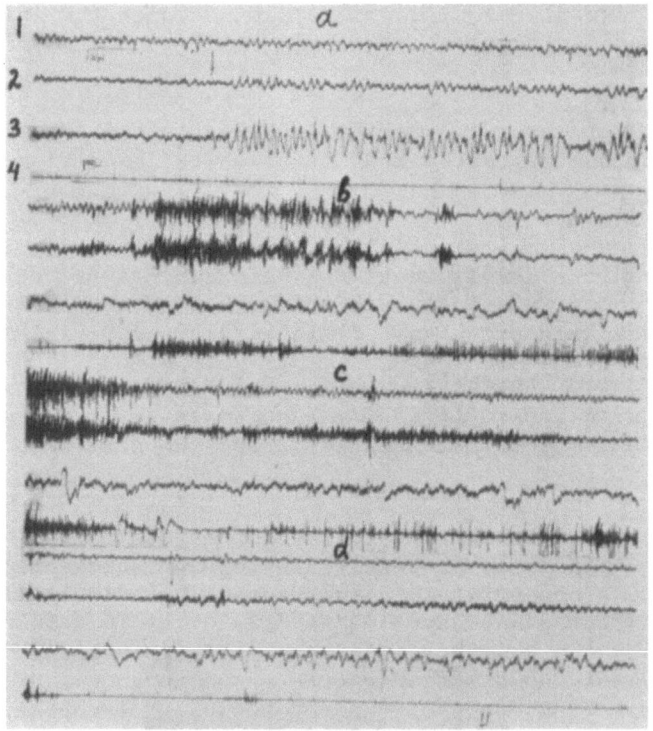

Figure 1.   EEG during the clonic-tonic state of a seizure: (a) background activity before the beginning of the seizure (the arrow indicates the moment of presentation of a flashing light); (b) beginning of the seizure; (c) end of the tonic stage; (d) reaction to
the light stimulus after the end of the seizure (a weakening of
the reaction to the light stimulus is clearly seen).   Designations:
(1)EG of the hippocampus; (2) EG of the medulla oblongata; (3) EG
of visual cortex; (4) electromyogram.

Whereas in the medulla oblongata and other subcortical structures oscillatory bursts of high frequency and amplitude as well as "spike-and-wave" complexes are recorded during a seizure, in the cortex the excitatory process proves to be of a considerably less pronounced character.

Most probably, the increase of the cortical bioelectrical activity is due to irradiation of the excitatory process from the subcortical structures.

According to the data obtained by A.P. Steshenko, the animal's motor response to the action of a strong stimulus during one and one-half minutes is accompanied by a sharp increase of the blood pressure (up to 160-200 mm of the mercury column) and by its subsequent decrease (to 70-90 mm) in spite of the continuing sonic stimulation.

Electrocardiograms recorded during seizures revealed marked disturbances of cardiac activity, such as disappearance of complexes of cardiac contractions, violation of the cardiac rhythm, bradycardia. It is noteworthy that ECG disturbances after a convulsive seizure caused by an acoustic stimulus are more profound and stable than after a seizure provoked by electric current (15). (Figure 2).

II. <u>Multiple seizures arising under the intermittent action of an acoustic stimulus</u>.

These seizures arise when after an audiogenic seizure provoked by a sonic exposition during one and one-half minutes, short ten-second sonic stimulations are continued with ten-second intervals between them.

In this case sharp excitation develops in response to the application of the acoustic stimuli, and this excitation sometimes ends in convulsions. In the original strain of rats that was raised by us and which manifested almost in all cases audiogenic seizures during an initial one and one-half minute application of an acoustic stimulus, a subsequent ten-second sonic stimulation usually revealed the animal's full refractivity. However, by way of selection it proved possible to raise a line of rats which after a seizure of audiogenic epilepsy still exhibited seizures of strong pathological excitation in response to each subsequent application even of short acoustic stimuli (33). These data show that multiple bursts of excitation arising as a result of frequently repeated sonic expositions have a specific genotypic foundation.

Characteristic of the state of rats which manifest multiple fits of excitation in response to a frequently repeated sonic

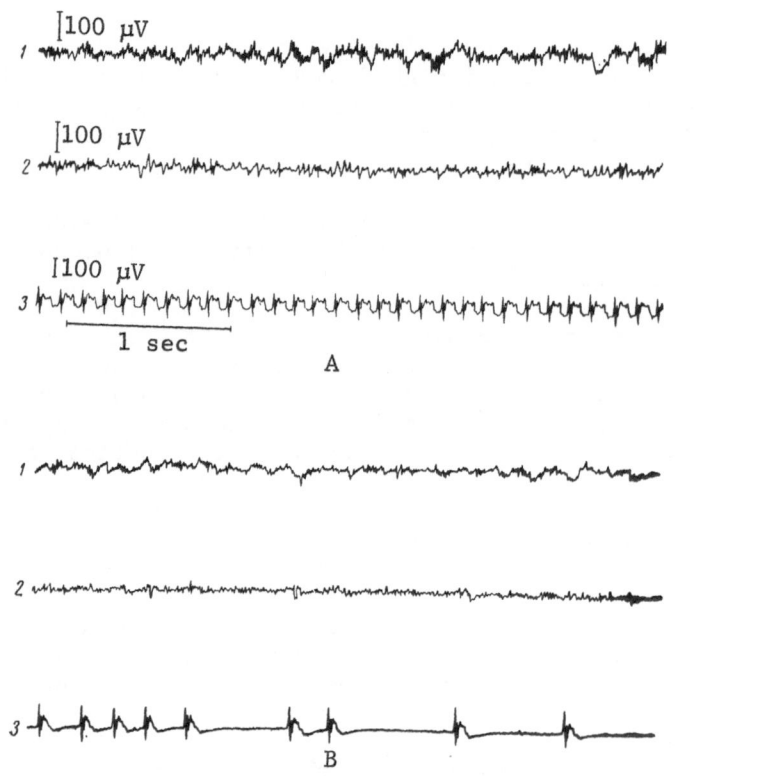

Figure 2. Bioelectrical activity of the rat's brain and heart during audiogenic seizures. A - prior to sonic stimulation; B - after the convulsive seizure. (1) EG of the anterior hypothalamus; (2) EG of the posterior hypothalamus; (3) EKG.

exposition is a gradual increase of the excitability of their nervous system with the application of acoustic stimuli. This is testified to by a reduction of the latency: whereas the latency of the beginning of motor excitation in response to the application of an acoustic stimulus is measured in seconds (usually five to seven seconds), in the case of a preceding sonic exposition it decreases to 0.12 - 0.08 sec. (60).

Against the background of the rising excitability of the nervous brain structures one can distinctly observe all phases of parabiotic, transmarginal inhibition described by N.E. Wedensky in 1901 (84) and I.P. Pavlov in 1938 (84), as well as some new relationships between the force of the stimulus and the intensity of the response which were never previously described.

The following relationships between the force of the acoustic stimulus and the intensity of the motor excitation could be revealed (39,46):

(a) Force relationships - conformity between the intensity of the acoustic stimulus and the magnitude of the response.

(b) Equalizing relationships - equality between the response to strong acoustic stimuli (120 - 100 dB) and that to weak ones (80-70 dB).

(c) Paradoxical relationships - a more powerful reaction to weak stimuli than to strong ones.

(d) Protracted excitation, i.e., excitation not only during the presentation of stimuli, but also in the intervals between their application.

(e) Inverted relationships - when a stimulus applied during protracted excitation weakens or fully eliminates this excitation. When the stimulation is discontinued the excitation reappears.

As shown by the investigations of L.V. Krushinsky, A.P. Steshenko and L.F. Fursina (48), the above-described phases of parabiotic inhibition can be clearly recorded not only during motor excitation, but also in cases of vascular reactions (changes of the blood pressure) in normal and curarized animals (Figure 3).

According to the findings of A.P. Steshenko, the role of trans-marginal inhibition consists in creating conditions under which the blood pressure does not rise above and does not fall below definite limits.

The succession of the phases of parabiotic inhibition clearly manifests itself also in electrograms of different parts of the brain (experiments performed by A.F. Semiokhina).

Pathological modifications of the EEG, which are characteristic of a sonic exposition during one and one-half minutes, show a steady

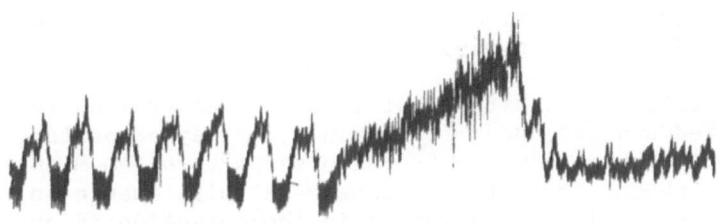

Figure 3. Blood pressure changes under the action of acoustic stimuli in a curarized rat; 130 - a strong bell; 95 - a weak bell.

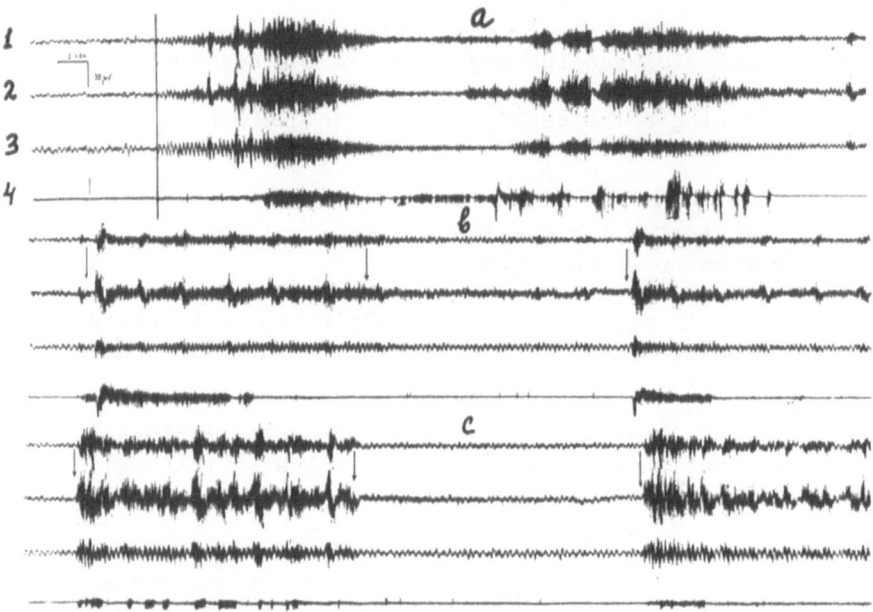

Figure 4. EEG during multiple seizures: (a) a clonic-tonic convulsive seizure; (b) force relationships (the arrows indicate the
beginning and end of the sonic stimulation; strong sound-interval-
weak sound; (c) paradoxical relationships (weak sound - interval -
strong sound). (1) EG of the amygdala complex; (2) EG of the medulla
oblongata; (3) EG of the motor cortex; (4) electromyogram.

increase against the background of the continuing sonic stimulation
when the conformity between the force of the stimulus applied and
the intensity of the rat's response becomes gradually violated
(paradoxical relationships) (Figure 4).

It was shown on a specially created mathematical and electronic
model that the afore-mentioned phases of parabiotic inhibition are
determined by a quantitative correlation between the processes of
excitation and inhibition (66) (Figure 5).

Experiments demonstrated that if the transmarginal inhibition
is intensified and extended (for example, by way of prolonging the
action of the strong stimulus from ten seconds to two minutes during
the inverted phase) there takes place a definite change in the state
of the rat's central nervous system in the direction of normalization:
the protracted excitation ceases and not infrequently force relationships prove to be established between the stimulus and the motor
reaction (32).

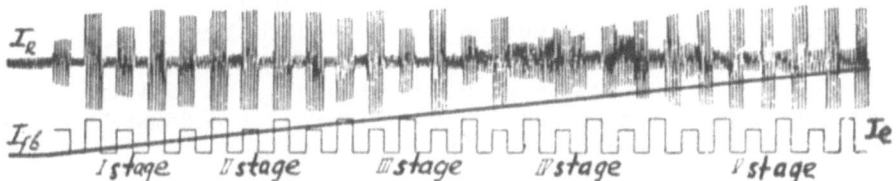

Figure 5. Oscillogram obtained on a model. I_R – impulses of the stimulus; I_{fb} – signal of positive feedback; I_e – impulses of excitation. I – force; II equalizing; III paradoxical; IV – inverted relationships; V – protracted excitation.

These data lead to the conclusion that the multiple seizures are limited by parabiotic, transmarginal inhibition which plays a protective role after the exhaustion of active inhibition during the first minute of sonic stimulation.

In contradistinction to single audiogenic seizures, which almost never result in any complications, multiple seizures often cause severe pathological disturbances not infrequently leading to a lethal end. Most pronounced among these disturbances is a shock-like state with an acute derangement of the blood circulation.

Macroscopic and microscopic investigations established hemorrhages in different parts of the brain in almost all rats which died as a result of the aforesaid disturbances. (Figure 6).

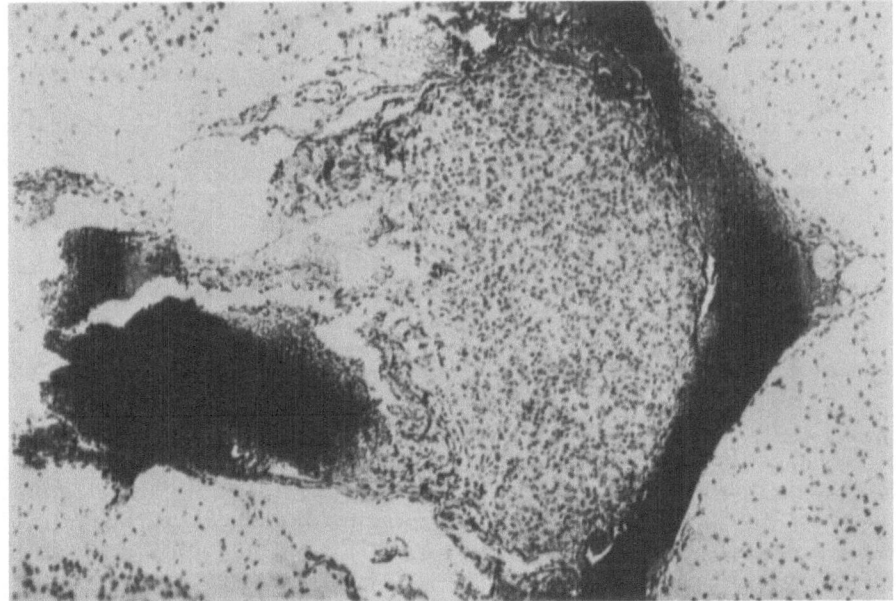

Figure 6. Hemorrhages around epiphysis; staining by hematoxylin-eosion (from ref. 30).

    The blood vessels manifested an exfoliation of the walles plas-
morrhage and sometimes a disruption of the endothelium.

    More often the phenomenon of blood stasis and a swelling of
the endothelial nuclei accompanied by perivascular oedemata were
observed in the blood vessels (30,58).

    These investigations disclosed considerable changes in the
cortical cells; the neurons of the medial geniculata body, optic
thalamus and auditory nuclei of the medulla oblongata proved to be
most affected.   These changes are expressed in marked (sometimes
total) chromatolysis, as well as in a swelling and deformation of
the nucleus (Figure 7).   In the motor region of the cortex binuclear cells are encountered.   In the myelinated fibres of the
internal capsule segmental demyelinization is observed: it is
often accompanied by a disintegration of the myelin.

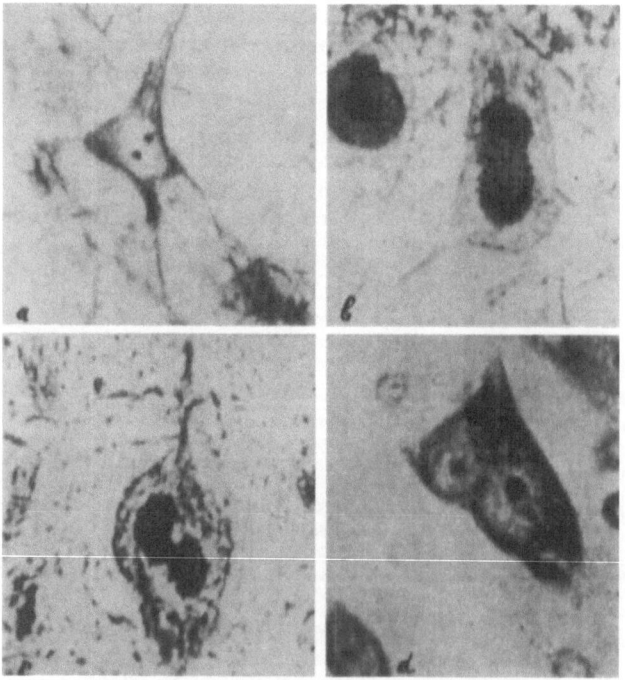

Figure 7.   Alterations of the nuclei and nucleoli of nervous cells.
(a) Binucleolar cell layer III of motor cortex area.   Staining by
the method of Golgi-Deineka.   (b) Dumb-bell-like nucleus; layer III
of auditory area of cortex.   Staining by the method of Cajal; (c)
Segmentlike nucleus; layer III of auditory area of cortex.   Staining by the method of Cajal; (d) Binuclear nervous cell of reticular
substance.   Immersion staining by the method of Nissle.

G.N. Krivitskaya (30) points out that the morphological changes of the brain taking place in rats after multiple seizures of reflex epilepsy are similar to those which are characteristic of a traumatic shock.

Investigations revealed some of the physiological conditions which contribute to an increase of lethality (36). One of such essential conditions is the state of excitability of the central nervous system before the beginning of the sonic stimulation. The administration of caffeine (Caffeinum natriobenzoicum) at the dose of 10 to 20 mg per 100 g of bodyweight increases the number of fatal cases almost threefold. Extirpation of the parathyroid glands, which weakens the inhibitory process and thereby intensifies the excitability of the brain increases the number of fatal cases in a group of animals subjected to such an operation more than two fold in comparison with a control group. Other glands of internal secretion likewise exert considerable influence on the outcome of multiple seizures. It proved that the rate of lethality among male rats is twice as high as among female rats. Emasculation of male rats results in a reduction of lethality following a sonic exposition (40).

Particularly great influence is exerted by the functional state of the thyroid glands (13,42). A preliminary administration of thyroid extract to rats increased the lethality rate as a result of multiple seizures (out of 108 control rats only 9 died, while in a group of 103 hyperthyroidized rats the number of lethal cases was 69).

Extirpation of the thyroid glands or their functional blocking by administering methylthiouracil caused a decline of the lethality rate. There is every reason to assume that hyperthyroidism leads to an increase of this rate, owing to a change in the level of blood pressure. It proved that those individuals predominantly died which possessed a higher initial level of blood pressure due to hyperthyroidism (76).

Investigations carried out on rats with renal hypertension (ischemia of one or two renal arteries provoked with the help of a metallic spiral) fully accord with the aforementioned data. The number of lethal cases in rats with renal hypertension (average level of arterial pressure - 160.2 mm) caused by multiple seizures as a result of sonic stimulation was five times higher than in intact rats subjected to a similar stimulation. It is essential to note that among the hypertensive rats it is individuals with a higher initial excitability of the nervous system which proved to die most frequently (47).

These data indicate that initial heightened excitability of the nervous system, heightened arterial pressure (of different

aetiology) as well as the presence of the male sex hormone are predisposing physiological conditions for lethality as a result of
multiple audiogenic seizures. A study of factors which prevent
such lethality showed that inhalation of carbonic acid exerts an
appreciable protective action (45). The number of lethal cases
in rats which were kept in an atmosphere containing seven percent
carbon dioxide during the whole period of traumatization by means
of an acoustic stimulus was three times smaller than in control
rats subjected to sonic stimulation in a normal atmosphere. These
rats did not manifest such a strongly pronounced decline of blood
pressure and all symptoms of a shock which were observed in control
rats.

III. Myoclonic convulsions

A characteristic feature of these convulsions is that they
never develop during the first two or three sonic exposures and
emerge only after ten to twelve such exposures on the average.
They are so stable that they prove to persist even after an interval of four to seven months (55).

At first the myoclonic convulsions affect only the facial
musculature (the eyelids, ears and lips), then the musculature
of the neck and of the extremities(fore and hind - in succession).
In their external manifestation they are similar to the first phase
of a seizure provoked by metrazol (29).

As a rule, myoclonus arises during the period of inhibition
between two waves of motor excitation; but it may also emerge at
the end of the first wave and at the beginning of the second.
Our investigations established an obvious connection and causal
dependence between the emergence of myoclonus and the inhibitory
functions of the nervous system. It proved that the exhaustion
of inhibition as a result of its systematic over-exertion under
a repeated action of the acoustic stimulus is predominantly responsible for the emergence of myoclonic convulsions, although in
this case the excitability of the nervous system declines (54).
Experiments performed by L.V. Krushinsky, L.N. Molodkina and N.A.
Levitina (43) showed that in order to recover the inhibitory
process (which must be strong enough to develop an inhibitory
phase) it is necessary to make a definite interval between two
sonic expositions, varying this interval for different rats. In
the case of insufficiently long intervals the myoclonus manifested
itself three times more often than in the case of intervals which
ensure a recovery of the inhibition.

The significance of the exhaustion of the inhibitory process
for the development if myoclonus is testified to also by experiments with the administration of sodium bromide to rats, which
reduced the number of myoclonic reactions twofold. Thus, the

well-known effect of bromide on the intensification of the inhibitory function (which was demonstrated by a number of I.P. Pavlov's
co-workers) indicates also a connection between the emergence of
myoclonic convulsions and a chronic exhaustion of the inhibitory
process (44). This is also confirmed by experiments with roentgen
irradiation. A single irradiation of rats at a dose of 250 roentgen, which led to a weakening of the inhibitory process, contributed to an earlier development of myoclonic convulsions (69).

Investigations of the bioelectrical brain activity during
myoclonic convulsion revealed in the cortex and subcortical structures pathological discharges of the type of "spike-and-wave" complexes with an amplitude of 500-800 mev (25,67,83).

A.F. Semiokhina (67-70) showed that these pathological discharges first arise in the subcortical division of the auditory
analyser - in the medial geniculate body; in the cortex they prove
to be recorded at the moment of generalized myoclonic convulsions
(Figure 8). These investigations indicate that the excitatory
process extensively irradiates along the whole brain, which takes
place during myoclonic convulsive seizures, and that the cerebral
cortex is involved in the pathological process.

Investigations carried out by B.I. Kotlyar (27,28) also demonstrated a distinct connection between the development of myoclonic
convulsions and the cortex. These convulsions were eliminated
after extirpation of the cortex (up to ninety percent of its whole
mass) and did not reappear even after 40-60 additional sonic expositions. Experiments with partial decortication showed that after
removal of the motor or cutaneous-kinesthetic zone of the rat's
cortex (in which the bulk of the big pyramidal cells is concentrated) the myoclonic convulsions were considerably weakened, while
after a bilateral removal of the auditory and visual zones of the
cortex they did not manifest any changes.

A functional decortication performed by the method of spread-
ing depression, with the help of a twenty-five percent solution of
potassium chloride, markedly weakened or fully eliminated the myoclonic convulsions (68).

A number of facts have now been obtained showing that the
limbic system participates in the genesis of myoclonus (21,86,88).
It was established that myoclonic convulsions are reproduced as a
result of electrical and chemical (metrazol, bemegride) stimulation
of the limbic structures, and do not develop as a result of stimulation of other structures, including those which are connected
with the genesis of audiogenic myoclonus. Besides, audiogenic
myoclonus can be eliminated by the injection of meprobamate and
librium, which suppress convulsive activity in the limbic system.

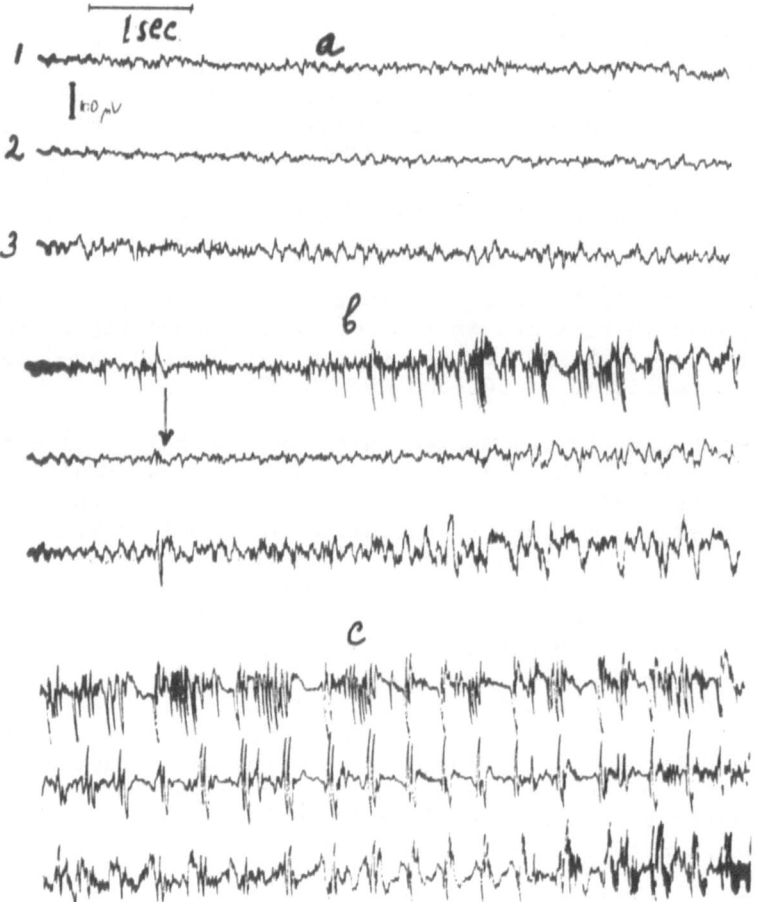

Figure 8. EEG during the myoclonic convulsions. (a) Background activity, (b) beginning of convulsions (the arrow indicates the presentation of sound and the beginning of myoclonus), (c) generalized myoclonic convulsions. (1) medial geniculate body; (2) hippocampus; (3) auditory cortex.

It is assumed that the gradual intensification of convulsive activity in the hippocampus with the repetition of sonic expositions, which was pointed out by K.G. Guselnikova (25) is one of the causes of audiogenic myoclonus.

All the above-mentioned investigations of the complex of pathological reactions which arise in rats under the action of a strong acoustic stimulus were performed on rats of the K-M line (Krushinsky-Molodkina) (33,49). By way of selection, which was conducted since 1946, the percentage of animals reacting to the sound of a bell with convulsive seizures was increased from 10-15 to 98-99 percent. The following symptoms were strictly considered during the process of selection:

(a) Short latency of the seizure, (b) Absence of the inhibitory phase, (c) Tonic seizure, (d) Protracted excitation (motor excitation sometimes lasting more than 20-30 minutes after the action of the acoustic stimulus is discontinued).

In spite of the genotypic conditionality of the audiogenic reaction, the precise picture of its hereditary transmission remains uncertain. The results of crossing with non-selected rats belonging to a population with a low level of sensitivity as well as with rats of the Wistar line manifested incomplete domination of this reaction.

We tried to analyse the character of the hereditary transmission of separate components entering this complex, i.e. of the protracted excitation and myoclonus. Our experiments gave grounds to conclude that the predisposition to protracted excitation is determined by two recessive genes with incomplete manifestability.

Selection according to the rate of emergence of myoclonic convulsions during sonic expositions allowed us to obtain two sub-lines which greatly differed as regards the rate of formation of such convulsions.

In the seventh generation the myoclonus developed in more than eighty percent of animals of the first sub-line after five to ten applications of the acoustic stimulus, while in ninety-one percent of animals of the second sub-line - not earlier than after the thirtieth application. Such a rapid divergence of these sub-lines and such a high hereditability of this feature (0.6, 0.4 and 0.65; 0.35 respectively) testify to the genotypic conditionality of this pathology (65).

CONCLUSION

Thus, the investigations carried out in our laboratory showed that the action of sonic stimulation may lead to some severe pathological disturbances, such as audiogenic convulsive seizures, myoclonic hyperkinesias, shock-hemorrhagic states and cerebral hemorrhages which often are lethal. The investigations also found that the noxious effect of sonic stimulation is to a considerable degree determined by a genotypic predisposition of the organism.

From the very outset of our research we could establish that the inhibitory - protective mechanisms of the nervous system are of great significance for the development of convulsive seizures and for the character of their course. The development of different types of epileptic seizures in rats under the action of one and the same acoustic stimulus is due to changes taking place in the state of the inhibitory-protective mechanisms during the

experiments.   A study of the role played by the inhibitory-
protractive processes made it possible to exert influence on the
development and character of the seizures; it contributed to a
physiological comprehension of such a complex process as emergence
of different stages of parabiotic inhibition.   Mathematical modeling of this process showed that the very fact of emergence and the
successive development of the stages of transmarginal inhibition
described in physiological experimentation may be obtained from
the interaction of the curves that show the quantitative growth
of the excitatory and inhibitory processes in the rat during a
sonic exposition.   We believe that these data may shed certain
light on the question (which is still unsolved in physiology) of
the phasic changes of excitability during the development of inhibitory states.

The significance of a functional derangement of the inhibitory-
protective system as a leading cause of convulsive seizures manifested itself with particular clarity in our investigations of
myoclonic convulsions.   The emergence of focal cortical epileptiform discharges during myoclonic convulsions is due to a weakening
of the inhibitory-protective mechanism of the brain.   Only in
conditions of a functional derangement of this system the transmission of epileptiform impulses from the medulla oblongata to the
cortex proves possible.   Whereas, according to the EEG, convulsive
seizures provoked by the action of an acoustic stimulus during one
and one-half minutes, as well as multiple seizures may be regarded
as seizures of a purely brain stem origin, myoclonic convulsions
develop with the indispensable participation of the cortex.   There
takes place a distinct corticalization of the pathological process.
The effectuation of generalized myoclonic convulsions is possible
only when discharges of excitation irradiate from the subcortical
structures to the cortex.

W. Penfield and H. Jasper (60) showed that the emergence of
cortical focal discharges during the impulsation of the optic
thalamus in cats and monkeys depends on the "susceptibility" of
the cortex to paroxysmal activity.   It is possible that in rats
the epileptiform activity which irradiates from the medulla oblongata
and leads to myoclonic convulsions is blocked in the cortex itself.
The "susceptibility" of the cortex to paroxysmal activity is
accounted for above all by a derangement of the inhibitory-protective
function of this brain division.   The state of the cerebral inhibitory-protective function is a leading factor determining the outcome
of the convulsive seizure.   Our research work established that the
nervous system possesses not less than two different "lines of defense" which prevent in rats convulsive seizures and their death
as a result of sonic stimulation.

The first "line of defense" is formed by active inhibition
which rapidly comes into operation in the case of a very strong

excitatory process in the brain.   In audiogenic seizures of rats
the action of this inhibition manifests itself in a temporary
stoppage of the pathological excitation.   Being rapidly mobilized,
it proves simultaneously instable and easily exhaustible.   After
its exhaustion it is transmarginal parabiotic inhibition which
assumes the protective role; during a considerable period it
strictly limits the excitation of the brain neurons.   However,
when certain conditions eliminating this kind of inhibition are
created in the nervous system (such conditions can be created also
in the course of experimentation), severe shock states accompanied
by cerebral hemorrhages develop in the animal which often end in
its death.   No matter how powerful is the protective role of transmarginal inhibition in this second cerebral "line of defense", the
lethal outcome is to a considerable degree determined by the physiological state of the organism at the moment of the sonic exposition.
Of great importance is the state of the endocrine glands (especially
of the thyroid gland) as well as the initial level of blood pressure.

On the whole, on the basis of our research the development of
convulsive seizures which were subjected by us to investigation
can be pictured as follows.   An epileptogenic focus of excitation
arising in the medulla oblongata under the action of an acoustic
stimulus irradiates along specific and non-specific pathways to
the subcortico-stem structures of the brain resulting in a convulsive audiogenic seizure.   Pathological impulsation does not reach
the brain cortex since the latter is probably protected by an
active inhibitory process after acute exhaustion of this inhibition,
the parabiotic inhibition begins to limit the pathological excitation.   In this case the character of the seizures changes.   They
are usually expressed in bursts of excitation arising in response
to the applied stimulus.   During this period various distortions
of the correlation between the force of the given stimulus and the
intensity of excitation are observed (parabiotic stages).

When the inhibitory process is chronically exhausted, myoclonic
convulsions appear during an audiogenic seizure.   They are the
result of a "break" of the excitatory process from the sub-cortico-
stem divisions of the brain to the brain cortex.

Thus, the most important cause of the emergence of various
epileptiform seizures is a change in the state of the inhibitory
brain functions under the influence of pathological impulsation
coming from an epileptogenic focus.   A change of the inhibitory-
protective brain mechanisms creates, as it were, the "pattern of
the pathological process" which can be observed and examined during
experiments with animals.

REFERENCES

1. Adams, I.S. and W.I. Griffiths, Jr., 1948.   The effect of
   tridione on audiogenic fits in albino rats.   Journal of Comparative Physiology and Psychology 41:319-326.

2. Anthony, A. and E. Ackermann, 1955.   Effect of noise on the
   blood eosinophil levels and adrenals of mice.   Journal of
   Acoustical Society of America 27:1144-1149.

3. Ballantina, E.E., 1963.   The effect of gamma-aminobutyril acid
   on the audiogenic seizure.   Colloques Internationaux de Centre
   National de la Researche Scientifique(Paris) 112:447-451.

4. Beach, F.A. and T.H. Weaver, 1943.   Noise induced seizures in
   the rat and their modification by cerebral injury.   Journal of
   Comparative Neurology 79:379-392.

5. *Beletsky, V.,. and A.A. Khachaturyan, 1959.   A histophysio-
   logical analysis of changes in the central nervous system in
   epilepsy.   Collected articles: Problem of Epilepsy, Medgiz,
   pp. 331-348.

6. Bures, J., 1953.   External inhibition of reflex epilepsy in
   rats and mice.   Czechoslovak Physiology 2:28-37.

7. Bures, J., 1953a.   Effect of the exclusion of analysers on the
   convulsive readiness in rats and mice.   Czechoslovak Physiology
   2:284-293.

8. Bures, J., 1953b.   Experiments on the electrophysiological
   analysis of the generalization of an epileptic fit.   Czechoslovak
   Physiology 2:347-356.

9. Bures, J., 1963.   Electrophysiological and functional analysis
   of the audiogenic seizure.   Colloques Internationaux de Centre
   National de la Researche Scientifique(Paris) 112:165-179.

10. Busnel, R.G., A. Lehmann and M.C. Busnel, 1958.   Etude de la
    crise audiogène de la Souris comme test psychopharmacologique,
    son application aux substances du type "tranquilliseur".
    Pathologie et Biologie (Paris) 9 - 10: 749-762.

11. Busnel, R.G. and A. Lehmann, 1960.   Nouvelles donnees pharma-
    codynamiques relatives à la crise audiogène de la Souris.
    Journal de Physiologie (Paris) 52:37-38.

12. Chocholova, L., 1959.   The effect of repeated acoustic stimulation

*in the Russian.

and of conditioned reflexes on audiogenic epilepsy in rats.
Physiologia Bohemoslovaca (Praha) 8:61-68.

13. *Dobrokhotova, L.P., 1957.   Effect of methyl-thyrouracil on
    shock-haemorrhagic states developing as a result of a nervous
    trauma.   Papers, of the USSR Academy of Sciences 114:1320-1321.

14. *Dobrokhotova, L.P., 1958.   Effect of hyperthyroidization on
    the functional state of the central nervous system during the
    development of shock-haemorrhagic states in animals as a result
    of nervous traumas.   Problems of Endocrinology and Hormonal
    Therapy 4:12-21.

15. Dobrokhotova, L.P., A.F. Semiokhina and M. Bonfitto, 1969.
    Some data concerning the regulation of cardiac activity in
    convulsive states.   Papers of the Higher School of Biological
    Sciences 4:43-48.

16. Dolina, S.A., 1963.   Contribution to the physiological analysis
    of the state of convulsive readiness.   Journal of Neuro-
    pathology and Psychiatry 63:867-873.

17. Donaldson, H.H., 1924.   The Rat, Henry Herbert Donaldson (Ed.),
    Memoirs of the Wistar Institute of Anatomy and Biology, No. 6,
    Philadelphia, p. 469.

18. *Fless, D.A., 1957.   Influence of factors changing the state
    of the excitatory and inhibitory processes on the phases of
    reflex epilepsy.   Abstracts for a scientific conference on
    problems of experimental pathophysiology and therapy of the
    animal's higher nervous activity, pp. 97-98.

19. *Fless, D.A. and A. Lehmann, 1960.   Comparative studies on the
    action of several tranquilizers on excitation and inhibition
    processes of Central Nervous System.   Abstracts of Proceedings
    of XIX Conference on Problems of the Higher Nervous Activity,
    Leningrad, 1960: 140-142.

20. *Fless, D.A. and A. Lehmann, 1961.   Action of tranquilizers on
    the excitatory and inhibitory processes in reflex epilepsy of
    rats.   Papers of the Higher School. Biological Sciences
    2:102-106.

21. *Fless, D.A. and Z.A. Zorina, 1965.   The role of the hippo-
    campus in the genesis of audiogenic seizures of the myoclonic
    type.   Bulletin of Experimental Biology and Medicine 10:43-45.

22. *Fless, D.A., Z.A. Zorina and S.A. Zinina, 1969.   Some data
    concerning the connection between the electrical excitability

*in the Russian.

of the hippocampus and the character of an audiogenic epileptiform
seizure in rats.   Journal of Higher Nervous Activity, 14 (in press

23. Griffiths, W.J., Jr., 1942.   The effect of dillantin on convulsive
    seizures in the white rat.   Journal of Comparative Psychology
    33:291-296.

24. *Guselnikova, K.G., 1959.   Contribution to the electrophysio-
    logical characteristics of the white rat's medulla oblongata
    during an audiogenic epileptiform seizure.   Papers of the
    Higher School of Biological Sciences 3:**101-106.**

25. *Guselnikova, K.G., 1959.   Study of the bioelectrical activity
    of the rat's olfactory brain during an epileptiform seizure.
    Collected articles "Problem of epilepsy", Medgiz, 270-276.

26. Kesuer, R.P., 1966.   Subcortical mechanisms of audiogenic
    seizures. Experimental Neurology 15:192-205.

27. *Kotlyar, B.I., 1958.   The role of the brain cortex in the
    development of motor pathological reactions.   Papers of the
    Higher School of Biological Sciences 4:98-101.

28. *Kotlyar, B.I., 1959.   On the localization of clonic convulsions
    of an epileptic seizure.   Papers of the Higher School of Bio-
    logical Sciences 2:73-76.

29. *Kotlyar, B.I. and D.A. Fless, 1962.   Concerning the mechanism
    of the action of metrazil on the central nervous system.   Papers
    of the Higher School of Biological Sciences 2:98-102.

30. *Krivitskaya, G.N., 1964.   The action of a strong sound on the
    brain.   Meditsine, Moscow, p. 158.

31. *Krushinsky, L.V., 1959.   New factors in the study of experi-
    mental epilepsy and of its physiological mechanisms.   Achieve-
    ments of Modern Biology 28:74-93.

32. *Krushinsky, L.V., 1954.   Study of the interrelation of excita-
    tion and inhibition in normal and pathological conditions by the
    method of sonic stimulations.   Achievements of Modern Biology
    37:74-93.

33. *Krushinsky, L.V., 1959.   Genetic investigations in the field
    of experimental pathophysiology of the higher nervous activity.
    Bulletin of the Moscow Society of Naturalists (section of Bio-
    logy) 64:105-117.

34. *Krushinsky, L.V., 1959a.   Investigation of the physiological

*in the Russian.

mechanisms of seizures in reflex epilepsy.   Collected articles,
Problem of Epilepsy, Medgiz, pp. 245-249.

35. *Krushinsky, L.V. 1960.   Animal behavior its normal and ab-
    normal development, Izd. M.G.U. (translated in English by J.
    Wortis, Consultants Bureau, New York).

36. Krushinsky, L.V. 1962.   Study of pathophysiological mechanisms
    of cerebral haemorrhages provoked by reflex epileptic seizures
    in rats.   Collected articles Epilepsia (Netherlands) 3:363-380.

37. Krushinsky, L.V., 1963.   Etude physiologique des differents
    types de crises convulsives de l'épilepsie audiogène du rat,
    Colloques Internationaux de Centre National de la Researche
    Scientifique(Paris) 112:71-92.

38. *Krushinsky, L.V., D.A. Fless and L.N. Molodkina, 1950.
    Analysis of the physiological processes lying at the base of
    experimental reflex epilepsy.   Journal of General Biology
    2:104-119.

39. *Krushinsky, L.V., D.A. Fless and L.N. Molodkina, 1952.   Study
    of transmarginal inhibition by the method of sonic stimulations.
    Bulletin of Experimental Biology and Medicine 33:13-16.

40. *Krushinsky, L.V., L.P. Pushkarskaya and L.N. Molodkina, 1953.
    Experimental study of cerebral haemorrhages as a result of
    nervous traumas.   Herald of the Moscow State University 12:
    25-44.

41. *Krushinsky, L.V., M.Y. Sereysky, L.P. Pushkarskaya and G.I.
    Fiodorova, 1955.   Experimental study of a new antiepileptic
    preparation. Journal of Higher Nervous Activity 5:892-900.

42. *Krushinsky, L.V. and L.P. Dobrokhotova, 1957.   Influence of the
    thyroid gland on the mortality rate in shock-haemorrhagie states
    provoked by strong acoustic stimuli.   Bulletin of Experimental
    Biology and Medicine 8:46-49.

43. *Krushinsky, L.V., L.N. Molodkina and N.A. Levitina, 1959.
    Time and conditions of rehabilitation of an exhausted inhibitory
    process during the action of acoustic stimuli.   Journal of
    Higher Nervous Activity 9:566-572.

44. *Krushinsky, L.V. and L.N. Molodkina, 1960.   A new experimental
    model of a chronic disease of the nervous system.   Journal of
    Higher Nervous Activity 10:779-785.

45. *Krushinsky, L.V., A.P. Steshenko and L.N. Molodkina, 1960.

*in the Russian.

The preventive action of carbonic acid inhalation in the development of shock-haemorrhagic states.   Papers of the Higher School
of Biological Sciences 2:73-77.

46. *Krushinsky, L.V., D.A. Fless and N.V. Dubrovinskaya, 1960.
Specific features of the development of parabiotic stages during
pathological states of the brain, provoked by the action of sonic
stimulations.   Proceedings of a scientific conference dedicated
to the memory of N. Wedensky, Vologda, pp. 221-227.

47. *Krushinsky, L.V., A.Kh. Kogan, L.P. Dobrokhotova and N.P.
Modenova, 1963.   Significance of the re-excitation of the
central nervous system in the genesis of an experimental hyper-
tonic stroke.   Pathological Physiology and General Therapy
6:40-43.

48. *Krushinsky, L.V., A.P. Steshenko and L.V. Fursina, 1964.
Study of the dynamics of arterial pressure in rats.   Tenth
Congress of the USSR Pavlov Physiological Society, Abstracts
of Scientific Communications, 2, issue 1:431.

49. *Krushinsky, L.V., L.N. Molodkina and R.G. Romanova, 1968.
Genetic investigations in the field of experimental pathophysiol-
ogy of the nervous activity.   Collected articles, "Neurophysiology and Genetics, Meditsina, pp. 186-199.

50. Lehmann, A. and R.G. Busnel, 1959.   Etude sur la crise audiogène.
Archives of Sciences Physiologiques (Paris) 8:193-225.

51. Lehmann, A. and D.A. Fless, 1962.   Etude de l'action de drogaes
psychotropes sur les processus d'excitation et de inhibition du
S.N.C. par le test de la crise audiogène du rat.   Psychopharmacologia 3:331-343.

52. Lehmann, A. and R.G. Busnel, 1963.   A study of the audiogenic
seizure.   In: R.G. Bushnel (Ed.) Acoustic Behaviour of Animals,

53. Mircier, I., 1963.   Modification de la crise audiogène du rat
albinos sous l'influence de divers agents pharmakodynamiques.
Compte Rendue de Biologie de Centre National de la Researche
Scientifique (Paris) 112:331-392.

54. *Molodkina, L.N., 1962.   State of excitability as a result of
systematic affecting on the nervous system with sonic stimulation.   Papers of the Higher School of Biological Sciences
2:94-97.

*in the Russian.

55. *Molodkina, L.N., 1966. Influence of a long interval in the systematic action of an acoustic stimulus on myoclonic hyper- kinesis provoked by this stimulus. Papers of the Higher School Biological Sciences 1:62-64.

56. Niaussat, M.M., 1963. Etude électrocardiographique de la crise audiogene de la souris, Colloques Internationaux de Centre National de la Researche Scientifique (Paris) 112:199-215.

57. Niaussat, M.M. and P. Laget, 1963. Etude électroencephalo- graphique de la crise audiogène de la souris, Colloques Inter- nationaux de Centre National de la Researche Scientifique (Paris) 112:181-197.

58. Nitschkoff, S.T. and G. Kriwizkajag, 1968. Lärmbelastung. Akustischer Reiz und Neurovegetative Störungen. Veb. Georg Thieme Verlag. Leipzig.

59. *Pavlov, I.P., 1938. Twenty years of objective study of the higher nervous activity (behaviour) of animals "Blomedgiz". Moscow-Leningrad.

60. Penfield, Wilder and Herbert Jasper, 1954. Epilepsy and the functional anatomy of the human brain. Little, Brown and Co., Boston.

61. Plotnikoff, N.P., 1958. Bioassay of potential tranquilizers and sedatives against audiogenic seizures in mice. Archives Internationales de Pharmacodynamie et de Therapie 116:130-135.

62. *Portnov, R.A. and D.D. Fedotov, 1959. Contribution to the question of the reflex mechanisms of epilepsy. Collected articles, Problem of Epilepsy, Medgiz, 5-12.

63. *Prokopets, I.M., 1958. Experimental investigation of the protective-rehabilitative role of a functional cataleptoid state. Papers of the Higher School of Biological Sciences 3:84-89.

64. *Prokopets, I.M., 1958. An experimental cataleptoid state. Pathological Physiology and Experimental Therapy 4:29-33.

65. *Romanova, L.G., 1969. The heritability of audiogenic myo- clonic seizures. Genetics 4:135-140.

66. *Savinov, G.V., L.V. Krushinsky, D.A. Fless and R.A. Valerstein, 1964. Study of the interrelation of the excitatory and in- hibitory processes by means of mathematical modeling. Collected

*in the Russian.

Articles: <u>Problems of Cybernetics</u> 2:11-24.

67. *Semiokhina, A.F., 1958.   Electrophysiological investigation
    of the auditory and motor analysers on a model of experimental
    motor neurosis.   Journal of Higher Nervous Activity 8:278-285.

68. *Semiokhina, A.F., 1959.   Study of the bioelectrical activity
    of the auditory and motor analysers during a reflex epileptic
    seizure.   Collected Articles: <u>The Problem of Epilepsy</u>,
    pp. 259-270.

69. *Semiokhina, A.F., 1962.   Study of the action of ionizing
    radiation on a model of reflex epilepsy.   Radiobiology 12,
    issue 7:69-74.

70. *Semiokhina, A.F., 1969.   Concerning the cortico-subcortical
    inter-relations during a spreading depression of the neocortex.
    Journal of Higher Nervous Activity, 19, issue 1: 143-148.

71. *Servit, Z., 1955.   Contribution to the question of the inter-
    relation of the excitatory and inhibitory processes in the
    patho-physiology of an epileptic seizure.   Journal of Higher
    Nervous Activity 5, issue 4: 474-479.

72. Servit, Z., 1960.   The role of subcortical acoustic centres
    (colliculi inferiores, laminae quadr.) in seizure susceptibility
    to an acoustic stimulus and in symptomatology of audiogenic
    seizures in the rat.   Physiologia Bohemoslovaca 9:42-47.

73. *Steshenko, A.P., 1959.   Physiological analysis of shock-
    haemorrhagic states.   Papers of the Higher School of Bio-
    logical Sciences 1:74-88.

74. *Steshenko, A.P., 1959a.   Physiological analysis of the pre-
    ventive action of carbonic acid against shock-haemorrhagie
    states in rats.   Papers of the Higher School of Biological
    Sciences 3:107-111.

75. *Steshenko, A.P., 1960.   The role of initial arterial pressure
    in the development of a shock-haemorrhagic state.   Bulletin
    of Experimental Biology and Medicine 4:33-35.

76. *Steshenko, A.P., 1961.   Dependence of the mortality rate as
    a result of nervous traumas on the level of blood pressure in
    hyperthyroidized rats.   Papers of the Higher School of Bio-
    logical Sciences 1:77-79.

77. Swinyard, E.A., 1963.   Some physiological properties of audio-
    genic seizures in mice and their alteraction by drugs.

*in the Russian.

Colloques Internationaux de Centre National de la Researche Scientifique (Paris) 112:405-427.

78. Valette, G. and G. Raynaud, 1963. Action des derives de la phenotherazine sur la crise audiogène de la Souris. Colloques Internationaux de Centre National de la Researche Scientifique (Paris) 112:311-329.

79. *Van, Bin, 1958. Conditioned reflexes in some functional and organic lesions of the brain cortex. Candidate Dissertation. Moscow University.

80. *Vasiliev, G.A., 1924. Concerning the mechanism of the "Parfenov reaction". Russian Physiological Journal 6 issue 4-6:74-81.

81. *Vasilieva, V.M., 1957. Modification of electrical activity in the cortical division of the motor analyser in white rats during a reflex epileptiform seizure. Bulletin of Experimental Biology and Medicine 1:57-61.

82. *Vasilieva, V.M., 1958. Electroencephalographic investigation of the motor area of the cerebral cortex in white rats during epileptiform reactions. Journal of Higher Nervous Activity 8 issue 4:602-610.

83. *Vasileva, V.M., 1960. Modifications of the electrogram of the cerebral cortex in white rats during epileptiform seizures complicated by experimental motor neurosis. Papers of the Higher School of Biological Sciences 2:69-72.

84. *Wedensky, N.E., 1901. Excitation, inhibition and narcosis. Complete Works, Saint Petersburg 4:5-146.

85. Weiner, H.M. and Morgan, C.T., 1945. Effect of cortical lesions upon audiogenic seizures. Journal of Comparative and Physiological Psychology 38:199-208.

86. *Zorina, Z.A. and D.A. Fless, 1968. Concerning the origin of the inhibitory phase in audiogenic epilepsy. Papers of the Higher School of Biological Sciences 9:34-38.

87. *Zorina, Z.A. and D.A. Fless, 1969. Pharmacological analysis of the mechanism of myoclonic seizures in rats. Bulletin of Experimental Biology and Medicine 1:48-50.

88. *Zorina, Z.A., 1969. Concerning the mechanism of the effect of meprobamate and bromide on experimental convulsions in rats. Pharmacology and Toxicology 3:272-275.

*in the Russian.

ACOUSTIC PRIMING OF AUDIOGENIC SEIZURES IN MICE

Kenneth R. Henry and Robert E. Bowman

University of Wisconsin
Regional Primate Research Center
Madison, Wisconsin   53706

The audiogenic seizure is a classic example of an extra-
auditory physiological effect of audible sound.   This distinctive
behavior has been studied in animals in the hope that knowledge of
its mechanism might aid in the understanding of both sensory-
induced and spontaneous seizures in man.   Recently, it has been
discovered that prior intense auditory stimulation can induce or
enhance susceptibility to audiogenic seizures in mice.   This
"acoustic priming" effect represents a new phenomenon which may
be useful in seizure research, especially if a similar effect is
observed in man.   This chapter describes a series of on-going experiments which have been designed to examine the priming effect.

AUDIOGENIC SEIZURES

The audiogenic seizure syndrome is a unique chain of behaviors
which occur in response to a loud sound.   After the stimulus onset,
there is a latency of a few seconds before the subject exhibits the
first of four successive stages of this syndrome.   The first stage,
wild running, is unique to sound-produced convulsions, and easily
differentiates it from other sensory-induced convulsions.   As
wild running progresses, the gait of the subject changes to a series
of stiff-legged bounds and, as the animal falls on its side, the
second stage is initiated.   This myoclonic (commonly called "clonic") convulsive stage is characterized by a series of rapid jerks
by all four limbs.   This stage ends as the hind limbs stiffen and
are drawn to the body so that the feet almost touch the mouth.
The next stage, myotonic ("tonic") seizure, begins as the hind limbs
start a slow descent to a fully extended position.   The fourth
stage, death, results from respiratory failure.   In a maximally

185

severe seizure, all four stages of the audiogenic seizure syndrome
will be seen; if the seizure is less severe, only wild running followed by a clonic seizure may be seen.

These stages are, of course, only the overt symptoms which reveal the presence of a series of rapidly developing physiological
events. But they are easily and reliably observed and, because
of their successive nature, provide a simple way of quantifying
the severity of the seizure. This laboratory typically reports
the incidence of these components in the primary reports so that
a direct comparison can be made with results from other laboratories. But our statistical analyses are usually performed on a
single composite score which reflects the severity of seizures
for each subject. By considering each of the four successive
stages as contributing equally (25%) to this composite severity
score, the appearance of only wild running would signify a score
of 25, while a tonic seizure would rate a score of 75, since the
animal first went through wild running and clonic seizures. This
simple score meets the requirements (1) of the higher order statist-
ics which are employed, and avoids the ambiguity (2) and bias which
can result from merely analyzing a single stage of the syndrome.

Although earlier investigators preferred the laboratory rat,
the inbred mouse has recently become the subject used in most
studies of audiogenic seizures. In addition to their economy,
more convenient size, and ease of handling, inbred mice have a
definite genetic advantage in the study of this syndrome. The
inbred mouse exhibits a very high degree of genetic uniformity
within a single strain, and a large number of inbred strains are
available for comparisons. Since large differences in susceptibility to audiogenic seizures have been observed between different
strains of mice, strain comparison studies have been performed as
a first step in determining the genetic determinants of this behavior. Of perhaps greater importance, physiological and biochemical
correlations have been made between "susceptible" and "nonsusceptible" strains in an attempt to identify the underlying mechanisms
of the seizure. While providing much valuable information, this
procedure has an inherent flaw which can result in a great deal of
confounding. Although strains of inbred mice all share the majority of their genes in common, comparisons between any two strains
may reveal differences resulting from one of the many alleles
which is unique to each particular strain. Therefore, any biochemical and physiological difference between "susceptible" and
"nonsusceptible" strains is merely a correlation, and stands a
good chance of not reflecting a causal mechanism of audiogenic
seizures. For the past three years, however, we have used the
acoustic priming phenomenon to circumvent this difficulty. This
technique produces a "behavioral phenocopy" by inducing susceptibility to audiogenic seizures in an otherwise nonsusceptible strain.
Even if this phenocopy results in a physiological alteration which

is not identical to that seen in nonprimed susceptible strains, it
still offers a valuable experimental tool in the study of sound
induced convulsions, per se.   Therefore, priming appears to have
distinct advantages over the correlative strain comparison approach,
since this experimental manipulation allows comparison of susceptible and nonsusceptible subjects within the same inbred strain by
merely comparing primed and nonprimed mice.

SPECIFICITY OF ACOUSTIC PRIMING

Subjecting a juvenile mouse to a loud sound for a few seconds
will prime it, making it susceptible to audiogenic seizures when
it is tested a few days later.   A 103 db electric bell provides
the acoustic source, on the assumption that noise will excite a
greater number of neurons than would a pure tone.   It should be
emphasized that the priming and testing stimuli are identical 30-
second exposures to the same bell.   The only difference is behavioral; "nonsusceptible" mice appear very nonreactive to the priming
stimulus, but convulsions terminating in death often result when
the test stimulus is presented a few days later.

In order to determine the mechanism of priming, it was necessary to obtain information concerning its specificity.   If acoustic
priming would increase the sensitivity of mice to all types of
seizures, one could look for some resultant general effect, such
as increased neural excitability.   If only audiogenic seizures
would be affected, a more specific effect might be the causal
mechanism.   Since the priming phenomenon was first reported in
the C57BL/6J inbred mouse (3), it was also desirable to determine
whether this effect would be observed in other genotypes.   Both
questions were answered in a study (4) in which mice of six different genotypes were primed and tested for three types of seizures.
Although priming did not enhance the already maximal susceptibility
of the DBA/2J strain, the five "nonsusceptible" genotypes (Table 1)
showed a dramatic increase in susceptibility to sound-produced convulsions.   However, none of the genotypes (Table 2) showed an increase in sensitivity to electroconvulsive or chemoconvulsive
stimuli.   These results showed that acoustic priming was not
specific to a single genotype, and suggested that it was highly
specific to audiogenic seizures.   The rank order of the severity
of seizures for each genotype also showed that strains that were
most susceptible to audiogenic seizures after priming were also
most susceptible to electroconvulsive seizures, whereas no such
correlation existed between chemoconvulsive (Metrazol) seizures
and either audiogenic or electroconvulsive seizures.   This suggested a nonauditory mechanism which was shared by both electroconvulsive and audiogenic seizures, although it revealed nothing
of the mechanism involved in priming per se.   Because Levine (5)
had shown that mild physiological stress could decrease the

Table 1. Incidence of audiogenic seizures for primed and nonprimed mice of 6 genotypes. Twenty mice of each genotype were primed at 16 days of age and tested 5 days later (4).*

	Wild Running		Clonic Seizure		Tonic Seizure		Death	
	P	N	P	N	P	N	P	N
DBA/2J	20	20	19	18	19	18	12	14
B6D2F$_1$	20	2	20	0	20	0	13	0
C57B1/6J	18	2	17	0	17	0	12	0
BcB6F$_1$	19	1	17	0	6	0	4	0
BALB/cJ	19	0	17	0	0	0	0	0
D2BcF$_1$	19	1	17	0	5	0	5	0

*P = primed. N = not primed.

Table 2. Effects of acoustical priming on three types of convulsions in 16 day old mice. Twenty mice of 6 genotypes were tested for either audiogenic, electroconvulsive or chemoconvulsive seizures when 21 days old (4).*

	Clonic-tonic seizures			Deaths		
	Primed	Nonprimed	$p <$ **	Primed	Nonprimed	$p <$ **
	Audiogenic seizures					
DBA/2J	19	18	---	12	14	–
B6D2F$_1$	20	0	1.3×10^{-11}	13	0	1.3×10^{-5}
C57B1/6J	17	0	2.6×10^{-8}	12	0	4×10^{-5}
BcB6F$_1$	6	0	2.0×10^{-2}	4	0	–
BALB/cJ	0	0	---	0	0	–
D2BcF$_1$	5	0	4.6×10^{-2}	1	0	–
	Electroconvulsive seizures					
DBA/2J	19	17	---	8	11	–
B6D2F$_1$	18	14	---	14	11	–
C57B1/6J	12	14	---	2	4	–
BcB6F$_1$	5	2	---	0	0	–
BALB/cJ	3	0	---	0	0	–
D2BcF$_1$	1	0	---	0	0	–
	Metrazol seizures					
DBA/2J	10	8	---	7	4	–
B6D2F$_1$	6	11	2.2×10^{-2}	0	1	–
C57B1/6J	12	9	---	5	7	–
BcB6F$_1$	3	7	---	2	4	–
BALB/cJ	6	5	---	1	4	–
D2BcF$_1$	5	7	---	1	1	–.

*The numbers in the cells give the number of mice out of 20 that responded as indicated; total = 720.
**Fisher exact probability test, two-tailed.

threshold to electroshock convulsions in rats, we wished to determine whether stressing agents would have a similar effect on audiogenic seizures in mice. We used two severe stressing agents,
electric shock and hypothermia, on juvenile mice, but found that
these agents did not induce susceptibility to audiogenic seizures.
In addition, attempts to prime the mice with electroconvulsive
shock (ECS) or chemoconvulsions did not increase their future susceptibility to these two agents. The evidence suggested, therefore, that acoustic priming was specific to audiogenic seizures,
that priming did not operate via the physiological stress mechanism,
and that priming was not a general feature of all seizures.

STABILITY OF PRIMING

     To determine the physiological events necessary for effective
acoustic priming, the physiological states of mice were varied in
a number of ways. Surprisingly, priming was found to be effective
(Table 3) even when it was performed on mice which were anesthesized
with sodium pentobarbital (Nembutal) or with ether (6). This indicated that any psychological or physiological process which required wakefulness could not be responsible for the priming effect,
ruling out the importance of learned events such as conditioned fear.

Table 3. Effects of priming 16 day old C57BL/6J
mice under anesthetic conditions. Fifteen 21 day
old mice were tested after priming under one of the
6 conditions listed below (6).

Treat- ment	Seizure behavior (%)			
	Wild run- ning	Clonic seizure	Tonic seizure	Death
Nembutal	0	0	0	0
Ether	0	0	0	0
None	0	0	0	0
Nembutal plus bell	93	87	87	53
Ether plus bell	100	100	100	67
Bell only	100	80	80	60

Because the barbituate, pentobarbital sodium, appears to have a
selective inhibitory effect on the short interneurons of the midbrain reticulum (7), it was also surmised that active participation
of the reticular activating system was not necessary for the initiation of priming.

After priming has occurred, testing must be delayed for several days before maximal sensitivity to audiogenic seizures is seen.
This raised the possibility that some active process akin to consolidation might be occurring, and suggested that any agent which
could disrupt this process might reveal evidence concerning its
mechanism. Agents which are capable of producing retrograde
amnesia in the mouse were therefore administered in an attempt to
alter this "consolidation process." Electroconvulsive shock administered at times varying from immediately after priming to one
day after priming had no effect. Nor did severe hypothermia,
which reduced the brain temperature to 17 degrees C (Table 4).

Table 4. Effects of electroconvulsive shock (ECS) and hypothermia
on acoustically primed audiogenic seizures. Sixteen day old C57BL/
6J nice were acoustically primed for 30 seconds, followed by either
hypothermia (immersion of body up to nose in 2 degrees C. water for
5 minutes) or ECS (800 ms. of 7.65 ma A.C. current via transcorneal
electrodes). ECS was applied either 30 seconds (Time 1), 1 hour
(Time 2), 2 hours (Time 3), or 24 hours (Time 4) after priming.
Hypothermia was initiated either 56 seconds (Time I), 3.75 minutes
(Time II), 15 minutes (Time III), or 60 minutes (Time IV) after
priming. All mice were subsequently tested for audiogenic seizures at 21 days of age.

Time of Treatment	Number of Mice	Seizure Severity	Standard Error
ECS			
1	20	76.25	7.318
2	19	80.25	4.635
3	16	64.0	9.415
4	17	78.0	6.398
Hypothermia			
I	20	78.75	6.603
II	20	73.75	8.403
III	20	71.25	6.855
IV	20	76.25	7.318

Subsequent pilot studies have also shown that increasing endogenous
levels of acetylcholine by the administration of physostigmine,
and inhibition of brain protein synthesis by intracranial injections
of cycloheximide were equally ineffective in preventing or reducing
the effects of acoustic priming.   It may therefore be assumed that
priming is a very stable phenomenon, which suggests that its mechanism has some more passive component.   For example, priming might
initiate some highly specific damage which requires several days
to be fully manifested.

ONTOGENETIC AND GENETIC STUDIES OF PRIMING

     Priming exhibits characteristics in common with phenomena
studied by embryologists and developmental psychologists in that
a seemingly minor event (priming) has a major effect (audiogenic
seizure) later in the life of the organism.   To determine the
developmental characteristics of priming, C57BL/6J mice were
acoustically primed at one of 19 discrete ages ($\pm$ ½ day) and
tested for audiogenic seizures under an X-21, X-28, or X-35 schedule.   (The X designates the age at priming and the last number
designates the age at testing.)   This experimental design, re-
quiring 680 mice (20 for each unique combination of priming and
testing), was created to tease out the three interdependent factors
of age at priming, age at testing, and the interval between priming
and testing.   Because testing at 35 days resulted in less severe
seizures than did testing at either 28 or 21 days of age, the
severity scores of all these schedules were transformed so that the

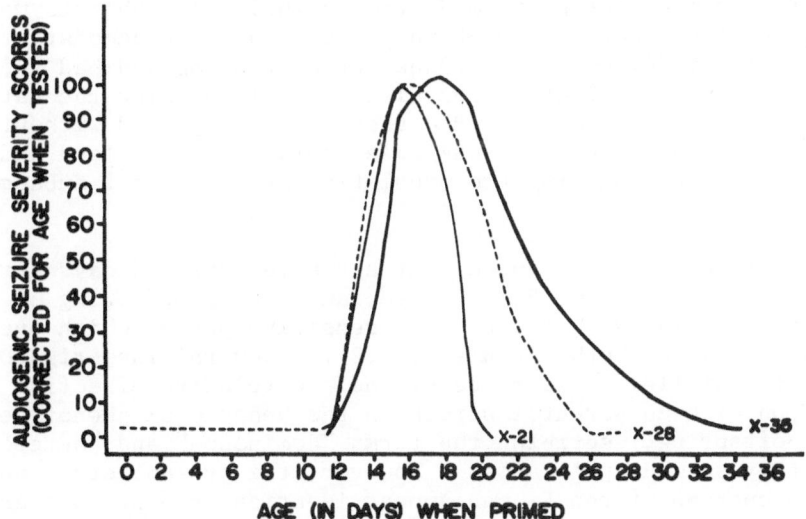

Figure 1. Effects of testing 21, 28, or 35 day old C57BL/6J mice
for audiogenic seizures after priming at one of several different
ages.

age which showed the most severe seizures for each schedule was
given a score of 100.   It was found (Fig. 1) that priming was
ineffective if it occurred before 14 days of age, regardless of
when testing occurred.   This corresponded to the age at which
the auditory startle reflex first appeared, and was probably coincidental with the age at which the auditory system becomes
functional (8).   Priming was also maximally effective in producing
susceptibility to audiogenic seizures if it occurred between the
ages of 16 and 19 days, regardless of when the mice were tested.
But the number of days during which priming was effective increased
as the animals were tested at an older age.   Had it not been for
this latter fact, it could confidently be said that priming had met
the rigid embryological requirements for having a "critical period."
Although it is possible that testing the mice at a still later age
would not result in a further increase of this age span, this has
not been investigated, so it can only be concluded at present that
a "sensitive period" exists during which priming is maximally effective.   These data suggest that priming requires a functional auditory system and that it operates by influencing some neurological
process which is most vulnerable between the ages of 16 and 19 days.
This vulnerability could have occurred because the underlying process
was undergoing its most rapid neurological development at that time.

The next step (1) in the research program was to combine the
ontogenetic approach described above with the simple 1 x 1 diallel
cross, which has often been used in genetic investigations of behavioral phenotypes.   Because one of the parental strains of mice,
the (DBA/2J), was very susceptible to audiogenic seizures even without acoustic priming, we performed priming while all subjects were
anesthetized with ether.   Use of anesthesia has been demonstrated
to have no effect (6) on the seizures during testing and had the
desired property of preventing convulsions while priming the naturally susceptible DBA/2J mice.   Had the mice been primed without
anesthesia, the more sensitive DBA/2J's would have convulsed and
died during priming, biasing the population available for subsequent
testing.

The sensitive period extended over a greater span of days for
the DBA/2J than for the C57BL/6J mouse, and the F_1 offspring of these
two parental strains showed a bimodal sensitive period which overlapped the curves of both parents (Fig.2).   Several investigators
have used the diallel cross to determine the relative effective
contribution of each parental strain to the behavioral phenotype
of the F_1 offspring, ascribing the terms "dominance" and "heterosis"
whenever they seemed applicable.   However, the present study showed
that the phenotype of the F_1 can appear identical to, or more extreme than either parental strain, as a function of the age at
priming.   This was illustrated by the way the F_1 curve overlapped
the curves of the DBA/2J and C57BL/6J in Figure 2.   Therefore,

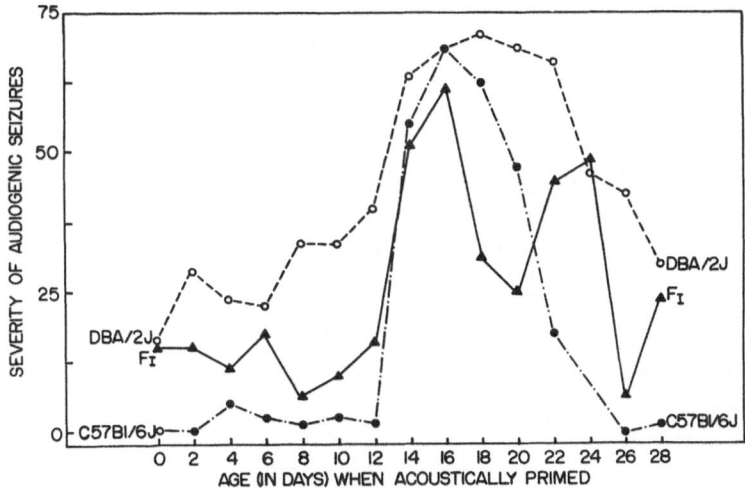

Figure 2. Behavioral genetic comparison of audiogenic seizure severity scores. Nine hundred mice of 3 genotypes were acoustically primed at one of 15 discrete ages and tested at 28 days of age (1).

the penetrance (degree to which the phenotype expressed the genotype) can vary radically as a function of age at priming, invalidating (Table 5) the use of such static terms as "dominance" and "heterosis." The pattern of inheritance of such a dynamic behavior obviously cannot be determined from mice which were primed at a simple, arbitrarily chosen age. The biomodality of the severity scores of the F_1 may have reflected the unequal or out of phase contribution of the two parental strains.

PSYCHOPHYSICAL AND NEUROLOGICAL INVESTIGATIONS

Acoustic priming undoubtedly produces some long term neural change. Psychophysics, neurophysiology, and neurology have derived basic relationships between stimulus parameters and neural or behavioral events which have aided in understanding the mechanisms underlying these events. We have recently initiated a program of stimulus manipulations in order that the resultant changes in audiogenic seizures may reveal its underlying neural properties.

C57BL/6J mice were acoustically primed with a 103 db electric bell for durations ranging from 5 seconds to 5 minutes (Fig. 3). Five days later, when they were 21 days old, they were tested for 30 seconds. In this way, the threshold for stimulus duration (that amount of priming necessary to produce 50% of maximal seizure severity score) was determined as being 20 seconds. If the individual stages of the seizure syndrome were used as dependent measures, 13 seconds of priming was the threshold of stimulus duration for

Table 5.   Comparison of audiogenic seizure severity scores of 28
   day old mice after acoustic priming at one of 15 ages (1).

Age when primed	DBA/2J scores C57BL/6J scores	F_1 scores deviate from midparent in direction of the DBA/2J parent	F_1 scores deviate from midparent in direction of the C57BL/6J parent	F_1 scores more extreme than either parent strain
28	.001	ns	ns	ns
26	.001	ns	.05	ns
24	.001	.05	ns	ns
22	.001	ns	ns	ns
20	.01	ns	.001	.01
18	ns	ns	.001	.001
16	ns	ns	ns	ns
14	ns	ns	ns	ns
12	.001	ns	ns	ns
10	.001	ns	ns	ns
8	.001	ns	ns	ns
6	.001	ns	ns	ns
4	.001	ns	ns	ns
2	.001	ns	ns	ns
0	.001	ns	ns	ns

wild running, and 28 seconds of priming was the stimulus duration
threshold for deaths.

   Binaural (spatial) summation has also been observed during

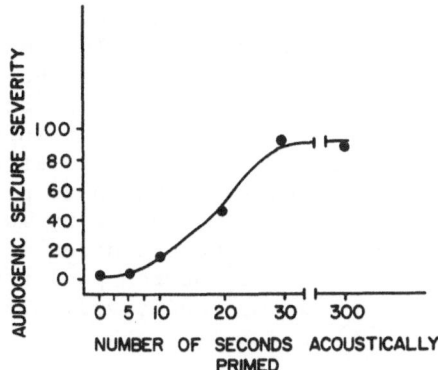

Figure 3. Audiogenic seizure severity as a function of the duration of priming. Eighty 16 day old C57BL/6J mice were primed for one of six different durations and tested for audiogenic seizures when 21 days old.

priming (9) by utilizing the glycerine plugging technique which was devised by Fuller and Collins (10). After verifying that plugging both ears of C57BL/6J and SJL/J mice with glycerine blocked 90% of the priming effect, 320 mice of each strain were primed with either one, both, or neither ears plugged with glycerine. Testing was also performed under one of these same conditions of ear plugging.

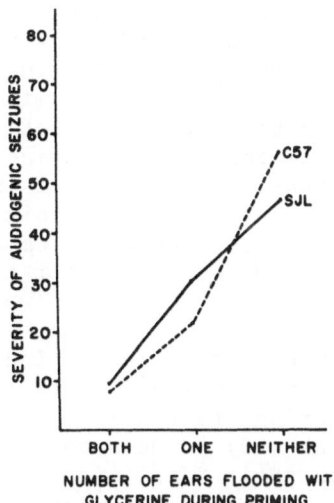

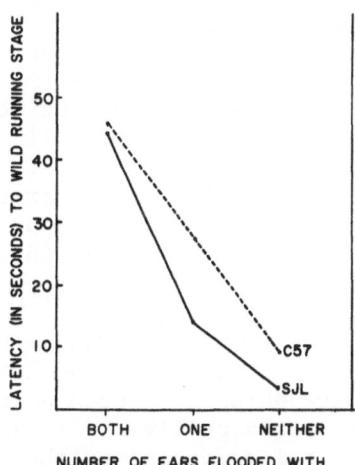

Figure 4. Binaural (spatial) summation of acoustic priming. Maximal priming of both ears (no glycerine plug) result in more severe seizures than maximal priming of only one ear (one ear flooded with glycerine) (9).

Figure 5. Binaural (spatial) summation of priming, shown by decreased latency as more ears were maximally primed (9).

The severity of seizures during testing decreased linearly (Fig. 4) as more ears were plugged during priming. In other words, priming two ears (no glycerine plug) resulted in twice the seizure severity during testing as did priming only one ear (one ear plugged). The latency observed between stimulus onset and the initiation of the wild running phase showed (Fig. 5) the opposite effect, being half as long when two ears were primed as compared to priming only one ear. These spatial summation effects were very large, accounting for 22-80% of the total variance of both indices in the two strains. It may be concluded that some common neuron pool responsible for the audiogenic seizures syndrome was additively affected by the input from the two ears.

It was also shown (11) that the number of ears plugged with glycerine during testing had little effect on seizures if the mice had been effectively primed. In other words, glycerine very effectively (90%) blocked the effects of priming, but had little efficacy in blocking the convulsions of primed mice when they were tested. This suggested that priming results in either an increased auditory sensitivity, or in an increased reactivity of structures which initiate or propagate the seizure.

After demonstrating nearly perfect binaural (spatial) summation, more C57BL/6J mice were tested on the same 16-21 schedule to determine whether temporal summation could also be observed. The experimental subjects were primed for 10 seconds followed by 60 seconds of silence, another 10 seconds of priming followed by another 60 seconds of silence, primed for a third 10 second interval and then removed from the chamber. The massed control mice were primed for 30 consecutive seconds, and spent the same total amount of time (3 minutes) in the chamber as the mice receiving distributed priming. Under these conditions, nearly perfect (98%) temporal summation was observed. When priming occurred at six 5 second intervals equally spaced over 150 minutes, temporal summation was 74% effective for the C57BL/6J mice and 70% effective for the SJL/J mice.

The glycerine technique has also been used to compare mice which have been primed and tested in the same ear (ipsilaterally) with mice which have been primed in one ear and tested in the other ear (contralaterally). Fuller and Collins (10) found a very large ipsilateral component to the seizures when their SJL/J mice were primed at 21 days of age and tested 2 days later; i.e., mice primed in one ear did not convulse when tested in the other ear. We were able to replicate their experiment and found an ipsilateral component, although its magnitude was only 20% of that seen by Fuller and Collins (10). Surprisingly, this ipsilateral effect was entirely absent if either C57BL/6J or SJL/J mice were primed at 16 days of age and tested 5 days later. Another experiment was performed to determine why an ipsilateral component should be present on a 21-23

Table 6. Audiogenic seizure severity scores and latencies to wild running of SJL/J mice primed with glycerine plugging one ear and tested with glycerine in either the same (Ipsilaterals) or the opposite (Contralaterals) ear. In Series A, each group contained 40 mice; in Series B, 20. Probabilities were determined by Fisher's t test, 1-tailed. Ages are expressed in days.

Age Primed	Age Tested	Severity Score (± S.E.)		Probability of I > C	Ipsilaterals (I)	Contralaterals (C)	Probability of I > C
		Ipsilaterals (I)	Contralaterals (C)				
				Series A			
16	21	30.0 ± 0.8	30.0 ± 0.7	N.S.	3.4 ± 0.5	19.0 ± .91	.0005
18	21	30.6 ± 0.8	22.5 ± 0.9	.0005	7.3 ± 0.6	32.3 ± 1.1	.0005
21	24	11.3 ± 0.6	4.4 ± 0.6	.0005	34.4 ± 1.4	54.1 ± 0.8	.0005
21	26	25.0 ± 0.8	13.8 ± 0.7	.0005	7.9 ± 1.1	41.7 ± 1.1	.0005
				Series B			
16	21	27.5 ± 2.5	28.8 ± 3.3	N.S.	5.3 ± 3.1	23.6 ± 4.5	.01
16	24	21.3 ± 2.1	17.5 ± 4.1	N.S.	11.1 ± 4.7	38.8 ± 5.4	.005
16	27	21.3 ± 2.1	17.5 ± 4.1	N.S.	10.9 ± 4.7	43.6 ± 4.5	.0005
21	24	13.8 ± 2.9	6.3 ± 2.5	.1	29.4 ± 6.4	51.8 ± 4.3	.01
21	27	13.8 ± 2.9	8.8 ± 2.7	N.S.	28.7 ± 6.5	52.4 ± 3.3	.025
21	30	18.8 ± 2.5	7.5 ± 3.7	.01	18.2 ± 5.6	58.1 ± 1.1	.0005

schedule, but absent on a 16-21 schedule.   These two schedules differed in three interdependent factors; age at priming, age at testing, and the interval between priming and testing.   By independently
varying these three factors, it was found (Table 6) that seizures
were bilaterally influenced by priming at 16 days of age, but an
ipsilateral effect became more pronounced if priming occurred at a
later age.   This effect was independent of the age at testing and
the interval between priming and testing.

        This effect could be explained by assuming that priming pro-
duces a change in a neural locus which has both bilateral and ipsilateral components.   Since priming is maximally effective at around
16 days of age, the same age at which the bilateral component is
most pronounced, auditory stimulation at this time may affect primarily the bilateral regions which are at the peak of their sensitive period.   The unilaterally ennervated regions are probably most
susceptible to acoustic priming at a later age.

POSSIBLE IMPLICATIONS OF THIS RESEARCH

        Thus far, all our studies have been confined to the inbred
mouse.   However, sound produced convulsions have also been observed
in several species of higher mammals, including man.   The demonstration of a sensitive period for acoustic priming which occurs
just after audition becomes functional suggests how investigations
of other species might efficiently proceed.   The initial appearance of the startle reflex (8) correlates well with the beginning
of this sensitive period in the mouse (6).   If the acoustic startle
reflex can also be used as an index of the beginning of this sensitive period in other mammals, one should examine the effects of
priming after 15 days postpartum in the cat, 25 days after birth in
the dog, and at (or shortly before) birth in man (8).   The unex-
plained presence of musicogenic seizures in man may possibly arise
in this fashion.   We are presently determining the frequency de-
pendency of priming in mice.   This will reveal whether seizures
can occur if the test stimulus is a different frequency than the
priming stimulus.   If the two auditory stimuli have to be of the
same frequency, this may be the first step in creating the uniquely
human musicogenic seizures in the mouse.

        The human environment is becoming increasingly polluted by high
intensity sounds which may occur at regular intervals, such as sonic
booms from regularly scheduled SST flights.   The data described in
this chapter suggest that any detrimental physiological effects of
audible sound might be additive over time.   If acoustic priming in
mice provides an index of the effects of intense environmental noise
on humans, the demonstration of temporal summation suggests that the
neuropathy is a positive function of the total time exposed to the
sound, but a negative function of the interval between these exposures

We have shown that priming is not a characteristic of all convulsions, but it is possible that it is a characteristic of all sensory induced convulsions. A first guess would suggest that those sensory modalities which are most sensitive in the organism would be more likely to show priming. The mouse can be classified as being primarily auditory and olfactory, whereas primates are auditory and visual mammals. The human primate shows both sound and light produced convulsions, and photic driven epilepsy has recently been demonstrated in the baboon (12). These same investigators have shown that 95% of their subjects convulsed with the second photic stimulation, whereas the first stimulus presentation produced only 50% convulsions. This is suggestive of a priming effect, although it may be more easily explained by other mechanisms.

The ratio of speculation to fact is very high when we discuss species and phenomena which have not been investigated in our laboratory. But these findings may guide others in their search for extra-auditory physiological effects of audible sound in man.

SUMMARY

Acoustic priming is a technique which, by merely exposing a juvenile mouse to 30 seconds of intense sound, can induce a high degree of susceptibility to sound produced convulsions. This technique allows biological investigations of this syndrome which do not have the confounding effects that are inherent in strain comparison studies. Acoustic priming appears to be specific to audiogenic seizures, having no effect on electroconvulsions or chemoconvulsions. The evidence suggests that priming does not work by the physiological stress mechanism, and that priming is not a general feature of all convulsions. Processes which require wakefulness or active involvement of the reticular activating system do not appear to be involved in priming. Once the 30 second initial exposure to the acoustic stimulus has occurred, none of our physiological manipulations have been able to reduce the susceptibility to audiogenic seizures, raising the possibility that priming initiates a passive destructive process, rather than an active process.

Priming is not effective if the mouse is exposed to the sound source before 14 days of age, corresponding to the age at which the auditory system becomes functional. Priming is maximally effective between 16 and 19 days postpartum, but the number of days during which priming is effective varies as a function of the age at which testing occurs. It was hypothesized that priming affects some developing neural structure during this sensitive period. The sensitive period also varies as a function of the genetic background of the mouse.

Priming shows both a stimulus threshold and spatial summation. The neural elements involved in priming are primarily bilaterally ennervated in the 16 day old mouse, but an ipsilateral component becomes more pronounced if priming occurs at later ages, giving further evidence to the hypothesis that elements undergoing rapid neural differentiation are involved. The possible relevance of these data to sensory induced seizures in humans was discussed.

REFERENCES

1. Henry, K.R. and R.E. Bowman, 1970.   Behavior genetic analysis of the ontogeny of acoustically primed audiogenic seizures in mice.   Journal of Comparative and Physiological Psychology 70:235-241.

2. Frings, H., M. Frings and M. Hamilton, 1956.   Experiments with albino mice from stocks selected for predictable susceptibilities to audiogenic seizures.   Behavior 9:44-52.

3. Henry, K.R., 1966.   Effects of Manipulations of Dietary Phenylalamine, Tyrosine, Thiamine and Ascorbic Acid on Audiogenic Seizures in Naive and Primed Inbred Strains of Mice.   Undergraduate honors thesis submitted as partial requirement for A.B. degree, Univ. of North Carolina, Chapel Hill.

4. Henry, K.R. and R.E. Bowman, 1969.   Effects of acoustic priming on audiogenic, electroconvulsive, and chemoconvulsive seizures. Journal of Comparative and Physiological Psychology 67:401-406.

5. Levine, S., 1962.   Psychophysiological effects of infantile stimulation.   In: E.L. Bliss (Ed.), Roots of Behavior, Hoeber-Harper, New York.

6. Henry, K.R., 1967.   Audiogenic seizure susceptibility induced in C57BL/6J mice by prior auditory exposure.   Science 158: 938-940.

7. French, J.D., M. Verzeano and H.W. Magoun, 1953.   A neural basis of the anesthetic state.   A.M.A. Archives of Neurology and Psychiatry 69:519-540.

8. Fox, M.W., 1966.   Neuro-behavioral ontogeny: A synthesis of ethological and neurophysiological concepts.   Brain Research 2:3-20.

9. Henry, K.R., R.E. Bowman, V.P. English and K.A. Thompson, Audiogenic seizures as a function of differential priming, in preparation.

10. Fuller, J.L. and R.L. Collins, 1968.   Mice unilaterally sensi-
    tized for audiogenic seizures.   Science 162:1295.

11. K.R. Henry and R.E. Bowman, 1969.   Audiogenic seizures in in-
    bred mice after priming and testing with maximal acoustic stimu-
    lation to one, both, or neither ears.   The Physiologist 12,
    No. 3 (abstract).

12. Killiam, K.F., 1967.   Talk given at Neurosciences Lecture pro-
    gram, Univ. of Wisconsin at Madison, 14 October.

GENETIC AND TEMPORAL CHARACTERISTICS OF AUDIOGENIC SEIZURES IN MICE

JOHN L. FULLER and ROBERT L. COLLINS

The Jackson Laboratory
Bar Harbor, Maine

Convulsive seizures induced by exposure to loud sounds have been described in rats, house mice, deermice and rabbits. Strain variation in susceptibility to audiogenic seizures has been described in all these species. The literature is particularly abundant for house mice which are excellent subjects for investigations in mammalian genetics (1,3,4,9,12,13,15). Among these reports one can find data used to support single factor, two factor, and multiple factor hypotheses for genetic determination of susceptibility. The multiple factor hypothesis is usually combined with a threshold concept. Genes are considered to affect some underlying physiological variable which is continuously, and perhaps normally, distributed. Above some threshold value for this variable, seizure risk is high; below the threshold, it is low. Actually the threshold is usually considered to be a range of values within which susceptibility fluctuates. At the extremes of the postulated continuous underlying variable, however, animals are clearly either susceptible or not susceptible.

Regardless of their attraction to monogenic or polygenic hypotheses, the majority of experimenters in this field have assumed that a particular genotype is either susceptible or not susceptible to audiogenic seizures. To be sure, it was recognized that susceptibility varied with age (6,14). It was also noted that some "susceptible" genetic lines seemed to require a warm-up trial before responding with a seizure (11). But it was not until Henry (10) demonstrated that a so-called resistant strain, C57BL/6J, could be made highly susceptible to audiogenic seizures by exposing mice to sound stimulation at a sensitive age when they were young that the arbitrariness of this dichotomous classification became apparent.

Although this paper will eventually deal with genetics, we shall begin by reviewing some of the temporal aspects of vulnerability to audiogenic seizures. There is a variety of such effects, some of which seem at this time to be related primarily to genetically programmed schedules of development. Others are clearly related to exposure to sound. The duration of such experientally modified changes in susceptibility may be a few seconds or several months.

An example of age dependence of audiogenic seizure susceptibility is found in Vicari's (14) study of the DBA/2 strain. Prior to 19 days of age no seizures were observed on first exposure to sound in 45 individuals. From the third through the sixth week of age the incidence of seizures on first exposure ranged from 60 to 90 per cent. From 50 to 59 days susceptibility fell to seven per cent and was zero from eighty days onward. Since mice can hear, as judged by the pinna reflex, from about 12 days of age, it is clear that the onset of audiogenic susceptibility is not simply associated with maturation of the auditory system. And since convulsions can be elicited in mice of advanced age by electroshock or metrazol, it is equally clear that the decline of audiogenic seizure susceptibility is not due to loss of capacity to convulse.

Fuller and Sjursen's (6) study covered a smaller range of ages, but employed eleven inbred strains which were tested at seven day intervals from three to six weeks of age. They retested their subjects rather than employing independent samples like Vicari. Thus the effects of maturational changes were confounded with the sensitization phenomenon of Henry. Nevertheless, the data demonstrate great interstrain variability in the relationship of age to susceptibility to audiogenic seizures and to their lethality. One cannot divide strains into two groups, one comprised solely of susceptible mice and the other of resistant mice. Genetic investigations must take account of this fact.

Other ways of modifying susceptibility involve exposure to sound, usually the same sound that is used to induce a seizure, although this does not appear to be necessary. Short-term effects of subconvulsive doses of sound on seizure induction were reported by Fuller and Smith (5). Five and ten-second bursts of sound enhanced seizure susceptibility for up to one minute. The term "priming" was applied to this phenomenon. The authors hypothesized that the priming stimulus initiated a process which continued for a time after discontinuance of the sound. In contrast, exposure to sound for twenty seconds resulted in a reduction of susceptibility for periods up to five minutes.

The short-term effects of priming stimulation in susceptible mice can thus be either enhancement or inhibition. Nothing is

known of the physiological basis for either phenomenon, although
Fuller and Smith pointed out a formal resemblance between them and
the effects of faradic stimulation of the cerebral cortex.   These
effects of priming are necessarily demonstrated in animals with
fairly high incidence of audiogenic seizures.   Fourteen years
elapsed before the long term effects of priming were demonstrated
in animals which were considered to be resistant (10).

The SJL/J strain is highly suitable for studing sensitization
of so-called resistant animals since the incidence of seizures on
first trials is consistently low (near 1 per cent) and the proportion of fatalities is low enough to permit repeated testing if this
is required by an experimental design.   The temporal parameters of
the process have been reported by Fuller and Collins (7).   Further
experience with this strain has shown that the absolute values of
these parameters vary more widely than we believed at the time of
our earlier publication, but the broad outlines remain unchanged.

To be effective as a priming stimulus a sound must have sufficient intensity (a doorbell emitted sound at 95 db above 0.0002
dyn/cm is ample) and sufficient duration.   One second of bell-
ringing is too short, 5 seconds produces a measurable effect, and
20 seconds is probably maximally effective.   The age of sensitivity in SJL/J males begins at about 18 days of age and declines
after 35 days of age.   However, we have found no sharp cut-off
of the sensitive period.   Fifteen out of 43 mice tested over an
age span of 22 through 28 weeks became convulsive after one or more
exposures.   The sample was a mixture of mice which had never been
primed and those which were primed at three weeks of age but failed
to sensitize.   These two groups did not differ in proportion of
subjects sensitized (Table 1).

Table 1. Convulsion history of resistant mice tested weekly.

SJL/J males, 22 to 28 weeks at start.

	Convulse with four trials	No convulsion in four trials	Total
Primed -- did not take	7	11	18
Unprimed	8	17	25
Total	15	28	43

Chi square = 0.218 (not significant)

The process of sensitization in 3 week old SJL/J male mice
requires a median time of about 30 hours.   In some samples the
rate of sensitization may be faster.   In one experiment, one-
fourth of the subjects convulsed eighteen hours after priming,
but five per cent would be more generally expected.   It is clear
that sensitization is slower in older mice of the same strain, but
we have not studied this matter quantitatively.   We have attempted
to modify the course of sensitization by drugs, hormones, and electroshock, but thus far without success.

Once an SJL/J mouse is sensitized it remains so for many weeks.
Mice primed at three weeks of age and placed in groups in an animal
room for from 19 to 25 weeks before retesting have a moderate to
high convulsive risk, while unprimed controls are resistant (Table
2).   The efficiency of priming in this study was less than usual
for unknown reasons.   SJL/J males in groups are vicious fighters
and it is conceivable that the continued stress of group living
may have had an effect on convulsibility.

The duration of induced convulsibility appears to differ among
strains.   In contrast with prolonged vulnerability of SJL/J mice,
C57BL/6J mice remain susceptible for only about one week after
sensitization.   Thus genetic variation in onset and duration of
convulsibility is manifested with induced as well as spontaneous
susceptibility.

When priming stimulations are repeated at six or twelve-hour
intervals the induction of susceptibility to audiogenic seizures
is impaired rather than assisted.   When they are eighteen hours
apart there is no interference with convulsibility.   Fuller and
Collins (7) suggested that the sensitization period could be divided into two phases, an early one which was reversible and a

Table 2.   Convulsions on first trial; 19 to 25
            weeks after priming at 3 weeks.

SJL/J males.

	Convulsion	No convulsion	Total
Primed	12	18	30
Unprimed	0	25	25
Total	12	43	55

Chi square = 23.7, d.f.= 1   P < .001

later one (commencing between 12 and 18 hours after priming) which
was irreversible.   Implicit in this formulation is the concept
that the interposed stimulation somehow resets the sensitization
process towards its beginning.

We now believe that this interpretation was in error.   Three
groups of SJL/J male mice were primed at zero hours.   Group 0-30
was retested after 30 hours, Group 0-12-30 was exposed to the bell
at 12 hours during the postulated reversible phase of sensitization.
If a reset mechanism is involved these mice should show reduced
convulsibility.   Group 0-18-30 were exposed at 18 hours in the
postulated irreversible phase when interference should not occur.
The results shown in Table 3 demonstrate that the hypothesis was
not confirmed.   In fact, the outcome is precisely opposite to
that predicted by the reset model.   It now seems clear that interference is an inappropriate name for the phenomenon, since it does
not seem to depend upon disruption of the sensitization process
but is rather a form of proactive inhibition or a post-stimulation
refractory state.

In addition to the study of temporal characteristics of audiogenic seizure susceptibility, two problems have been our particular
concern recently.   Both of us have been concerned with the anatomical and physiological basis of the difference between audiogenic
seizure susceptible and resistant mice.   Collins has in addition
investigated the genetic loci which contribute to these differences.

The most provocative finding is that sensitization can be
localized on one side of a mouse (8).   If an SJL/J mouse is primed
with one ear open and the other blocked by a few drops of glycerine,
it will convulse 48 hours later when again stimulated by bell ringing

Table 3. Test of reset versus proactive inhibition hypotheses for
                         interference.

Hours of exposure to bell	Convulsion	No convulsion	Percent
0---------------------30	31	8	79.5
0----12----------------30	28	9	75.7
0---------------18----30	10	19	34.5

Subjects convulsing prior to 30 hours were excluded from the tabu-
lation.
Losses were: Group 0-30, 1 at 0 h.; Group 0-12-30, 2 at 0 h, 2 at
12 h.; Group 0-18-30, 2 at 0 h., 9 at 18 h.

through the previously open ear, but not if stimulated through the
previously blocked ear.   In order to control for possible persistent effects of glycerine upon hearing, it was placed in the
open ear a few seconds _after_ priming.   The seizures in such unilaterally susceptible mice are similar in form to those in mice
which are susceptible on both sides.   Subsequently Collins (2)
found that post-stimulation inhibition could also be restricted to
one side.

   We cannot be sure that the difference between a convulsible
SJL/J mouse (made so as a result of priming) and a nonconvulsible
SJL/J mouse is the same as that between a convulsible DBA/2 mouse
(which required no priming) and a nonconvulsible SJL/J mouse.
However, there are no differences between the seizures in animals
with induced and those with spontaneous susceptibility.   The in-
ference is that the anatomical locus for audiogenic susceptibility
is very closely associated with the auditory receptor and its
primary afferent connections, and this locus may be the same in
mice with induced or genetically produced susceptibility.

   The determination of the genetic loci which influence audiogenic seizure susceptibility may seem less relevant to the theme
of this symposium than the study of effects mediated through experience.   The two are, however, related.   Recognition of the
occurrence of sensitization in the C57BL/6J strain and its hybrids
led to the recognition of a gene _asp_ (for audiogenic seizure prone)
which in homozygous state makes mice susceptible to audiogenic
seizures without need of prior sensitization (1).   The gene has
been located in the 8th linkage group of the mouse genone.   Dr.
Collins has extended his studies of the genetics of seizure susceptibility to a variety of strains with a view to determining how
many independent genetic factors may be involved.   The preliminary
results are encouraging.   It should be emphasized that the acuity
of such genetical analysis is increased by the recognition of the
temporal characteristics of audiogenic seizure susceptibility.
In certain situations it may be appropriate to classify groups as
convulsible or nonconvulsible, but any general genetic theory must
take account of the effects of age and of experience at specific
ages.   The identification of specific loci may lead to a genetic
factoring of the physiological complex underlying this unusual
response to high-intensity sound.

   What is the general significance of the audiogenic seizure
syndrome?   Does the induction of long- or short-term susceptibility
in otherwise resistant genotypes have any counterpart in man?
There is no evidence of such, but the seizures had been studied in
mice for twenty years before sensitization was clearly recognized.
Looking at SJL/J mice being primed for twenty seconds one would
say that they were only slightly responsive to the sound.   I
think it would be very difficult to detect any comparable phenomenon

in human infants by retrospective study made in late childhood.
We should not, therefore, exclude the possibility of similar sensi-
tization of infants and young children by strong sensory stimulation.
Caution in permitting such exposures makes good sense.

SUMMARY

Susceptibility to audiogenic seizures in mice varies with age
and with previous exposure to sound at a critical age.   Induced
seizure susceptibility in some strains may last for months.   Thus
a brief event in the immediate post-weaning stage has a major influence upon resistance to stress for a substantial portion of the
lifespan.   Genetic studies must take into account the temporal
pattern of audiogenic seizure susceptibility and the difference
between induced and natural susceptibility.   Although many genes
influence the trait, it has been possible to identify a specific
locus which plays a major role in producing natural susceptibility.

ACKNOWLEDGEMENT

This research was supported by grant MH-11327 from The National
Institute of Mental Health.

REFERENCES

1. Collins, R.L., and J.L. Fuller, 1968.   Audiogenic seizure prone
   (asp): a gene affecting behavior in linkage group VIII of the
   mouse.   Science 162:1137-1139.

2. Collins, R.L., 1970.   Unilateral inhibition of audiogenic
   seizures in mice.   Science 167:1010-1011.

3. Frings, H., and M. Frings, 1953.   The production of stocks of
   albino mice with predictable susceptibilities to audiogenic
   seizures.   Behavior 5:305-319.

4. Fuller, J.L., C. Easler, and M.E. Smith, 1950.   Inheritance of
   audiogenic seizure susceptibility in the mouse.   Genetics 35:
   622-632.

5. Fuller, J.L., and M.E. Smith, 1953.   Kinetics of sound induced
   convulsions in some inbred mouse strains.   American Journal of
   Physiology 172:661-670.

6. Fuller, J.L., and F.H. Sjursen, 1967.   Audiogenic seizures in
   eleven mouse strains.   Journal of Heredity 58:135-140.

7. Fuller, J.L., and R.L. Collins, 1968. Temporal parameters of sensitization for audiogenic seizures in SJL/J mice. Developmental Psychobiology 1:185-188.

8. Fuller, J.L., and R.L. Collins, 1968. Mice unilaterally sensitized for audiogenic seizures in mice. Science 162:1295.

9. Ginsburg, B.E., and D.S. Miller, 1963. Genetic factors in audiogenic seizures. Colloq. Int. Centre Nat. Rech. Sci. Paris 112:217.

10. Henry, K.R., 1967. Audiogenic seizure susceptibility induced in C57BL/6J mice by prior auditory exposure. Science 158:938-940.

11. Huff, S.D., and J.L. Fuller, 1965. Audiogenic seizures, the dilute locus, and phenylalanine in DBA/1 mice. Science 144:304-305.

12. Lehmann, A., and E. Böesiger, 1964. Sur le determinisme genetique de l'epilepsie acoustique de Mus musculus domestique (Swiss/Rb). Compte Rendu Acadamie Sciences (Paris) 258:4858-4861.

13. Schlesinger, K., W. Boggan, and D.X. Freedman, 1965. Genetics of audiogenic seizures: I. Relation to brain serotonin and norepinephrine in mice. Life Science 4:2345-2351.

14. Vicari, E.M., 1951. Fatal convulsive seizures in the DBA mouse strain. Journal Psychology 32:79-97.

15. Witt, G., and C. S. Hall, 1949. The genetics of audiogenic seizures in the house mouse. Journal of Comparative and Physiological Psychology 42:58-63.

INFLUENCE OF AGE, AUDITORY CONDITIONING, AND ENVIRONMENTAL NOISE

ON SOUND-INDUCED SEIZURES AND SEIZURE THRESHOLD IN MICE

GREGORY B. FINK, Department of Pharmacology
Oregon State University, Corvallis, Oregon

W. B. ITURRIAN, Department of Pharmacology
University of Georgia, Athens, Georgia

Experimental seizures may be evoked in animals by electrical, chemical, or sensory technics. Sound is the most frequently employed sensory stimulus. With sufficient stimulus intensity, all the technics produce a maximal seizure characterized by a sequence of tonic flexion and tonic extension; the pattern of submaximal seizure activity varies with the type of evoking stimulus and its intensity (1,2). Because of their simplicity and reliability, electroshock and chemoshock tests have been widely employed to study the convulsive process and to evaluate drug activity. Several tests, e.g., maximal electroshock seizure and low frequency electroshock seizure tests, are well standardized and have a reasonable neurophysiological explanation (1,2).

The physiological and/or psychological mechanisms underlying susceptibility to sound-induced seizures are not well defined. The etiology of audiogenic seizures has been ascribed variously to conflict-induced behavior, fright escape reaction, and specific sound-induced convulsions. Audiogenic seizures are apparently produced by an exaggeration of a normal element (sound) in the animal's environment, whereas other experimental seizures are produced by foreign stimuli. All mice are susceptible to convulsions produced electrically or chemically. However, sound produces a high incidence of convulsive behavior in only certain strains of mice. Many factors have been found to determine audiogenic seizure susceptibility, e.g., genetic constitution, age, hormones, nutrition, audition, environmental influences, and behavioral situations (3,4). Hence, sound-induced seizures have not been standardized and quantitative duplication of experiments is often difficult.

Most mice of genetically predisposed strains, e.g., DBA/2 strain,

211

exhibit seizures when first subjected to sound provided they are
of proper age (5,6).   Sound resistant strains of mice, such as CF
#1, exhibit very little seizure activity upon first exposure to
sound, a condition independent of age (2,7).   When mice of various strains and age are subjected to repeated exposure to sound
the reported seizure incidence is variable and conflicting; the
discrepancies in the literature are probably due to different testing procedures (3-11).   We have found that exposing CF #1 mice
to sound more than once greatly increases seizure susceptibility
provided proper selection of age and the interval (days) between
initial and subsequent sound exposure is made (7,12).   The initial
exposure (audio-conditioning or acoustic prime) at a critical early
age induces sensitivity to the test sound.   Seizure incidence and
severity in CF #1 and other sound-resistant strains is dependent
upon the temporal aspects of the sensitization process.   (7,12-14).
This response of sensitized mice has been designated the "Audio-
Conditioned Convulsive Response" of "ACCR" to distinguish it from
genetically controlled audiogenic seizures and from non-auditory
sensitizing procedures (12).

    The research potential of audiogenic seizures for investigat-
ing the action of drugs or maturation of the central nervous system
has limitations due to complexities involved in maintaining animals
of predictable susceptibility.   Furthermore, death frequently follows maximal audiogenic seizures in the susceptible strains.   The
ACCR offers seizures of predictable incidence and severity without
the use of genetically susceptible strains, special diets, chemicals
or surgical manipulation.   The CF #1 strain was selected for study
since it is a widely used strain known not to be susceptible ordinarily to sound-induced seizures and to have a low death rate following maximal electroshock seizures (2).   This report is concerned
with the characterization of the ACCR, some factors that influence
it, and the use of electroshock and chemoshock procedures for comparison and interpretation.

                   AUDIO-CONDITIONED CONVULSIVE RESPONSE

    Mice (CAW:CF#1-SW from Carworth Farms) were bred to produce
dated pregnancies and examined for offspring with such frequency
that six hours was the maximum variability in recorded age; all mice
were maintained in a controlled environment (12,15,16).   At selected days of age, individual mice were placed in a glass chamber (25
cm diameter and 15 cm deep) and subjected to 60 sec of sound (95 db
relative to .0002 dynes/cm^2) produced by an electric door bell (7,12).

    As a sound resistant strain, CF #1 mice exhibit a very low
seizure incidence upon first exposure to sound.   However, upon
a second exposure to sound 1 to 5 days later, the incidence of
convulsive behavior may be markedly altered.   As illustrated in

Table 1. Effect of age and prior auditory stimulation
    upon convulsions in CF #1 mice. (Mean ± S.E.)

Age (days) when Conditioned	CONVULSIONS (%) First exposure to sound	CONVULSIONS (%) Second exposure 3 days later	Number of Animals
12	0	0	53
20	4	90 ± 4 (tonus 65 ± 7)	63
30	8	31 ± 3 (tonus 3)	61
45	5	3	92
60	2	2	97

Table 2. Profile for seizure susceptibility among CF
#1 mice subjected to an initial bell-sound at 18 days
of age and tested 1 to 6 days later.   (Mean ± S.E.)

Condition Test Interval	Number Animals Exposed	Incidence of Convulsions (%) Maximal Response	Total
0 days	416	1	3 (±1)
1 day	23	26 (±1)	65 (±10)
2 days	208	57 (±6)	88 (±4)
3 days	36	47 (±12)	58 (±9)
4 days	74	35 (±9)	43 (±9)
5 days	54	0	7 (±4)
6 days	21	0	5 (±5)

Table 1, seizure incidence and severity greatly depends upon the
age of initial sound exposure (audio-conditioning or sensitization).
Mice audio-conditioned at 20 days of age exhibited 90% seizure activity upon presentation of the test sound 3 days later, whereas
mice audio-conditioned at 12 or 45 days showed little response upon
test.   As illustrated in Table 2, the time between audio-conditioning and the test sound (conditioning-test interval) is as important
as age.   The greatest incidence of maximal seizures occurs at a 2
or 3 day condition-test interval and the audio-sensitive period
lasts about 4 days.   The period of audiosensitivity and the convulsive response induced by the test sound have been characterized
in detail at various days of age with total seizure incidence and,
in particular, susceptibility to maximal seizures delineated (12).
Mice audio-conditioned at 12 days of age or less did not exhibit
seizures at any conditioned test interval.   Convulsive behavior
upon test sound began at about 14 days of age and reached a peak
incidence at 18 to 23 days of age.   The highest total incidence
of seizures and of maximal seizures occurred with a 2 or 3 day
condition-test interval in mice audio-conditioned at 18 or 20 days
of age.   A high incidence of maximal seizures first appeared in
mice audio-conditioned at 15 days of age; few maximal seizures were
observed in mice audio-conditioned after 22 days of age.   Neither
sex nor season of the year affected seizure susceptibility.   Death
may follow a maximal seizure and the death rate was approximately
11% for all of the animals tested.   The highest death rate occurred
in subjects tested at 20 days of age.   Mortality was higher with
a 1 or 5 day condition-test interval than with an interval of 2 or
3 days.   Susceptibility to seizure and capacity for survival appear
to involve independent mechanisms (17, 18).

It should be noted that repeated exposure to the sound stimulus prolongs audiosensitivity.   Some seizure susceptibility persisted for several weeks if the mice were tested at 2 days intervals
after audio-conditioning at age 16 or 18 days (11, 12).   Audiosensitization at age 18 days will induce 90% seizure incidence upon test
at 20 days but only 5% if test is at day 30.   However, if tested
at day 20 and also day 30, the third exposure incidence is 42%.
Thus, an overt convulsion elicited by sound prolongs seizure susceptibility.   The fact that seizure susceptibility persists once on
animal experiences a convulsion indicates that audio-sensitivity
and convulsive behavior may reflect separate mechanisms.

Previous auditory stimulation was found to be absolutely essential for the genesis of convulsions.   Other experimental conditioning procedures were not effective in inducing seizure susceptibility
(18).   With proper selection of age and of conditioning-test interval,
the ACCR offers seizures that are reproducible and of predictable incidence and severity (12).   Seizure activity begins after
a latency of several seconds with an explosive burst of wild running
for 3 to 5 seconds.   The running is succeeded by a convulsive spasm,

clonic convulsion, or clonic-tonic (maximal) convulsion.

MAXIMAL SEIZURES

The patterns of maximal seizures induced by sound, electrical and chemical stimuli in CF #1, Frings, and O'Grady mice are very similar at any given age (1,2,17,19). The Frings and O'Grady strains of albino mice are genetically predisposed and exhibit maximal audiogenic seizures (MAS) when first exposed to sound, whereas the CF #1 strain is sound resistant and must be audio-conditioned at a critical age before exhibiting a maximal audio-conditioned convulsive response (ACCR).

In all three strains, the incidence of maximal seizures is age dependent and the effect of maturation on the development of the seizure pattern has been studied (2,12,19.). In Frings strain, MAS first occurred on day of age 13 with peak incidence (96%) by day 20, and in the O'Grady strain day 16 with peak (86%) by day 20-24. By day 32 and day 60, 86% and 80% of Frings mice still exhibited MAS but only 49% and 29% of O'Grady mice responded maximally. In CF #1 mice, maximal ACCR first occurred on day 16 with peak incidence (65%) by day 22-23; however, incidence was less than 10% by day 32. The effect of maturation on the development of maximal electroshock seizures (MES) was similar in all 3 strains of mice and resembled that of MAS, except that all mice respond maximally to electroshock throughout life. Thus, ontogenesis of maximal seizure susceptibility in audiogenic susceptible (first trial) strains of mice and of ACCR susceptibility follow a remarkably similar time course. The ability of CF #1 mice to respond to sound with a maximal ACCR persists for only a very brief period (unless a previous convulsion has occurred) whereas the ability of O'Grady and, in particular, Frings mice to exhibit MAS persists for weeks or months.

Similarities and differences in the pattern of maximal seizures induced by various stimuli in young and adult mice are shown in Table 3. The data in the table are a compilation of experiments reported previously (1,2,17). The pattern of MES is tonic-clonic, whereas maximal pentylenetetrazol seizure (MPS) pattern is clonic-tonic. The maximal ACCR and MAS pattern are the same and appear to be a combination of the previous two in that they are clonic-tonic-clonic.

The pretonic phase is absent in MES, but consists of periods of latency and clonus in MPS. In contrast, the pretonic phase of MAS is characterized by a latent period of several seconds, a running component, and a period of clonus. The tonic phase is remarkably similar in all maximal seizures and relatively independent of the stimulus employed. Duration of the components do change with age;

TABLE 3. Maximal seizure patterns in mice.

SEIZURE (Strain, Age)	Mean duration of seizure components, seconds					
	LATENCY	RUNNING	CLONUS	HINDLEG TONIC FLEXION	EXTENSION	CLONUS
Audiogenic (Frings, Adult)	2.9	2.8	2.7	1.0	12.9	yes
Audiogenic (O'Grady, Adult)	3.0	4.0	2.0	1.9	11.1	yes
Electroshock (CF#1 Adult)		a	a	1.8b	12.8	yes
Pentylenetetrazol (CF#1 Adult)	3.4	a	2.5	0.7	12.8	no
Audio-Conditioned (CF#1, 20 day)	6.	3.2	3.0	1.8	15.1	yes
Electroshock (CF#1, 20 day)		a	a	1.6b	17.0	yes
Pentylenetetrazol (CF#1, 20 day)	3.1	a	3.0	1.8	20.4	yes

a Component absent
b Includes latency

e.g., the duration of electroshock tonic-flexion increases until
about 30 days of age, whereas tonic extension decreases until about
45 days of age (17,18).

A period of terminal clonus follows the tonic phase of electroshock and sound-induced seizures.   Although terminal clonus in
adult mice is absent after MPS, it is frequently observed in younger
mice.   A cataleptic state follows terminal clonus of sound-induced
convulsions, and occasionally electroshock in younger mice.   This
cataleptic state has been observed to persist for as long as 3 to 5
minutes.   Recurrent convulsions frequently follow MPS since time
is necessary for concentration of the drug to decline to the level
that does not cause seizure activity.   In young and adult mice,
recurrent seizures are not observed subsequent to supramaximal
electroshock or to sound if the stimulus is discontinued; but, if
the auditory stimulus is continued, seizures of a running or clonic
nature occur.

The mortality rate after maximal seizures is influenced by age.
Adult DF #1 mice seldom die following MES, but over 50% die in tonic
extension following intravenous injection of 38 mg/kg pentylenetetrazol.   About 20% of young CF #1 mice (20 day) die after maximal
seizures induced by electroshock and sound whereas death occurs in
almost all the mice injected with pentylenetetrazol.   In the Frings
and O'Grady strains, about 20% of young mice (20 day) die after MAS
and MES (2,19), but adults usually survive the maximal seizure.

Certain features of the maximal seizure pattern and its sequelae
depend on the parameters of the stimulus used to initiate the convulsion.   Supramaximal electrical stimulation fulfills the requirements for a maximal convulsion almost instantaneously; hence, the
MES pattern is tonic from the start.   On the other hand, auditory
stimulation is not immediately supramaximal and three distinct components precede the tonic phase of the seizure.   The maximal seizure
pattern is virtually an all-or-none phenomenon and represents maximal activity of the spinal cord, the final common pathway of seizure
discharge.   Since electroshock, chemoshock, and sound stimuli appear
to excite the spinal cord maximally, the distinguishing characteristics of sound-induced seizure susceptibility probably involves supra-
spinal mechanisms.   In contrast to electroshock and chemoshock procedures, sound induced seizure procedures have not been standardized
and the parameters of the sound stimulus have not been sufficiently
defined.   An electric bell is generally employed as the sound stimulus, and it is assumed that the physical properties of the stimulus
are not critical determinants provided it is sufficiently intense
and of proper frequency range (3,4).   However, the characteristics
of the sound that induces audiosensitivity and seizures in CF #1
mice does influence the incidence and severity of the convulsive
response.   Mere measurement of sound intensity may not be sufficient
for standardization from bell to bell as other characteristics are

apparently important in determining the seizure pattern. After extensive use a bell apparently changes frequency without a change in intensity which results in a decreased incidence of maximal seizures (17,18). The latency of the different seizure components was also altered as the bell changed tone. Therefore, each bell must be bioassayed periodically to provide a reference for responses throughout an investigation. The criteria for standardization were seizure incidence of approximately 90% with 60% maximal seizures among 20 day old CF #1 mice tested 48 hours after audiosensitization. The duration of the initial auditory stimulus to induce sensitization is not critical, as 10 seconds of bell is nearly as effective as 60 seconds (14,16,18). Prolonged duration of the initial sound or repeated brief auditory stimulation (less than 12 hours apart) will block the development of audiosensitivity and subsequent convulsions upon test (14,16). Also environmental noises of sufficient intensity, such as Galton whistle, fire alarm, clanging metal, will elicit audiosensitivity (vide infra).

SEIZURE THRESHOLD

The experimental techniques employed to induce maximal or minimal electroshock seizures may be modified to measure quantitatively the seizure threshold. Maximal electroshock seizure (MES) threshold or low frequency (minimal) electroshock seizure (l.f.ES) threshold are measured by the intensity of stimulus required to evoke a maximal seizure (tonic extension) or a minimal seizure (stun response or minimal clonus) in 50% of mice (2,17). A neurophysiological explanation of these seizures (2) states that in order for a minimal seizure to be evoked, it is necessary for a substantial number of neurons to discharge over a finite period of time; this collection of neurons has been designated the "oscillator" to distinguish it from the seizure focus or exogenous stimulus which serves to trigger the oscillator. With l.f.ES the oscillator is excited with maximum efficiency, but the discharge spreads very little to adjacent areas and maximal spread is never obtained. With MES the oscillator discharges maximally, the discharge spreads over the brain, and a maximal seizure occurs. The l.f.ES threshold may be used to measure the relative susceptibility of the animal to seizures (or the level of brain excitability), and the MES threshold to measure the relative ease in which seizure discharge spreads to other neuronal circuits.

Audiogenic seizure susceptible animals have been reported to be more susceptible to electroshock, caffeine, nicotine, strychnine and pentylenetetrazol induced seizures than animals that are resistant to sound-induced seizures (4). The thresholds for l.f.ES in both submaximal-audiogenic and maximal-audiogenic-susceptible O'Grady mice were significantly lower than that for resistant CF #1 controls; the threshold for MES was lower in only maximally-audiogenic-susceptible O'Grady mice (2). This suggests that mice which display a

maximal audiogenic seizure are more prone to oscillator discharge
and to spread of the oscillator discharge; those that display only
submaximal seizures are more sensitive than the control group to
oscillator discharge (more susceptible to seizures) but have essentially normal resistance to seizure spread (exhibit only minimal
seizures). Minimal and maximal pentylenetetrazol seizure thresholds
were essentially the same in O'Grady audiogenic seizure mice and CF
#1 controls (2).

The effect of maturation on the development of electroshock
and chemoshock seizure thresholds in CF #1 mice has been studied
(17,20). The MES threshold did not change after 20 days of age,
but the l.f.ES threshold markedly decreased from 20 to 30 days of
age and reaches adult level about day 45 (Table 4). Minimal and
maximal pentylenetetrazol seizure threshold also decreased from 20
to 30 days (17,18). The electroshock thresholds determined at any
given age were not altered by previous determinations.

Audio-conditioning (a single 60-second sound stimulation) produced profound changes in the ontogeny of l.f.ES (Table 4) and chemoshock thresholds (Fig. 1) but did not affect maximal seizure threshold

Table 4. Effect of age, repeated low-frequency electroshock seizures (l.f.ES) and audio-conditioning upon low frequency electroshock
threshold.

		Mean Threshold (Volts)			
		Age in Days			
18	20	22	25	30	45
Bell	52*	38*	31*	29*	25
87	66	49	39	34	25
—	63	48	40	34	25
—	—	53	40	32	25
—	—	—	Bell	34	25
—	—	—	—	34	25
—	—	—	—	—	25

*p < .05

(17).   Maturation of MES pattern was not altered by electroshock
alone.   However, in sound sensitized mice, the seizure pattern
to second and third MES was altered in that the extensor component
was lengthened (Fig. 1).   Also, supramaximal electroshock in 18
day old mice does not substitute for audio-conditioning and produce
susceptibility to sound-induced convulsions.   The thresholds for
minimal (clonic) and maximal (tonic) pentylenetetrazol were increased for about 6 days after audio-conditioning.   In Table 4, it
may be seen that l.f.ES thresholds in mice audio-conditioned at 18
days of age were significantly lower than for control animals until
30 days of age.   Audio-conditioning mice at 25 days of age does
not alter l.f.ES threshold (Table 4) nor does it produce ACCR susceptibility.   Thus, the audiosensitive period in CF #1 mice is
accompanied by changes in seizure threshold.   During the audio-
sensitive period, the mice are more sensitive to oscillator discharge, as indicated by lowered l.f.ES threshold, and hence more
susceptible to seizures.   Changes in seizure threshold are longer
lasting than audiosensitivity.   Determinations of seizure threshold
appears useful to evaluate the effect of postnatal exogenous stimuli,
such as sound (12,18).

ANIMAL HOUSE NOISES

Following initial auditory stimulation, CF #1 mice exhibit a
transitory audiosensitivity that lasts about 4 days (Table 2).

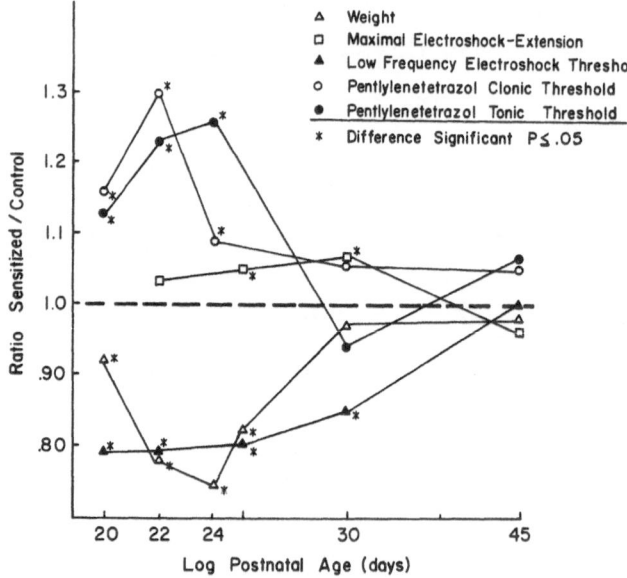

Figure 1.   Effect of audio-conditioning upon threshold for pentylenetetrazol and low frequency electroshock seizures, weight gain,
and duration of maximal electroshock seizure tonic extension.

Table 5. Modification of seizure susceptibility by environmental noise. The mice were exposed to sound initially at 18 days of age and tested for seizure 5 days later.

Intertrial stress age 20 days	Number animals exposed	Convulsions (%) 2nd exposure to bell	
		Maximal response	Total
Controls	54	0	7
Concrete drill noise	14	21	93
Galton whistle	14	57	86
Banging can-lid	10	50	90
Barking dog	12	75	75
Barking dog	11	45	64
Bell	43	51	92

Table 6. Effect of noises in the animal room in eliciting seizure susceptibility. The duration of the initial sound was approximately one minute.

Source of initial sound	Number animals exposed	Convulsions (%) exposure to bell 2 days later	
		Maximal response	Total
None	21	0	5
Galton whistle			
setting 8	11	9	27
setting 16	11	37	37

This period of audiosensitivity (as measured by susceptibility to maximal seizures) can be profoundly influenced by environmental situations that may occur while housing the mice (16).   If mice audio-conditioned at age 18 days and tested 5 days later are exposed to environmental noise at age 20 days, markedly different results are obtained.   As shown in Table 5, various noises such as occur from hammering metal or barking dogs are most effective in prolonging seizure susceptibility, even though they may be brief in duration and not sufficiently intense to produce seizure activity in the 20 day old mice.   An electric drill employed in the animal room or a garbage can lid carelessly banged results in over 90% seizure incidence rather than the 7% expected at the 5th day condition-test interval.   Extraneous noise in the animal room can also serve as an audio-conditioning stimulus in a manner similar to the initial exposure to the bell.   The intensity of the extraneous conditioning stimulus may influence the incidence and severity of seizures produced by the test-bell (Table 6).   A Galton whistle (95 db) adjusted to two different frequencies demonstrated different effectiveness and differences in latency of the convulsions.

Finally, it should be noted that a single 60-second audio-conditioning exposure to the bell has a marked effect upon the growth of weanling mice even though no overt seizures occur (Figure 1).   The effect is most apparent 4 days after audiosensitization; the gross appearance of the mice changes and the coat becomes unkempt.   Such mice recover normal weight, and mate and reproduce in an apparently normal manner.

These results emphasize the importance of environmental conditions in the study of sound-induced seizures.   Unless the environment is rigidly controlled while housing the mice as well as during testing, the experimental results are not reproducible. This is particularly important in the study of the ACCR but no doubt influences experiments in mice genetically predisposed to audiogenic seizures.

ALTERATION OF SEIZURE RESPONSE BY DRUGS

Experimental seizure technics are very useful to analyze the convulsive process and to evaluate drug activity.   Electroshock and chemoshock procedures are the most popular because of simplicity and reliability.   Since sound-induced seizure technics are variable and not standardized, the results of drug studies reported in the literature are conflicting (1-4,6,21).   Drugs from numerous pharmacological classes can modify the audiogenic seizure and considerable inconsistencies remain to be explained.   For example, reserpine has been reported to afford no protection, but to actually increase seizure severity and also to protect mice and rats from seizures.

We have confined our drug studies to maximal seizures in Frings strain of mice and to the ACCR in CF #1 mice. Drug activity was determined on the basis of ability to prevent hindleg tonic-extensor component of maximal audiogenic seizures (MAS tonic-extension test, the counterpart of the MES test), and to prevent the running component (MAS running test) which represents protection from any seizure activity. The Frings strain was chosen because adult mice will exhibit MAS repeatedly for long periods (19) and thus can be tested for a maximal response before and after drug administration for control purposes (1). Drugs which effectively modify maximal seizure pattern, trimethadione, phenobarbital, diphenylhydantoin, phenaglycodol, and meprobamate, are at least three times more effective against MAS than MES tonic extension; they also block MAS-running in lower doses than MES tonic extension. Chlorpromazine and other phenothiazine tranquilizers block MAS tonic extension, but not MAS running or MES tonic extension. Thus MAS appears to be more susceptible to modification by drugs than MES. Correlation of the ability of anticonvulsant drugs to modify MAS pattern and their mechanism of action suggests that effective drugs modify MAS by preventing spread of the seizure discharge (2).

Since previous auditory stimulation is essential for genesis of the ACCR, pharmacological protection from audio-conditioning would prevent the development of the convulsion produced by the second exposure to sound. Prototype drugs from various pharmacological classes were administered at their time of peak effect before audio-conditioning the mice which were then tested 2 or 3 days later. No drug was found to clearly protect the mice from the effects of audio-conditioning although changes in latency and pattern of the subsequent ACCR sometimes occurred. It should be remembered that only weanling CF #1 mice are audio-sensitive, that dramatic behavioral changes occur in this prejuvenile (15-26 days of age) period, and that immature mice may respond differently to drugs than adult mice. Hence, apparent inhibition of audio-conditioning by drugs may be due to residual drug activity affecting the subsequent seizure response upon test. The convulsion (ACCR) produced by the second exposure to sound is apparently modified by drugs in a similar manner as the MAS. For example, phenobarbital and diphenylhydantoin block the tonic-extension of the ACCR in low doses.

DISCUSSION

Convulsive disorders are common, represent a dramatic symptom of brain disfunction or disease, and are not completely controlled by drugs. The value of experimentally-induced seizures in animals to search for candidate anticonvulsant drugs has been firmly established. By itself, no single laboratory test can conclusively establish the presence or absence of anti-epileptic activity of a

drug, and hence a battery of assay methods has evolved. The MES
test has good predictive value for drugs effective against human
grand mal, the l.f.ES some for psychomotor and petit mal epilepsy,
and pentylenetetrazol seizures for drugs against petit and grand
mal. With the possible exception of MES, none of the methods are
particularly selective for specific clinical types of epilepsy;
e.g., petit mal, which is common in children. The causes of convulsions in children seem more diversified than among adults, and
genetic factors play a considerable role. Audiogenic seizures,
with hereditary aspects, physiological and behavioral implications,
correlation with maturation, and induced by an environmental stimulus, provide a potential test and also a unique model of convulsive
disorders. However, the multiplicity of factors that influence
audiogenic seizures has prevented satisfactory definition of the
underlying mechanisms and the standardization of a simple laboratory
anticonvulsant test. The research potential of audiogenic seizures
for investigating the action of drugs or maturation of the central
nervous system has limitations due to complexities involved in maintaining predictable susceptible animals. It has been demonstrated
that sound-induced convulsions can be obtained in ordinarily resistant strains of mice (e.g., CF #1, C57Bl/6, SJL, CBA) by proper
selection of age, audio-conditioning, and control of the temporal
aspects of the sensitization process. The audio-conditioned convulsive response, ACCR, offers seizures of predictable incidence
and severity without the use of genetically susceptible strains,
special diets, chemicals, or surgical manipulation if proper experimental precautions are taken. The overt appearance of the ACCR
closely resembles audiogenic seizures in genetically susceptible
strains and the ontogenesis of seizure incidence and severity
follows a similar time course. The important differences between
the types of sound-induced seizures are that prior auditory stimulation is essential, that duration of susceptibility to maximal
seizures is shorter, and that adaptation or prolongation of the
sensitization due to repeated stimulation or chronic noise occur
with ACCR. Hence, experimental conditions must be carefully controlled since environmental sound and animal house noise has a
profound effect on the incidence and severity of ACCR. Immature
mice are more sensitive to a variety of environmental factors than
mature mice and perhaps intense auditory stimuli in a critical
period disrupts neurological maturation. The ACCR may provide a
useful experimental model for the study of psychological, pathophysiological, and biochemical mechanisms in the developing central
nervous system. Sound-induced seizures might be added to the
battery of laboratory anticonvulsant tests with apparent usefulness
for evaluating drugs effective in the young. Drugs may affect the
audiosensitization induced by the initial sound or the seizure induced by the second exposure to sound. Thus, the ACCR has potential
for possible assay of drugs of several pharmacological classes and
as a model for studies in developmental pharmacology or physiology.

REFERENCES

1. Fink, G.B. and E.A. Swinyard, 1959.   Modification of maximal
   audiogenic and electroshock seizures in mice by psychopharmacologic drugs.   Journal of Pharmacology and Experimental
   Therapeutics.   127:318-324.

2. Swinyard, E.A., A.W. Castellion, G.B.Fink and L.S. Goodman,
   1963.   Some neurophysiological and neuropharmacological
   characteristics of audiogenic-seizure-susceptible mice.
   Journal of Pharmacology and Experimental Therapeutics.
   140:375-384.

3. Bevan, W., 1955.   Sound precipitated convulsions: 1947-1954.
   Psychological Bulletin.   52:473-504.

4. Fuller, J.L. and R.E. Wimer, 1966.   Neural, sensory and motor
   functions.   In: E.L. Green (Ed.), Biology of the Laboratory
   Mouse, McGraw-Hill, New York, pp. 609-629.

5. Vicari, E.M., 1951.   Fatal convulsive seizures in the DBA
   mouse strain.   Journal of Psychology.   32:79-97.

6. Suter, C., W.O. Klingman, O.W. Lacy, D. Boggs, R. Marks and
   C.B. Coplinger, 1958.   Sound-induced seizures in animals.
   Neurology. 8:117-120.

7. Iturrian, W.B. and G.B. Fink, 1967.   Conditioned convulsive
   reaction.   Federation Proceedings. 26:736.

8. Frings, H., M. Frings and A. Kivert, 1951.   Behavior patterns
   of the laboratory mouse under auditory stress.   Journal of
   Mammalogy. 32:60-72.

9. Frings, H. and M. Frings, 1952.   Audiogenic seizures in the
   laboratory mouse.   Journal of Mammalogy. 33:80-87.

10. Fuller, J.L. and F.H. Sjursen, Jr., 1967.   Audiogenic seizures
    in eleven mouse strains.   Journal of Heredity. 58:135-140.

11. Al-Hachim, G.M. and G.B. Fink, 1967.   Effect of DDT or para-
    thion on audiogenic seizures of offspring from DDT or parathion
    treated mothers.   Psychological Reports. 20:1183-1187.

12. Iturrian, W.B., and G.B. Fink, 1968.   Effect of age and con-
    dition-test interval (days) on an audio-conditioned convulsive
    response in CF #1 mice.   Developmental Psychobiology.
    1:230-235.

13. Henry, K.R., 1967.   Audiogenic seizure susceptibility induced

in C57BL/6J mice by prior auditory exposure.   Science.
158:938-940.

14. Fuller, J.L. and R.L. Collins, 1968.   Temporal parameters of
    sensitization for audiogenic seizures in SJL/J mice.   Devel-
    opmental Psychobiology. 1:185-188.

15. Iturrian, W.B. and G.B. Fink, 1968.   Comparison of bedding
    material: Habitat preference of pregnant mice and repoductive
    performance.   Laboratory Animal Care. 18:160-164.

16. Iturrian, W.B. and G.B. Fink, 1968.   Effect of noise in the
    animal house on seizure susceptibility and growth of mice.
    Laboratory Animal Care. 18:557-560.

17. Iturrian, W.B. and G.B. Fink, 1969.   Influence of age and
    brief auditory conditioning upon experimental seizures in
    mice.   Developmental Psychobiology. 2:10-15.

18. Iturrian, W.B., 1970.   Effect of noise on experimental seizures
    and growth of weanling mice.   IV International ILAR-ICLA
    Symposium - In: W.L. Gay (Ed.), Defining the Laboratory Animal
    in the Search for Health. National Academy of Sciences,
    Washington, D.C. In press.

19. Castellion, A.W., Swinyard, E.A. and L.S. Goodman, 1965.
    Effect of maturation on the development and reproducibility
    of audiogenic and electroshock seizures in mice.   Experimental
    Neurology. 13:206-217.

20. Vernadakis, A. and D.M. Woodbury, 1965.   Effects of diphenyl-
    hydantoin on electroshock seizure thresholds in developing
    rats.   Journal of Pharmacology and Experimental Therapeutics.
    148:144-150.

21. Plotnikoff, N., 1963.   A neuropharmacological study of escape
    from audiogenic seizures.   Colloquium Internationale Recherche
    Sciences (Paris), 112:429-443.

PSYCHOPHARMACOLOGY OF THE RESPONSE TO NOISE, WITH SPECIAL REFERENCE

TO AUDIOGENIC SEIZURE IN MICE

ALICE G. LEHMANN

Laboratoire de Physiologie Acoustique
Domaine de Vilvert
78 - Jouy-en-Josas, France

Troubles induced by intense noise can affect various physiological functions: cardiovascular or circulatory systems, sleep,
endocrines, reproduction, susceptibility to infection. One of the
most obvious harmful effects of noise is its influence on the nervous system. The first nervous system damages induced by noise
are located at its most peripheral level: the ear. Very loud and
long lasting noises can give auditory troubles going as far as total
deafness. The only protection against these injuries is to obstruct
the ears, as done in airfields, for example. But noises loud enough
to induce injuries at the auditory level can, however, be very harmful for the nervous system and can give troubles going from an abnormal excitability to loss of sleep and to a real nervous break-
down. The degree of nuisance of this overstimulation is felt
differently according to different individuals and is closely
related to the state of their nervous system; a defect in it can
sensitize to noise and lead to more or less important disorders.

As to defects, the role of the pharmacologist is to find drugs
that are able to specifically correct them and, if this is not
possible, at least to non-specifically protect against their consequences. The screening of these drugs has to be done on animals
in which similar disorders to those observed in man can be found or
artificially induced, as for any screening. One of these troubles,
known as the audiogenic seizure, exists in some rodents and can be
used as a screening test for psychopharmacologic drugs acting
against the harmful effect of noise. Most rodents, like normal
human beings, submitted to intense noise have just a startle response followed by a tendency to escape, but no other reactions.

227

However, some mutants possess a nervous system defect which modifies their usual behavior: submitted to a loud sound stimulation, after a startle reaction and a latent period of a few seconds, the animal runs wildly, and this running phase ends in clonic and tonic convulsions which can even lead to the animal's death (1). This reaction appears to be much stronger than the normal or even pathological human responses to noise, though, in some circumstances, convulsive states known as mucicogenic epilepsy can occur in man (2). This difference between rodent's and man's reaction is probably due to the smaller development of the cortex, and therefore to a lower level of nervous integration in the rodents.

The audiogenic seizure was discovered in 1924 in mice and rats (1). It was studied by many scientists involved in basic research in psychology, genetics, physiology and nutrition, who were interested by its similarity to human epilepsy. Since 1950, and after the discovery of many psychotropic drugs, it became necessary to find screening tests to verify their activity. The audiogenic seizure, being a nervous system reaction induced by a physiological stimulus, seemed to have the qualities required to be a good test for psychotropic drugs. Moreover, as it was a reaction induced by noise, it was used to study drugs active against the harmful effect of noise. At the same time, the actions of psychopharmacologic drugs on audiogenic seizure helped to clarify the mechanisms by which noise injured the nervous system.

The first psychopharmacological studies on the audiogenic seizure were done mostly on genetically unselected rats, which often presented poor and not fully reproducible responses. This explains the discrepancies of the results obtained, which were quite difficult to interpret. The only strain of rats selected for the audiogenic seizure is bred in Moscow by Krushinski; some pharmacological studies were done on it, and its results are in complete agreement with those obtained on mice (3). With mice the studies are more reliable because there are several highly inbred strains selected for their sensitivity to sound, and their homozygoty is high enough to provide a well known material for these studies; in these strains, the mutants sensitive to sound stimulation have a nervous defect, which first appeared in behavioral studies, showing a greater emotionality, some difficulties in learning ability and in conditioning (4, 5).

However, it must be pointed out that the abnormalities of the nervous system which lead mutants to react to sound stimulations by a seizure may be of different origins. In fact, in different strains of mice the genetic defect may be mono or polygenic, and there are several metabolisms involved in their reactions against noise, depending on the strain studied (6, 7).

Likewise, there are differences in the sensitivity of men's

nervous system to noise, and in the defects responsible for this
sensitivity.   For this reason we wish to distinguish between two
types of psychopharmacological drugs acting on audiogenic seizures,
one of which is non-specific and effective for all strains of mice
as for all human beings.   The other one acts on particular metabolisms and can be used only when the metabolic defect is known.   The
non-specific drugs tested are sedatives, minor tranquilizers and
neuroleptics derived from phenothiazines.   These drugs increase
the threshold of nervous system excitability by acting either on
the cortex or subcortex.   Because sensitivity to noise is due to
an exhausted nervous sysceptibility, almost all of these drugs
protect against the harmful effect of noise, as could be expected.

The study of drugs acting on <u>specific</u> metabolisms is much more
interesting and can help to clarify the biochemical metabolisms of
the nervous system involved in the noise reaction.   They are,
nevertheless, like those involved in many other different stresses,
electroshock, cold, etc.

There are several substances which may be considered as chemical mediators in the central nervous system, some of them being
at present more or less doubtful.   These are: gamma-amino-butyric
acid (GABA), norepinephrine (NE), serotonin (5-HT) and acetylcholine.   Drugs increasing GABA level give a strong effective protection against the audiogenic seizure.   On the contrary, substances
decreasing its level increase sensitivity to sound.   While, from
these results, it can be inferred that GABA metabolism is involved
in the effect of noise, it is not known, at present, if its effect
is due to a non-specific inhibitory action on the nervous system
or linked to its interference with the tricarboxylic acid cycle.

Studies with drugs acting on norepinephrine and serotonin
metabolisms give some good evidence of the involvement of these
two substances in the action of noise.   Experiments show that a
decrease in the brain level of these two biogenic amines increases
the severity of audiogenic seizures, but that an increase of their
brain level decreases the severity of seizures, and even gives complete protection.   For most of these drugs, biochemical studies
have defined the steps of biogenic amine metabolism on which they
are acting.   A thorough comparison of their actions on the audiogenic seizure and on biogenic amine metabolism gives a good idea
of one of the processes of the harmful effect of noise.   Moreover,
the actions of some drugs suggest that the amount of NE available
at the sympathetic nervous ganglion level is even more important
than NE brain level in the process of noise effect.   There seems
to be in this case, as in many others, an interaction between NE.
and acetylcholine metabolisms.   This study points out some of the
metabolisms involved in the harmful effect of noise and gives information about drugs that may help to protect against it.

I. DRUGS WITH NON SPECIFIC ACTION

1. Sedatives

The first drugs studied were standard sedatives of different types: barbiturates, hydantoines or dione derivatives. The results obtained with these drugs are given in Table 1 and show that all of them give full protection against sound stimulation, in mice of all strains, at small doses which are far from the LD 50, and even from any behavioral sedation.

2. Minor tranquilizers

Except for hydroxyzine, usual minor tranquilizers also efficiently protect mice against audiogenic seizures. Table 2 gives the results obtained with these drugs. They are efficient for all strains and the protective doses are also far from LD 50 and from doses modifying behavior.

3. Phenothiazine derivatives

The study of phenothiazine neuroleptics shows that their action on audiogenic seizures in different strains can be different, according to their molecular substitutions, although most of them more or less protect all strains of mice against the audiogenic seizure. The results obtained with these drugs are shown in Table 3. It must be noted that the two last drugs, Thioproperazine and Prochlorperazine, unlike other phenothiazine derivatives, do not protect the animals, but increase the death rate in seizures. The origin of this effect is not clear, but it must be outlined that, on one hand, these drugs easily induce catatonia, and that another drug inducing catatonia, haloperidol, has the same effect on the audiogenic seizure. On the other hand, it seems that the action of these drugs can also be related to their induced modification of the permeability of cells granules containing the catecholamines. However, the results obtained with all the drugs non-specifically decreasing nervous system excitability, show that most of them are effective against the audiogenic seizure.

II. DRUGS ACTING ON SPECIFIC METABOLISMS

1. Drugs acting on Gamma-amino-butyric acid (GABA) metabolism

GABA is considered by some authors as a chemical mediator of the nervous system (23); others consider it a non-specific nervous system inhibitor (24). However, whatever it may be, several studies showed that a decrease in brain GABA level, induced by hydrazides, sensitize to noise and to audiogenic seizures (25, 26, 27). On the contrary, it might be expected that an increase in GABA brain level would protect against sound stimulations.

Table 1. Effects of sedatives on audiogenic seizures.

	Protective doses mg/kg i.p.	LD 50	Author	Strain
Diphenylhydantoine	25 5-35 15	200 490	Suter (1958) Swinyard (1959) Halpern (1957)	DBA/2 (9) Frings (10, 11) Rb (8, 12)
Thiopental	8	200	Wilson (1959)	Rb (13)
Phenobarbital	4,6 - 20 10 8 - 16 275 (p.o.)	200 325	Swinyard (1959) Halpern (1957) Suter (1958) Plotnikoff (1958)	Frings (10, 11) Rb (12) DBA/2 (9) Swiss (14, 15)
Trimethadione	333 375-550	1800	Suter (1958) Swinyard (1959)	DBA/2 (9) Frings (10, 11)
Hydroxydione	5 - 20 25	50	Langer (1962) Lehmann	A_2G (16) Rb (unpubl.)

Table 2   Effects of minor tranquilizers on audiogenic seizures.

	Protective doses mg/kg i.p.	LD 50	Author	Strain
Meprobamate	**29–89**	1000	Swinyard (1959)	Frings (10, 11)
	60		Busnel (1958)	Rb (17)
	100		Foglia (1963)	Rockland (18)
	90–150		Plotnikoff (1963)	Swiss (19)
	4,5		Plotnikoff (1963)	DBA/1 (19)
Benactyzine	65	115	Busnel (1958)	Rb (17)
Methyl pentynol carbamate	50	400	Halpern (1956)	Rb (12, 20)
Hydroxizine	50 (low efficacy)	137	Foglia (1963)	Rockland (18)
	ineffective		Lehmann	Rb (unpubl.)
	140 (low efficacy)	515	Swinyard (1959)	Frings (10, 11)

Table 3. Effects of phenothiazine derivatives on audiogenic seizures.

	Protective doses mg/kg i.p.	LD 50	Author	Strain
Promethazine	40		Busnel (1958)	Rb (17)
Diethazine	45		Busnel (1958)	Rb (17)
Chlorpromazine	10	92	Halpern (1957)	Rb (12)
	2		Foglia (1963)	Rockland (18) (low efficacity)
	21 - 157	580	Swinyard (1959)	Frings (10, 11)
	4,8		Plotnikoff (1963)	DBA/1 (19)
	10		Plotnikoff (1961)	Swiss (14)
	inéffective		Plotnikoff (1961)	Swiss inbred (14)
	5-30		Raynaud (1963)	Rb (21, 22)
Promazine	24 - 235	485	Swinyard (1959)	Frings (10, 11)
	ineffective		Plotnikoff (1961)	Swiss inbred (14)

Table 3 continued

Aminopromazine	ineffective	Raynaud (1963)	Rb (21, 22)
Trifluoperazine	18 – 100	330	Swinyard (1959)	Frings (10, 11)
Trifluoperazine	ineffective	Plotnikoff (1961)	Swiss inbred (14, 19)
Alimemazine	40	Raynaud (1963)	Rb (21, 22)
Levomepromazine	5	Raynaud (1963)	Rb (21, 22)
Methopromazine	30	Raynaud (1963)	Rb (21, 22)
Thioproperazine	10	Raynaud (1963)	Rb (60% death in seizure) (21, 22)
Prochlorperazine	10	57	Raynaud (1963)	Rb (60% death in seizure) (21, 22)
	ineffective	Plotnikoff (1961)	Swiss inbred (14)

Table 4 shows the effect of substances increasing brain GABA
level.   It appears that, while GABA itself is inactive and does not
cross the blood brain barrier (23), all the other substances decrease
the severity of convulsions.   Some of them, like di-n-propylacetate
or gamma-glutamylhydrazide, even completely protect mice against the
effect of noise.   In agreement with these findings, our laboratory
showed that, when the blood brain barrier was by-passed by topical
application to the brain, GABA itself also had a protective effect
(28).

The action of GABA may be due either to its unspecific inhibitory action on the nervous system (35) or to an interference with
the tricarboxylic cycle.   Thus, GABA is a key member of a shunt
around the alpha-keto glutarate oxidase step of the tricarboxylic
acid cycle in the brain, which ends in the irreversible oxidation
of succinic semi-aldehyde to succinate (23, 30).   Up to now, there
is no clear evidence permitting a choice between these two explanations.   In favor of the last hypothesis are the facts that some
substances from the tricarboxylic acid cycle protect mice against
audiogenic seizures.

2. Drugs interfering with the tricarboxylic acid cycle

The first studies with these substances were done by Ginsburg
in DBA/2 DBA/2 and HS (originating from Frings stock) (29, 36 - 40).
Some other studies have been done in our laboratory with a strain
also originating from Frings mice, and by Schlesinger on DBA/2J (32).
Table 5 gives the results of the studies of substances related to
the tricarboxylic acid cycle, and which are active on audiogenic
seizure.   It shows that some substances interfering with the tricarboxylic cycle protect mice against audiogenic seizures.   Special
actions are those of glutamic acid and glutamine which act differently on different strains of mice according to their genetic background.   The data obtained with substances active on the tricarboxylic acid cycle or related with it, as GABA, suggest that the
action of noise may be related to the mechanisms controlling energy
turnover in the nervous system.

3. Drugs acting on the metabolism of biogenic amines: norepinephrine
(NE) and serotonin (5 HT)

a) Mechanism of action of drugs on biogenic amine metabolisms.
To understand these actions we must briefly describe these metabolisms.   NE metabolism is represented on Fig. 1.   NE is synthesized
inside the nerve cell by hydroxylation of tyrosine to DOPA.   DOPA
is decarboxylated and hydroxylated into NE.   NE is then stored in
special granules and released by nerve stimulation.   After its release, it is either destroyed inside the cell by monoamine oxidase
(MA), or released into the receptor sites.   In the synapse, NE
is either destroyed by catechol-o-methyl-transferase (COMT), or

Table 4. Effects of drugs which increase brain GABA levels upon audiogenic seizures.

Drugs	Effect on		Author	Strain
	Decrease of the % of death	Decrease of % of tonic seizure		
Gamma-aminobutyric acid	0	0	Ginsburg (1954)	DBA/1 (6, 29)
	0	65 (i.c.)	Lehmann (1963)	Rb (30)
	0	12	Trifaro (1964)	A_2G (31)
Gamma-glutamylhydrazide	18	16	Ginsburg (1954)	DBA/1 (6, 29)
	0	100	Lehmann	Rb (unpubl.)

Table 4 continued

Aminooxyacetic acid	100		Ginsburg–Lehmann	HS (unpubl.)
	79		Ginsburg–Lehmann	DBA/2 (unpubl.)
	10		Ginsburg–Lehmann	DBA/1 (unpubl.)
	protect		Schlesinger (1968)	DBA/2J (32)
	0	100	Lehmann (1963)	Rb (30)
Di-n-propylacetate	0	100	Simler (1968)	Rb (33)
Nicotinamide	0	16 – 40	Lehmann (1967)	Rb (34)
Hydroxylamine	0	93	Lehmann (1963)	Rb (30)
	50	22	Trifaro (1964)	A_2G (31)
5-methyl-5-phenyl-2-Pyr-rolidine	0	50	Lehmann	Rb (unpubl.)

Table 5. Effects on audiogenic seizures of substances related to the tricarboxylic acid cycle.

Drugs	Effect on		Author	Strain
	Decrease of the % death	Decrease of % of tonic seizure		
Alanine	15	15	Ginsburg (1954)	DBA/1 (6, 29)
Glutamic acid	43	43	Ginsburg (1954)	DBA/1 (6, 29, 36 – 40)
	decrease	decrease	Schlesinger (1968)	DBA/2J (32)
	0	0	Lehmann	Rb (unpubl.)
	0	0	Ginsburg (1950)	HS (6, 29, 36 – 40)
Lactic acid	100	50	Ginsburg (1954)	DBA/1 (6, 29)
Oxalacetic acid	22	22	Ginsburg (1954)	DBA/1 (6, 29)
Succinic acid	43	50	Ginsburg (1954)	DBA/1 (6, 29)
Glutamine	increase	increase	Ginsburg (1954)	DBA/1 (6, 29)
	decrease	decrease	Schlesinger (1968)	DBA/2J (32)

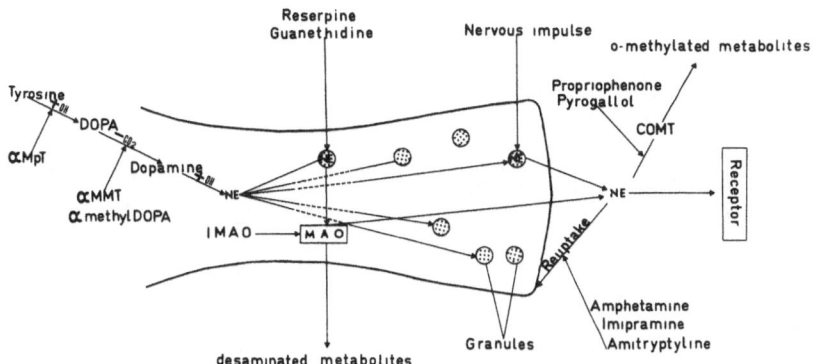

Figure 1. Norepinephrine metabolism.   Explanations in the text.

re-enters the nerve cell by reuptake (41).   This explains why there
can be an increase in the total brain level of NE or of its level
in the synaptic cleft only, according to the pathways on which drugs
act.

     Similarly, 5-HT is synthesized in the cell by hydroxylation of
tryptophane, and 5-hydroxytryptophane is decarboxylated to 5-HT.
After its release, 5-HT is oxidated by MAO to 5-hydroxyindolacetic
acid (5 HIAA) and eliminated (42).   As shown in Figure 1, drugs
acting on these metabolisms may decrease the level of the amines
by:
- inhibition of tyrosine or tryptophane hydroxylases: i.e. alpha-
  methyltyrosine.
- inhibition of DOPA decarboxylase: i.e. alpha-methyl-metatyrosine.
- impairment of storage in the granules leading to a destruction of
  the amines by MAO: i.e. reserpine.

     They can increase it:
- in the cell by inhibition of MAO: i.e. monoamine-oxidase inhibitors.
- into the receptor sites by - inhibition of COMT : i.e. pyrogallol.
                             or - inhibition of reuptake: i.e. thymo-
  leptics.

     We will successively examine the effect of drugs decreasing
or increasing the total amount of biogenic amines, the amount of
amine released into the receptor sites, or acting differently in
the central nervous system and at the periphery.

     b. Drugs decreasing norepinephrine (NE) and or serotonin (5-HT)
levels. Most of these drugs increase the severity of the seizure.
Table 6 shows the effect of drugs decreasing NE and 5-HT levels or
the level of only one of those amines, and increasing the severity

Table 6. Action of drugs decreasing NE and/or 5-HT brain levels and increasing the severity of audiogenic seizure.

Drugs	Doses increasing the severity of seizures, mg/kg i.p.	% death	Author	Strain	Amine involved
	1	75	Lehmann	Rb (14, 43)	NE + 5-HT
	0.25	28	Langer	A$_2$G (44, 45)	
			Trifaro	(46)	
Reserpine	0.1	19	Foglia	Rockland (18)	
	5	increases susceptibility	Bielec	Peromyscus (47)	
	1		Schlesinger	DBA/2J (32)	
	2	increases susceptibility	Schlesinger	C57BL/6J (32)	

Table 6 continued

Drug			Author	Reference	
Tetrabenazine	20	70	Lehmann	Rb (48)	NE + 5-HT
	5	increases susceptibility	Schlesinger	DBA/2J C57BL/6J (32)	
RO 4 - 1284	5	60	Busnel	Rb (49)	NE + 5-HT
Alpha-methyl-para-tyrosine	80	19	Lehmann	Rb (50)	NE
	80	30	Schlesinger	DBA/2J (32)	NE
Alpha-methyl-para-tyrosine	500	50	Lehmann	Rb (51)	NE
Metaraminol	3	60	Lehmann	Rb (52)	NE
6-hydroxydopamine	twice 34 + 68	100	Lehmann	Rb (52)	NE
Parachlorophenyl-lalanine	316	4 increases susceptibility	Lehmann	Rb (50)	5-HT
	320		Schlesinger	DBA/2J (32)	
Guanethidine	10	40	Langer	A$_2$G (45)	NE

Table 7. Action of drugs decreasing NE or 5-HT brain level and
         without effect, or protecting against audiogenic seizure.

Drugs	Doses mg/kg, i.p.	Effect on seizures	Amine involved
Alpha-methyl DOPA	400	no effect	NE
d-Amphetamine	15	protect	NE
Chlormethamphetamine	30	protect	5-HT
Fenfluramine	30	protect	5-HT

of seizures, as revealed by the death rate. It was observed that
all strains studied react identically to these drugs.

    To the contrary, however, some drugs that decrease NE or 5-HT
level protect mice, or are without effect on the audiogenic seizure.
The results obtained with these drugs are shown in Table 7. A
study of their mechanisms of action shows that this effect may be
due to their special way of releasing biogenic amines.

    c. Psychoanaleptics: Monoamine oxidase inhibitors (MAOI).
These drugs increase central nervous system activity and act as
anti-depressants in human therapy. Their action is generally
ascribed to an induced increase in total NE and 5-HT brain level.
Table 8 shows the effect of these drugs on the audiogenic seizure,
and it appears that, if some of them protect mice against the harmful effects of noise, some others are totally ineffective. It must
be noticed that the most active of the MAOI is tranylcypromine
which, at the same time, acts on monoamine oxidase inhibition and
on NE release (53). This suggest that the amount of the release
of free NE is more important than its total level. According to
this hypothesis Table 9 shows (54) that they all counteract the
increased severity of seizures induced by reserpine. In fact, in
this case, NE is released by reserpine, and leaves the cell without
being destroyed by monoamine oxidase.

    d. Thymoleptics and inhibitors of Catechol-o-methyl transfer-
ase (COMT). All these drugs increase NE release into the receptor
sites without increasing its total level, either by inhibiting reuptake, like imipramine and amitryptyline, or by inhibiting COMT,
like pyrogallol and propriophenone (41). Table 10 shows that these

Table 8. Effect of monoamine oxidase inhibition on audiogenic seizures.

Drugs	Number of animals	Dose mg/kg, i.p.	Reactions				Deaths	% of total protection	% partial protection (clonic seizure)
			=	X	C	0			
Tranylcypromine (54)	10	5	10						
	15	10	14	1					
	15	15	9	2	1	3		26	7
	16	25	4	3	2	7		56	13
	10	30				10		100	18
(55)		16,5						50	
Harmaline (54)	12	30	12						
	10	50	9	1					
	10	100	6				4	toxic	10
Phenelzine (55)		19							50
Nialamid (54)	12	20	11	1				20	8
	10	50	3	5	2			44	50
	9	100	3	2	2	2			22
(55)		8,7							50

table continues

Table 8 continued

Substance						
Etryptamine (55)		4,1				50
Isocarboxazid (54)	10	10	10			
	10	20	10			
	10	25	10			
	10	30	2	4	toxic	
Iproniazid* (54)	16	100	5	3	18	50
(32)	39	75–150	8			40
(55)		163				50
Pheniprazine (44, 56)	31	5				63
	34	10				83
	32	20				97,5
(55)		11				50
(32)		10				80
Pargyline	8	50	8	2	1	0
	17	75	15			10
	10	100	9			10

= tonic–clonic seizure   C running
X clonic seizure   0 no reaction
* 3 injections at 24 hours intervals

Table 9. Comparison between the action on audiogenic seizure of
reserpine when injected alone, and when preceeded by MAOI administration.

Drugs	Number of animals	Doses mg/kg, i.p. plus 1 mg/kg reserpine	Percentage of protection		% death in seizure
			Total	Partial	
Iproniazid*	32	100	18	43	12
Isocarboxazid	48	25	4	29	8
Tranylcypromine	40	25	75	12	12
Nialamid	28	20	0	16	8
Harmaline	24	30	0	0	0
Pargyline	10	75	10	0	10
Reserpine	156		0	0	78

* 3 injections at 24 hours intervals.

drugs decrease the severity of the seizures, thymoleptics giving
even total protection. The results represented in Tables 6 - 10
suggest that there are real relationships between biogenic amine
metabolism and the harmful effects of noise, as measured by the
severity of audiogenic seizures. It is obvious that the increase
or the decrease of the severity of the seizure is not closely related either to the total increase or decrease of the amines induced by these drugs per se, or to the specific amine on which they
act. To understand their actions we must not consider the crude
modifications of brain amine level, but how these drugs modify it.
This means knowing on which step of its synthesis or degradation
they act, and how they affect the available NE. Table 11 compares
the action on NE metabolism, on NE brain level and on audiogenic
seizure of the drugs studied. It appears that all drugs protecting
mice or decreasing the severity of the seizure increase NE level into
the receptor sites, and that all drugs increasing the severity of the
seizure decrease the NE level.

If we now consider drugs acting on 5-HT metabolism only, we
see that 5-HT depletion, due to parachlorophenylalanine (Table 6),
induced 4% of death only. It would suggest that NE depletion is
much more important than 5-HT depletion. On the other hand, for
chlormethamphetamine and fenfluramine which, at the same time,

Table 10. Action of thymoleptics and of COMT inhibitors on the audiogenic seizure.

Drugs	Number of animals	Doses mg/kg	Reactions =	X	C	O	Death	% of total protection	% of partial protection
Imipramine (57)	20	5	20					5	35
	20	15	12	7		1	1	20	30
	20	25	6	10		4			
	20	30				20		100	
(44, 56)	28	12,5							80
	32	25							85
	28	50							96
Amitryptyline (57)	20	5	14	3		3	2	15	15
	27	9	7	16		4	1	15	59
	28	10	3	13		12		46	42
	10	15				10		100	
Pyrogallol (57)	60	250	41	13		6		9	22
Propriophenone (U-0521)	10	150	5	3		2	2	20	30
	10	175	4	4	1	5		50	50

= tonic seizure   C running

X clonic seizure   O no reaction

Table 11. Drugs acting on biogenic amine metabolism: relationship between their mode of action and their effect on audiogenic seizures.

Drugs	Action on NE metabolism	NE total level	NE level into the receptor sites	Action on audiogenic seizure
Reserpine	impairment of storage	↘	↘	↗
Tetrabenazine	"	↘	↘	↗
RO 4-1284	"	↘	↘	↗
Alpha-methyl paratyrosine	inhibition of tyrosine hydroxylase	↘	↘	↗
6-Hydroxydopamine	destructs nerve endings	↘	↘	↗
Guanethidine		↘	↘	↗
Alpha-methyl metatyrosine	replaces NE by metaraminol	↘	↘	↗
Metaraminol	replaces NE by metaraminol	↘	↘	↗
Alpha-methyl DOPA	replaces NE by alpha-methyl NE	↘	Alpha-methyl NE acts as NE	0
D-Amphetamine	inhibits reuptake	↗	↗	↘
MAOI	inhibits MAO	↗	↗	↘
Pyrogallol	inhibits COMT	0	↗	↘
Propriophenone	inhibits COMT	0	↗	↘
Imipramine	inhibits reuptake	0	↗	↘
Amitryptyline	inhibits reuptake	0	↗	↘

decrease the 5-HT level and the severity of seizure (Table 7), it is well known that these drugs decrease 5-HT level without increasing its normal metabolite 5-hydroxyindole acetic acid (58). Thus, they modify 5-HT metabolism by an unknown pathway, and this may be related to their effect on the audiogenic seizure. This may be compared to the action of amphetamine, which (Table 11) depletes total NE level but, inhibiting reuptake, releases free NE into the receptor sites (41). If this is the case it would suggest that a release of 5-HT could be as efficient as NE release to protect against noise. But NE depletion increases the harmful effect of noise more than 5-HT depletion.

     e. Nervous pathways involved in the harmful effect of noise, as revealed by the action of drugs acting on NE metabolism. The above results already show that the effect of NE on the audiogenic seizure is related to the level of its release into the receptor sites. Now, this substance is a mediator in the central nervous system, as well as in the sympathetic nervous system, and if some drugs deplete NE in both systems, some deplete it in the peripheral sympathetic system only. With specific drugs, we may be able to find which part of the nervous system is involved in the audiogenic seizure, and thus one of the pathways by which noise becomes harmful.

Table 12. Action of drugs provoking NE central and/or peripheral
         depletion on the severity of the audiogenic seizure

Drugs	% of death in seizure	Central NE depletion	Peripheral NE depletion
Tetrabenazine	70	+	+
Reserpine	75	+	+
Alpha-methyl-meta-tyrosine	50	+	+
Alpha-methyl-para-tyrosine	19	+	+
6-hydroxydopamine	100	+	+
Metaraminol	60	0	+
Guanethidine	40	0	+
Immunosympathectomy	80	0	+

Table 12 shows that drugs inducing only peripheral NE depletion, as
well as these inducing central depletion, can increase the severity
of audiogenic seizures. If, for some of these drugs, we now plot

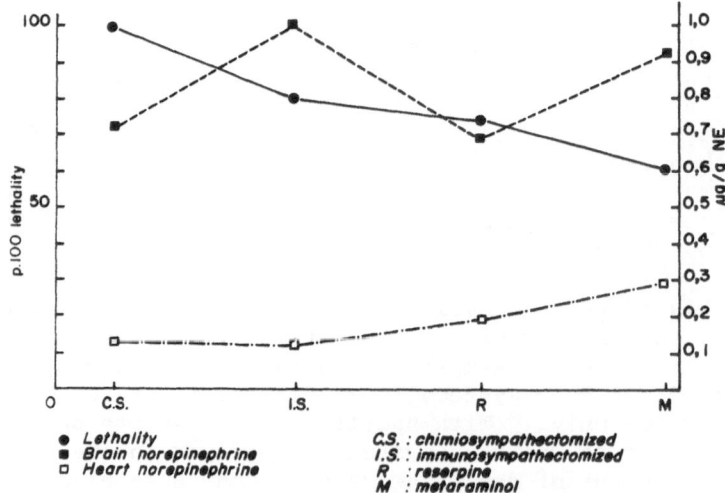

Figure 2. Comparison between lethality and NE levels in brain and hear

Table 13. Modifications of the severity of audiogenic seizures
           induced by drugs acting on sympathetic ganglia.

Drugs	Number of animals	Doses mg/kg i.p.	% of death in seizure
Nicotine	21	4.5	50
Nicotine (59)	46	2 - 5 (s.c.)	100
Pentamethonium	20	2.5	45
Atropine	20	0.4	55
Eserine	30	0.4	0
Arecoline	10	25	0

lethality rate in relation to the decrease of brain and heart NE,
the heart NE representing the peripheral tissue decrease, we can
see on Figure 2 that lethality closely follows heart NE depletion
and not at all brain depletion.   This indicates that lethality is
mostly due to peripheral amine depletion.   These results lead us
to the hypothesis that the increase in the severity of audiogenic
convulsions may be due to an impairment of ganglionic transmission.
In order to confirm it, we studied the effect of some cholinergic
blocking agents on one hand, and, on the other hand, some substances
increasing acetylcholine level.

Table 13 shows that, as expected, cholinergic blocking agents
increase the severity of audiogenic seizures, whereas parasympathet-
icomimetics such as arecoline or eserine, are without effect.
According to the interrelationship between NE and acetylcholine in
the sympathetic ganglia, both mediators seem to act in response to
noise (60).

DISCUSSION

The results presented above point out that many metabolic or
nervous pathways may be involved in the harmful effects of noise
on the nervous system.

The efficacy of all sedatives, minor tranquilizers and phenothiazine derivatives in protecting against audiogenic seizure clearly
shows that noise provokes an exhausted excitability of the nervous
system which may be decreased by sedation.

The effect of substances increasing the GABA level may also be related to a non-specific inhibitory action of this substance on the nervous system.   But some of the experiments with drugs acting on GABA metabolism suggest that reaction to noise is also connected at least in some strains of mice, with the tricarboxylic cycle, and, hence, with the energetic turnover of the nervous system.   It may therefore be suggested that in one way or another GABA metabolism is one of the specific metabolisms involved in the harmful effect of noise.

Studies performed with drugs acting on NE metabolism are even more fruitful.   The increasing or decreasing severity of audio-genic seizures, induced by drugs acting on NE metabolism, obviously shows the relationships existing between this metabolism and the harmful effect of noise.   These observations are in agreement with the decrease of brain NE, which appears after audiogenic seizures (61), as after other stresses (62, 63, 64).   Moreover, as we had the possibility to use drugs acting on the different steps of this metabolism, we demonstrated that the level of NE release into the receptor sites is the most important fact modifying reactions to noise.

At the same time, our experiments may help to define which nervous system receptors are primarily involved in the reaction to noise.   The increase in the severity of seizures, obtained by sympathetic nervous system blocking agents, suggests that the sympathetic nervous system is deeply involved in the harmful effect of noise, and supports the hypothesis that some stimulations, when reaching the lower brain stem, act by stimulating the sympathetic nervous system.   This stimulation of the sympathetic nervous system by noise is confirmed by some behavioral observations on audiogenic seizures, such as pupil dilation, post-convulsive brady-cardia, death due to respiratory failure preceeding heart failure, and by the biochemical findings showing a decrease of NE after seizures.

This hypothesis is supported by the results obtained with all the psychotropic drugs the effects of which are in close relation-ship with the free NE released into the receptor sites.   The level of free NE regulates the sympathetic nervous system activity; a high level facilitating its activity protects against noise, and a low level inhibiting it, increases the harmful effects of noise.   The specific relations found between noise stress and NE release in the sympathetic nervous system agree with findings on other stress situ-ations.   For instance, Leduc (65) shows that cold stress leads to a depletion of NE peripheral stores, and that the effects of cold are increased by ganglia blocking agents.   Likewise, Olivieri (66) shows that stress increases NE turnover in the mouse heart.

Of course, it is obvious that the audiogenic seizure is a

special phenomenon studied in mice and that it is out of the question to apply directly to man the findings exposed here.   As in any pharmacological study, and leaving aside the question of toxicity, the drugs used in this test have to be studied more thoroughly and in other animals species, before being used in clinic.

However, the results obtained by studying the effect of psychotropic drugs on the audiogenic seizure may help to define some pathways involved in the harmful effects of noise, support some hypotheses on the effect of this stress, and provide a preliminary screening of drugs that may be helpful against nervous disorders induced by noise.

SUMMARY

Noise affects the central nervous system of man as well as that of animals.   These harmful effects may even induce convulsions, but their physiopathology is not well known at present.   The audiogenic seizure in mice is a good test for the screening of drugs counteracting the harmful effects of noise on the nervous system.   Among the numerous efficient drugs, some are atypical nervous sedatives, and others act on specific metabolisms.   A study of the latter can lead to a better knowledge of mechanisms involved in the harmful effects of noise.

From our studies, it appears that the central nervous system and the sympathetic nervous system may both be affected by loud noises, and that serotonin, norepinephrine and gamma-aminobutyric acid metabolisms are involved in the reaction to noise.

REFERENCES

1. Lehmann, A. and R.G. Busnel, 1963.   A study of the audiogenic seizure.   In: R.G. Busnel (Ed.), Acoustic behaviour of animals, Elsevier, Amsterdam, pp. 244-274.

2. Forster, F.M., P. Hansotia, C.S. Cleeland and A. Ludwig, 1969. A case of voice-induced epilepsy treated by conditioning. Neurology 19:325-331.

3. Lehmann, A. et D.A. Fless, 1962.   Etude de l'action de drogues psychotropes sur les processus d'excitation et d'inhibition du S.N.C. par le test de la crise audiogène du rat.   Psychopharmacologia 3:331-343.

4. Cosnier, J., A.M. Brandon et A. Duveau, 1962.   Quelques particularités du comportement des souris sensibles à la crise audiogène. Compte Rendu Societe des Biologie (Paris) 156:2071-2074.

5. Cosnier, J., A. Duveau et E. Vernet-Maury, 1963.   Caractères
    différentiels du comportement des rats et des souris sensibles
    et réfractaires à la crise audiogène.   Journal de Physiologie
    (Paris) 55:130-131.

6. Ginsburg, B.E., 1954.   Genetics and the physiology of the ner-
    vous system.   Genetics and the inheritance of integrated neuro-
    logical and psychiatric patterns.   Proceedings Association
    Research Nervous and Mental Disease 33:39-56.

7. Lehmann, A. et E. Boesiger, 1964.   Sur le déterminisme génétique
    de l'épilepsie acoustique de Mus musculus (Swiss Rb).   Compte
    Rendu Academie des Sciences (Paris).   258:4858-4861.

8. Mouse News Letter, 1959.   21:42.

9. Suter, C., W.O. Klingmann, D. Boggs, O.W. Lacey, R. Marks and
    C.B. Coplinger, 1958.   Sound-induced seizures in animals.
    The efficiency of certain anticonvulsivants in controlling
    sound-induced seizures in DBA/2 mice.   Neurology 8, Suppl.
    1:121-124.

10. Swinyard, E.A., 1963.   Some physiological properties of audio-
    genic seizures in mice and their alteration by drugs.   In:
    R.G. Busnel (Ed.), Psychophysiologie, Neuropharmacologie et
    Biochimie de la crise audiogene, Editions du C.N.R.S., Paris.
    pp. 405-421.

11. Fink, G.B. and E.A. Swinyard, 1959.   Modification of maximal
    audiogenic and electroshock seizures in mice by psychopharma-
    cologic drugs.   Journal Pharmacology and Experimental Thera-
    peutics 127:318-324.

12. Halpern, B.N. et Lehmann, A., 1957.   Bases expérimentales de
    l'action thérapeutique d'une nouvelle médication sédative et
    anti-anxieuse, le carbamate de méthyl 3 pentyne 1 ol 3.
    Presse Médicale 27:622-625.

13. Wilson, C.W.M., 1959.   Drug antagonism and audiogenic seizures
    in mice.   British Journal of Pharmacology 14:415-419.

14. Plotnikoff, N.P., 1961.   Drug resistance due to inbreeding.
    Science 134:1881-1882.

15. Plotnikoff, N.P., 1958.   Bioassay of potential tranquilizers
    and sedatives against audiogenic seizures in mice.   Archives
    Internationales de Pharmacodynamie et de Therapie 116:130-135.

16. Langer, S.Z., J.M. Trifaro and V.G. Foglia, 1962.   Accion
    protectora de la hidroxidiona sobre las convulsiones audiogenas.

Revista de la Sociedad Argentina de Biologia (Buenos Aires) 38: 50-54.

17. Busnel, R.G., A. Lehmann et M.C. Busnel, 1958.   Etude de la drise audiogène de la souris comme test psychopharmacologique: son application aux substances du type "tranquilliseurs". Pathologie et Biologie (Paris) 1-10:749-762.

18. Foglia, V.G., R.P. Montanelli, S.Z. Langer et R. Epstein, 1963. Action des drogues psychotropes sur les convulsions audiogènes chez la souris.   Compte Rendu Société Biologie (Paris) 157: 1813-1814.

19. Plotnikoff, N.P., 1963.   A neuropharmacological study of escape from audiogenic seizures.   In: R.G. Busnel (Ed.), Psychophysiologie, Neuropharmacologie et Biochimie de la crise audiogene, Editions du C.N.R.S., Paris, pp. 429-443.

20. Halpern, B.N. et A. Lehmann, 1956.   Action protectrice du carbamate de methyl 3 pentyne 1 ol 3 (CMP) contre la crise convulsive audiogene.   Compte Rendu Societe Biologie (Paris) 150:1863-1866.

21. Raynaud, G. et G. Valette, 1963.   Action des dérivés de la phénothiazine et de l'halopéridol sur la crise audiogène de la souris.   Archives Internationales de Pharmacodynamie et de Therapie 142:425-439.

22. Valette, G. et G. Raynaud, 1963.   Action des dérivés de la phénothiazine sur la crise audiogène de la souris.   In: R.G. Busnel (Ed.), Psychophysiologie, Neuropharmacologie et Biochimie de la crise audiogene, Editions du C.N.R.S., Paris, pp. 311-324.

23. Roberts, E., C.F. Baxter and E. Eidelberg, 1960.   Some aspects of cerebral metabolism and physiology of gamma-amino-butyric acid.   In : D.B. Tower and J.P. Schade (Eds), Structure and function of the cerebral cortex, Elsevier, Amsterdam, pp. 392-404.

24. Kravitz, E.A., S.W. Kuffler and D.D. Potter, 1963.   Gamma-amino-butyric acid and other blocking compound in crustacea. III - Their relative concentrations in separated motor and inhibitory axons.   Journal Neurophysiology 26:739-751.

25. Balzer, A., P. Holtz und D. Palm, 1960.   Untersuchen über die biochemischen Grundlagen der Konvulsiven Wirkung von Hydraziden. Naunyn-Schmiedberg's Archive für Experimentelle Pathologie und Pharmakologie (Berlin) 239:550-552.

26. Buchel, L., A. Debay, J. Levy et O. Tanguy, 1961.   Conditions

de la sensibilisation des Rongeurs, par des hydrazides, à la crise convulsive audiogene. Etude des substances protectrices. Therapie 16:729-742.

27. Lehmann, A., 1963. Action des hydrazides convulsivants sur l'épilepsie acoustique dite crise audiogène de la souris. Journal de Physiologie (Paris) 55:282-283.

28. Ballantine, E., 1963. The effect of gamma-amino-butyric acid on audiogenic seizures. In : R.G. Bushnel (Ed.), Psychophysiologie, Neuropharmacologie et Biochimie de la crise audiogene, Editions du C.N.R.S., Paris, pp. 447-450.

29. Ginsburg, B.E., 1963. Causal mechanisms in audiogenic seizures. In : R.G. Busnel (Ed.), Psychophysiologie, Neuropharmacologie et Biochimie de la crise audiogene, Editions du C.N.R.S., Paris, pp. 228-237.

30. Lehmann, A., 1963. L'acide gamma-amino-butyrique est-il un inhibiteur du système nerveux central. Convulsions et acide gamma-amino-butyrique. Therapie 18:1509-1523.

31. Trifaro, J.M., E. Mikulio, H. Armendariz and V.G. Foglia, 1964. Accion del Acido gamma Aminobutirico (GABA) y de la Hidroxilamina sobre las convulsiones audiogenas del Raton. Acta Physiologica Latino-Americana 14:315-323.

32. Schlesinger, K., W. Boggan and D.X. Freedman, 1968. Genetic of audiogenic seizures. II-Effects of pharmacological manipulation of brain serotonin, norepinephrine and gamma amino butyric acid. Life Sciences 7, Part I :437-447.

33. Simler, S., H. Randrianarisoa et A. Lehmann, 1968. Effets du di-n-propyl-acetate sur les crises audiogenes de la souris. Journal de Physiologie (Paris) 60, Suppl. 2:547.

34. Lehmann, A., S. Simler et P. Mandel, 1967. Protection contre les crises audiogenes de la souris par la nicotinamide. Journal de Physiologie (Paris) 59:446-447.

35. Mitchell, J.F. and V. Srinivasan, 1969. Release of ^{3}H-gamma aminobutyric acid from the brain during synaptic inhibition. Nature 224:663-666.

36. Ginsburg, B.E., S. Ross, M.J. Zamis and A. Perkins, 1950. An assay method for the behavioral effect of L-glutamic acid. Science 112:12-13.

37. Ginsburg, B.E., S. Ross, M.J. Zamis and A. Perkins, 1951. Some effect of 1(+)glutamic acid on sound induced seizures in

mice.   Journal of Comparative and Physiological Psychology
44:134-141.

38.   Ginsburg, B.E. and E. Roberts, 1951.   Glutamic acid and central
nervous system activity.   Anatomical Record 111:492-493.

39.   Ginsburg, B.E. and J.L. Fuller, 1954.   A comparison of chemical
and mechanical alterations of seizure patterns in mice.   Journal of Comparative and Physiological Psychology 47:344-348.

40.   Miller, D.S., B.E. Ginsburg and M.Z. Potas, 1955.   The effect
of glutamine and glutamic acid on audiogenic seizures in mice.
Anatomical Record 122:438.

41.   Iversen, L.L., 1967.   The uptake and storage of noradrenaline
in sympathetic nerves, University Press, Cambridge.

42.   Page, I.H., 1958.   Serotonin.   Year Book Medical Publisher,
Inc., Chicago.

43.   Lehmann, A. et R.G. Busnel, 1963.   Le métabolisme de la
sérotonine cérébrale dans ses rapports avec la crise audiogène de la souris et ses variations sous l'influence de divers
composés psychotropes.   In : R.G. Busnel (Ed.), Psychophysiologie, Neuropharmacologie et Biochimie de la crise audiogène,
Editions du C.N.R.S., Paris, pp. 453-468.

44.   Langer, S.Z., 1962. Convulsiones audiogenas.   Thesis, Buenos-
Aires.

45.   Langer, S.Z., R. Epstein and V.G. Foglia, 1962.   Comparacion
de las acciones de la reserpina y la guanetidina sobre las
convulsiones audiogenas.   Revista de la Sociedad Argentina
de Biologie (Buenos Aires) 38:45-49.

46.   Trifaro, J.M., S.Z. Langer, A. Gallo and V.G. Foglia, 1962.
Estudio Farmacodinamico de las Respuestas Audiogenas mortales
en el Raton.   Actas de las II sesiones de Biologia, Univ.
Nac. Cordoba, Argentina, p. 97.

47.   Bielec, S. 1959.   Influence of reserpine on the behavior of
mice susceptible to audiogenic seizures.   Archives Internationales de Pharmacodynamie et de Therapie 119:352-357.

48.   Lehmann, A., 1964.   Contribution à l'étude psychophysiologique
et neuropharmacologique de l'épilepsie acoustique de la Souris
et du Rat.   Ph.D. Thesis, Agressologie 5:211-221 et 311-347.

49.   Busnel, R.G. et A. Lehmann, 1960.   Nouvelles données pharma-
codynamiques relatives à la crise audiogène de la Souris.

Journal de Physiologie (Paris) 52:37-38.

50.   Lehmann, A., 1968.   Modification de l'intensité de la crise
      audiogène par des substances actives sur le métabolisme des
      amines biogènes du cerveau de souris.   Compte Rendu Société
      Biologie (Paris) 162:24-27.

51.   Lehmann, A., 1965.   Rapports entre le taux cérébral de nora-
      drénaline et la léthalité au cours de la crise d'épilepsie
      acoustique de la Souris.   Journal de Physiologie (Paris) 57:
      646-647.

52.   Lehmann, A., 1969.   Relationship between sympathetic ganglion
      activity and audiogenic seizure in mice.   Life Sciences (in
      press).

53.   Golberg, L.I., 1964.   Monoamine Oxidase Inhibitors - Adverse
      reactions and possible mechanisms.   Journal American Medical
      Association 190:456-462.

54.   Lehmann, A. and R.G. Busnel, 1962.   A new test for detecting
      MAO-inhibitor effects.   International Journal of Neuro-
      pharmacology 1:61-70.

55.   Plotnikoff, N.P., J. Huang and P. Havens, 1963.   Effect of
      monoamine-oxidase inhibitors on audiogenic seizures.   Journal
      of Pharmaceutical Science 52:172-173.

56.   Langer, S.Z., J.M. Trifaro, A. Gallo and V.G. Foglia, 1962.
      Estudio comparativo de las acciones de la feniprazine y el
      imipramine sobre las convulsiones audiogenas.   Revista de
      la Sociedad Argentina de Biologia (Buenos Aires) 38:84-89.

57.   Lehmann, A., 1967.   Audiogenic seizures data in mice, support-
      ing new theories of biogenic amines mechanisms in the central
      nervous system.   Life Sciences 6, pt. I : 1423-1431.

58.   Le Douarec, J.C., H. Schmitt et M. Laubie, 1966.   Etude
      pharmacologique de la fenfluramine et de ses isomeres optiques.
      Archives Internationales de Pharmacodynamie et de Therapie
      161:206-232.

59.   Frings, H. and A. Kivert, 1953.   Nicotine facilitation of
      audiogenic seizures in laboratory mice.   Journal of Mammalogy
      34:391-393.

60.   Burn, J.H. and M.J. Rand, 1965.   Acetylcholine in adrenergic
      transmission.   Annual Review of Pharmacology 5:163-182.

61.   Lehmann, A., 1965.   Action de crises répétées d'épilepsie

acoustique sur le taux de noradrénaline des zones corticales et sous-corticales du cerveau de Souris.   Compte Rendu Société Biologie (Paris) 159:62-64.

62. Welch, A. and B.L. Welch, 1968.   Reduction of norepinephrine in the lower brainstem by psychological stimulus.   Proceedings of the National Academy of Sciences, U.S.A., 60:478-481.

63. Thierry, A.M., F. Javoy, J. Glowinski and S.S. Kety, 1968. Effects of stress on the metabolism of norepinephrine, dopamine and serotonin in the central nervous system of the rat.   I- Modifications of norepinephrine turnover.   Journal of Pharmacology and Experimental Therapeutics 163:163-171.

64. Bliss, E.L. and J. Zwanziger, 1966.   Brain amines and emotional stress.   Journal of Psychiatric Research 4:189-198.

65. Leduc, J., 1961.   Catecholamine production and release in exposure and acclimatation to cold.   Acta Physiologie Scandinavia 53: Suppl. 183.

66. Oliviero, A. and L. Stjarne, 1965.   Acceleration of noradrenaline turnover in the mouse heart by cold exposure.   Life Sciences 4:2339-2343.

NEUROCHEMICAL FACTORS IN AUDITORY STIMULATION AND DEVELOPMENT OF SUSCEPTIBILITY TO AUDIOGENIC SEIZURES

PAUL Y. SZE

Department of Biobehavioral Sciences
The University of Connecticut
Storrs, Connecticut 06268

Susceptibility to audiogenic seizures has been observed in several higher animal species, including man. Several inbred strains of mice are known that show either resistance or susceptibility to convulsive seizures during the presentation of an auditory stimulus, and these differences have been attributed to specifically defined genetic backgrounds (1, 2). Mice of the C57BL/6 strain are known to be highly resistant to such seizures (1, 2). In recent studies, Henry (3) and Jumonville (4) reported, independently, that the genetically seizure-resistant C57BL/6 mice can be induced to develop high susceptibility by prior exposure to auditory stimulation during a sensitive period of postnatal development. Similar effects of priming by sound in inducing susceptibility to audiogenic seizures have also been found in other resistant strains of mice (5, 6, 7). It has now been made clear that susceptibility to audiogenic seizures is not only determined by genetic differences but that it is also inducible by prior auditory input, at least in some strains of mice. The findings on this phenomenon of acoustic priming from several behavioral studies and their neurological implications have been reviewed and discussed in the article by K. R. Henry and R. Bowman in this volume.

Nothing is known of the physiological basis of priming auditory stimulation and of the subsequent development of seizure susceptibility, although a few suggestions have been made. The sensitive period for effective priming in the C57BL/6 strain was found to occur on day 19 post partum for animals subsequently tested for seizure susceptibility at 28 days of age (3). The existence of a sensitive period for maximally effective induction suggests that priming might involve some neural change inducible only in that specific period during ontogenesis of the central nervous system.

259

This neural change, once induced, appears to result in a long-term modification of behavioral response to sound. A simple explanation for this neural change would be that some specific damage, such as destruction of cells in some specific brain area, is caused by the intense auditory stimulation during priming. However, experiments by Fuller and Collins (5) indicate that repeated exposure to sound at 6-hour or 12-hour intervals following priming can reduce, not enhance, the effect of priming. Since it is difficult to conceive of repair being made by the same sound that caused damage, the neural change appears unlikely to be a lesion in the structural sense. Since priming by sound can occur even in animals that are under ether or sodium pentobarbital anesthesia, Henry surmised that neither the midbrain reticulum nor a conscious mechanism seems to be involved (3). Working with SJL/J mice, Fuller and Collins demonstrated that mice primed with one ear blocked are later susceptible to audiogenic seizures only if stimulated through the ear that was open at the time of priming (5). The unilateral nature of the "site of sensitization" led these investigators to suggest that the "site" might reside either in the ear or in those parts of the auditory system receiving auditory input solely from one side.

In order to better understand the neural basis of priming, it is desirable to elucidate some neurochemical events underlying the auditory stimulation that leads to the eventual development of seizure susceptibility. This article outlines a study of neuro-chemical changes that occur developmentally following priming auditory stimulation.

Several amino acids and biogenic amines have been suggested to be involved in the neurochemical mechanism of various types of convulsive seizures, including audiogenic seizures. (For review, see ref. 8.) Among these metabolic substances, γ-aminobutyric acid (GABA) has received particular attention because of its probable role as an inhibitory neurotransmitter in some parts of the mammalian central nervous system (9). It must be emphasized, however, that the problem posed in the present study is not the nature of the neurochemical basis of seizures or a seizure sus-ceptibility per se, but, rather, the neurochemical basis of the induction and development of seizure susceptibility. In fact, previous work in our laboratory using seizure-susceptible mutants arising from nonsusceptible C57BL/6 showed that differences in brain GABA level cannot be correlated in a simple quantitative way to susceptibility or resistance to seizures (1). However, differences in brain levels of GABA and its precursor, glutamic acid, do exist between susceptible and nonsusceptible animals at various stages of ontogenesis of the central nervous system. These changes in GABA metabolism could be significant especially when comparison is made between mutants and their otherwise isogenic control. My hypothesis is that GABA metabolism is in some way involved in gene-induced development of seizure susceptibility.

Mice used in the experiments reported here were from inbred C57BL/6 strain maintained in our own colony. After priming with an electric bell (101 db relative to 2×10^{-4} dyne/cm^2) on day 19 post partum, 73% of C57BL/6 mice showed clonic-tonic convulsion on day 28 when exposed to the same sound stimulus. Brain levels of GABA, glutamate and aspartate were determined at various developmental stages following priming exposure to sound on day 19. The animal was killed by immersion in liquid nitrogen. The amino acids were measured in the whole brain by the fluorometric methods of Graham and Aprison (10). As seen from the results (Table 1), the only significant change found during this period of development (day 19 to day 28) was an immediate decrease of brain GABA occurring 20 minutes after priming. This change in brain GABA was further verified and its time course is shown in Fig. 1. It appears that auditory stimulation caused a decline in brain level of endogenous GABA within 10-20 minutes, after which it was restored to the normal level. Both 5-hydroxytryptamine (5-HT) and norepin-

Table 2. Effect of auditory stimulation on brain levels of GABA, glutamate and aspartate. Amino acids are shown as umoles/g wet tissue ± S.D. Number in parenthesis indicates the number of mice determined. C: control mice. P: mice primed by sound on day 19 post partum. *Signicant difference from control (P < 0.02).

Time after priming		GABA	Glutamate	Aspartate
20 min	C	1.82 ± 0.06 (4)	9.00 ± 0.26 (4)	3.96 ± 0.05 (4)
	P	*1.64 ± 0.08 (4)	8.85 ± 0.25 (4)	4.21 ± 0.27 (4)
1 day	C	1.87 ± 0.02 (4)	9.12 ± 0.41 (9)	-----
	P	1.92 ± 0.05 (4)	9.19 ± 0.46 (9)	-----
2 days	C	1.84 ± 0.02 (5)	9.25 ± 0.20 (13)	3.44 ± 0.06 (5)
	P	1.85 ± 0.04 (5)	8.72 ± 0.60 (13)	3.48 ± 0.15 (5)
5 days	C	1.82 ± 0.09 (13)	9.43 ± 0.45 (11)	3.77 ± 0.23 (8)
	P	1.88 ± 0.07 (4)	9.25 ± 0.17 (4)	3.84 ± 0.33 (4)
7 days	C	-----	-----	-----
	P	1.84 ± 0.04 (4)	9.38 ± 0.35 (4)	3.55 ± 0.10 (4)
9 days	C	2.05 ± 0.12 (15)	10.11± 0.53 (14)	3.56 ± 0.17 (13)
	P	2.19 ± 0.09 (4)	10.35± 0.25 (4)	3.45 ± 0.20 (4)

Table 1. Effect of auditory stimulation on brain levels of GABA, glutamate and aspartate. Amino acids are shown as umoles/g wet tissue ± S.D.   Number in parenthesis indicates the number of mice determined. C: control mice. P: mice primed by sound on day 19 post partum.   *Significant difference from control (P<0.02).

Time after priming		GABA	Glutamate	Aspartate
20 min	C	1.82 ± 0.06 (4)	9.00 ± 0.26 (4)	3.96 ± 0.05 (4)
	P	*1.64 ± 0.08 (4)	8.85 ± 0.25 (4)	4.21 ± 0.27 (4)
1 day	C	1.87 ± 0.02 (4)	9.12 ± 0.41 (9)	-----
	P	1.92 ± 0.05 (4)	9.19 ± 0.46 (9)	-----
2 days	C	1.84 ± 0.02 (5)	9.25 ± 0.20 (13)	3.44 ± 0.06 (5)
	P	1.85 ± 0.04 (5)	8.72 ± 0.60 (13)	3.48 ± 0.15 (5)
5 days	C	1.82 ± 0.09 (13)	9.43 ± 0.45 (11)	3.77 ± 0.23 (8)
	P	1.88 ± 0.07 (4)	9.25 ± 0.17 (4)	3.84 ± 0.33 (4)
7 days	C	-----	-----	-----
	P	1.84 ± 0.04 (4)	9.38 ± 0.35 (4)	3.55 ± 0.10 (4)
9 days	C	2.05 ± 0.12 (15)	10.11 ± 0.53 (14)	3.56 ± 0.17 (4)
	P	2.19 ± 0.09 (4)	10.35 ± 0.25 (4)	3.45 ± 0.20 (4)

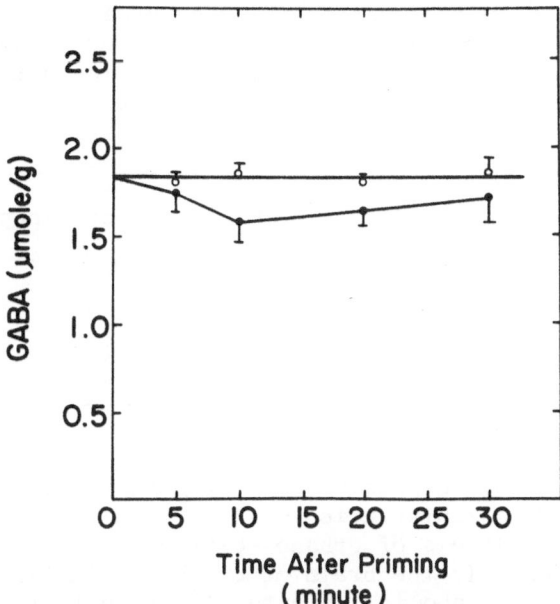

Figure 1. Effect of auditory stimulation on brain GABA. C57BL/6
litter-mates, 19-day-old, were divided into control (open circle)
and primed (solid circle) groups. Each point represents the mean
of 4-8 animals with the standard deviation indicated by the vertical bar. Significant difference from the control: $P < 0.05$ for
the 10-minute point; $P < 0.01$ for the 20 minute point.

ephrine, measured by the fluorometric method of Mead and Finger
(11) as modified by Wise (12), remained unchanged during this
period when change in GABA was observed (Table 2).

If this transitory reduction of brain GABA immediately following priming was indeed causally related to the sound-induced
development of later seizure susceptibility, then by blocking this
reduction of GABA, prevention of seizure development should be expected. This was found to be the case. As it is well known that
GABA does not penetrate the blood-brain barrier, manipulation of
brain level of GABA was made by the administration of amino-oxyacetic acid (AOAA), a selective inhibitor of GABA-α-ketoglutarate
transaminase activity in vivo (13). After subcutaneous injection
of AOAA (20 mg per kg), endogenous brain GABA in 19-day-old mice
was found to be greatly accumulated, reaching a five-fold increase
in five hours (Table 3). When priming was applied to mice with
brain level of GABA elevated at this high level by pretreatment
with AOAA five hours earlier, the effect of priming in inducing
later seizure development was markedly reduced (Table 4). Susceptibility to audiogenic seizures, as tested on day 28, was only
13% in the AOAA pretreated mice as compared with 73% in the un-

Table 2. Effect of auditory stimulation on brain levels of 5-HT
and norepinephrine. Amines are shown in ng/g wet tissue ± S.D.
Number in parenthesis indicates the number of mice determined.
Mice were primed on day 19 post partum, and sacrificed 20 minutes
after priming.

Treatment	5-HT	Norepinephrine
Control	558 ± 48 (5)	444 ± 76 (5)
Primed	568 ± 107 (6)	516 ± 126 (6)

treated mice. Previous extensive work by Baxter and Roberts showed
that in adult rats, administration of AOAA resulted in no significant changes in the levels of ethanol-extractable, ninhydrin-
reactive constituents in the brain other than GABA, indicating the
selectivity of this metabolic inhibitor in its effect on GABA
degradation (13). In the young mice used in priming, except also
for a slight decrease of aspartate, none of the other brain metabo-
lites examined, glutamate, 5-HT or norepinephrine, appeared to be
affected by AOAA, (Table 3), thus providing further support for
its selectivity of action in vivo.

Table 3. Effect of AOAA on brain levels of GABA, glutamate, aspartate, 5-HT and norepinephrine. The amino acids are shown in umole/
g wet tissue ± S.D.; the amines are shown in ng/g ± S.D. Number
in parenthesis indicates the number of mice determined. Mice, 19-
day-old, were given subcutaneous injection of AOAA (20 mg per kg)
five hours before sacrifice.

Metabolite	Control	AOAA treated
GABA	1.87 ± 0.05 (4)	10.45 ± 0.31 (5)
Glutamate	9.12 ± 0.41 (9)	10.23 ± 0.45 (5)
Aspartate	3.86 ± 0.10 (5)	3.01 ± 0.14 (5)
5-HT	471 ± 33 (12)	416 ± 78 (13)
Norepinephrine	474 ± 112 (13)	474 ± 135 (12)

Table 4. Effect of auditory stimulation on the development of audiogenic seizure susceptibility. Both untreated and amino-oxyacetic acid (AOAA) pretreated mice were primed by sound on day 19 <u>post partum</u> and tested for seizure susceptibility on day 28. Pretreated mice were injected subcutaneously with AOAA (20 mg per kg) five hours before priming. Each group consisted of at least 60 mice.

Treatment	Seizure Susceptibility (%)
None	73
Pretreated with AOAA	13

Earlier attempts by several investigators to prevent or reduce the effect of priming by the use of general anesthetics, anticonvulsant drugs and physostigmine, and by electroconvulsive shock and hypothermia, were unsuccessful (3, 5). In this study, by observing and then manipulating a neurochemical change, it was possible to markedly reduce the effect of priming. Since the change in GABA

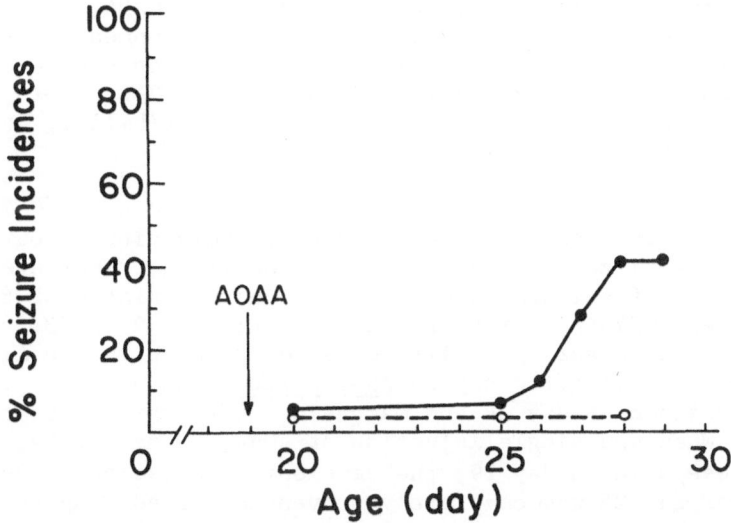

Figure 2. Effect of AOAA on genetic development of susceptibility to audiogenic seizures in Agouti mice. Mice were untreated (solid circle) or treated (open circle) with one single subcutaneous injection of AOAA (20 mg per kg) on day 19 <u>post partum</u>. Each point represents a group of at least 60 mice.

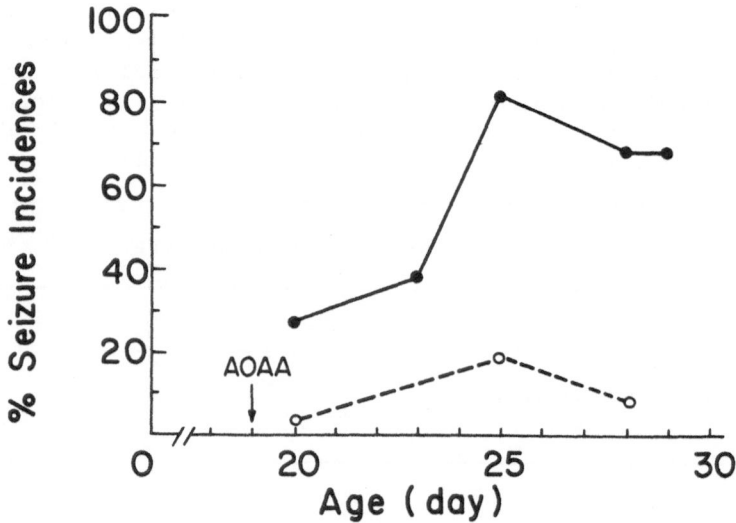

Figure 3.   Effect of AOAA on genetic development of susceptibility
to audiogenic seizures in C57BL/6s mice. Mice were untreated
(solid circle) or treated (open circle) with one single subcutaneous
injection of AOAA (20 mg per kg) on day 19 post partum.   Each
point represents a group of at least 60 mice.

level following priming was transitory and no further changes were
observed in later development, the involvement of GABA, if causal,
might be in a chain of events leading to the development of seizure
susceptibility. To test this possibility, experiments were carried
out to determine if similar manipulation of brain GABA could also
affect gene-induced development of seizure susceptibility, using
two mutants of the C57BL/6 strain.

The two seizure-susceptible mutants, designated as Ag(C57BL/6)*
and C57BL/6s, have been subjected to genetic selection in our laboratory and stabilized in the homozygous condition. The developmental profiles of seizure susceptibility characteristic of the
Ag(C57BL/6) and C57BL/6s mice are shown in Figures 2 and 3, respectively. Seizure susceptibility does not develop in the Ag
(C57BL/6) mice until after day 25 post partum, and normal levels
of 41% are reached on day 28 and maintained afterwards for at least
one month. When one single injection of AOAA, 20 mg per kg, was
given to these mice on day 19, the development of seizure susceptibility after day 25 was completely prevented for as long as the

*This mutant occurred in a pedigree that had also given rise to a
coat color mutation but the change in pigmentation was shown to be
independent of audiogenic seizures.   Since it is phenotypically
indistinguishable from the white-belly agouti, it is sometimes also
designated as C57BL/6A^W.

one month observed. A similar effect of AOAA was also demonstrated
in the C57BL/6s mice. In this mutant strain, maximal seizure susceptibility occurs at 25 days of age, but a stabilized level of
65% is reached after 28 days of age, as in the other mutant. After
one single injection of AOAA on day 19, the development of seizure
susceptibility in the C57BL/6s mice became markedly suppressed,
with the stabilized level after day 28 reduced to as low as 8%. It
is important to note that while elevation of endogenous GABA in the
19-day-old mouse brain was restored to normal level within three
days after the administration of 20 mg per kg of AOAA (14), suppression of audiogenic seizures in both mutants appeared to remain
for prolonged period.

These genetic differences support the suggested role of GABA
as a causal factor in the induction of long lasting seizure susceptibility by priming. Auditory stimulation associated with a
reduced level of brain GABA, when occurring in a sensitive period
of postnatal development, can produce a susceptible animal from a
genetically nonsusceptible one. In the reverse case, chemical inter-
vention associated with an elevated level of brain GABA, when occurring in the same period, can produce a nonsusceptible animal from a
genetically susceptible one. Whether or not a similar neural
mechanism underlies both sound-induced and genetically determined
audiogenic seizures is at present not known, but in both cases GABA
appears to be causally involved in the formation of the neural
mechanism.

The role of GABA, possibly in a chain of events, in inducing
long lasting neural change is at present only speculative. It
seems pertinent, however, to mention two recent findings on the
biochemical function of GABA. First, Tewari and Baxter (15) found
that synthesis of protein in a ribosomal system from immature rat
brain is stimulated by GABA. Secondly, Sze (14) found that glutamic
acid decarboxylase, the enzyme that synthesizes GABA, is probably
regulated by feedback repression by its own product in the developing mouse brain, and he and his colleagues (16) further demonstrated
that the enzyme activity that is repressed by GABA occurs mainly in
nerve endings. Although relevance of these findings to the inter-
vention of seizure development by GABA is far from being decided,
it may be surmised that biochemical events of this kind probably
play a role in the neural change leading to the development of
seizure susceptibility. Conceivably, a long lasting behavioral
change may be the result of several interacting biochemical alterations in the brain. It seems necessary, therefore, to obtain more
information about other biochemical parameters as well.

If induction of the development of seizure susceptibility by
sound is viewed in the broad sense as a long-term modification of
behavior by sensory input, the causal involvement of GABA following the auditory input is of particular interest. It supports the

hypothesis that modification of neural mechanisms, and hence behavior, by sensory input may be initiated or mediated by some
synaptic transmitter substance(s) responding to sensory input.
In investigating this possibility, sensitization to audiogenic
seizures by sound thus provides a simple behavioral system for
biochemical experimentation.

ACKNOWLEDGEMENT

This study was supported by research grant No. MH-03361 from
National Institute of Mental Health, and research grant No. 17-302
from the State of Illinois Department of Mental Health.

REFERENCES

1. Ginsburg, B.E., J.S. Cowen, S.C. Maxson and P.Y. Sze, 1969.
   Neurochemical effects of gene mutations associated with audiogenic seizures. In: A. Barbeau and J.R. Brunette (eds.),
   Progress in Neuro-Genetics, Excerpta Medica, Amsterdam, pp.
   695-701.

2. Ginsburg, B.E., 1967. Genetic parameters in behavioral research.
   In: J. Hirsch (Ed.), Behavior-Genetic Analysis, McGraw-Hill,
   New York, pp. 135-153.

3. Henry, K.R., 1967. Audiogenic seizure susceptibility induced
   in C57BL/6J mice by prior auditory exposure. Science 158: 938-
   940.

4. Jumonville, J.E., 1968. Influence of genotype-treatment inter-
   actions in studies of "emotionality" in mice. Doctoral dissertation, the University of Chicago.

5. Fuller, J.L., and R.L. Collins, 1968. Mice unilaterally sensitized for audiogenic seizures. Science 162:1295.

6. Fuller, J.L., and R.L. Collins, 1968. Temporal parameters of
   sensitization for audiogenic seizures in SLJ/J mice. Developmental Psychobiology 1:185-188.

7. Fuller, J.L., and F.H. Sjursen, Jr., 1967. Audiogenic seizures
   in eleven mouse strains. Journal of Heredity 58:135-140.

8. Lovell, R.A., 1970. Some neurochemical aspects of convulsion.
   In: A. Lajtha (Ed.), Handbook of Neurochemistry, Plenum Press,
   in press.

9. Roberts, E., and K. Kuriyama, 1968. Biochemical-physiological

correlations in studies of the γ-aminobutyric acid system. Brain Research 8:1-35.

10. Graham, L.T., and M.H. Aprison, 1966. Fluorometric determination of aspartate, glutamate, and γ-aminobutyrate in nerve tissue using enzymic methods. Analytical Biochemistry 15:487-497.

11. Mead, J.A.R., and K.T. Finger, 1961. Single extraction method for the determination of norepinephrine. Biochemical Pharmacology 6:52-58.

12. Wise, C.D., 1967. The fluorometric determination of brain serotonin. Analytical Biochemistry 20:369-374.

13. Baxter, C.F., and E. Roberts, 1961. Elevation of γ-aminobutyric acid in brain: selective inhibition of γ-aminobutyric-α-ketoglutaric acid transaminase. Journal of Biological Chemistry 236: 3287-3294.

14. Sze, P.Y., 1970. Possible repression of L-glutamic acid decarboxylase by γ-aminobutyric acid in developing mouse brain. Brain Research, in press.

15. Tewari, S., and C.F. Baxter, 1969. Stimulatory effect of γ-aminobutyric acid upon amino acid incorporation into protein by a ribosomal system from immature rat brain. Journal of Neurochemistry 16:171-180.

16. Haber, B., P.Y. Sze, K. Kuriyama and E. Roberts, 1970. GABA as a repressor of L-glutamic acid decarboxylase in developing check embryo optic lobes. Federation Proceedings 29:348 (abstract).

EFFECTS OF NOISE DURING SLEEP

GEORGE J. THIESSEN

Division of Physics
National Research Council of Canada
Ottawa 7, Ontario, Canada

The standard method of assessing the effect of noises in the nuisance region (below about 85 dB) is by means of the subjective judgement of juries. The subjectivity of such results is emphasized by the fact that they vary depending on what country carries out the experiment and also on whether the members of the jury are indoors or outside (1).

Ths use of sleeping subjects has two advantages. First, interference with sleep is traditionally regarded as the most serious of disturbances. Second, the subject is not reacting consciously to the sound stimulus and hence no judgement on his part is involved. The indications of our experiments are that much lower levels of noise disturb the sleep pattern of the subjects than was believed to be the case hitherto. Levels of 40 to 45 dBA have a 10 to 20 per cent probability of shifting the level of sleep for a subject picked at random, but individual differences are very great. Constant levels of sound on 12 successive nights show no adaptation to within the statistical error limits.

The noise used for our experiments was that of a passing truck. It was recorded at random intervals on magnetic tape, the total duration of the play-back being about six hours which included seven noises. The play-back levels were 40 to 70 dBA on different nights but were at a constant level on any one night.

The electroencephalograph, fed from frontal electrodes, was used as an indicator of the level of the subject's sleep, but besides the usual paper record of the brain waves a magnetic tape record was also obtained. This, when speeded up by a factor of 60 to 70, enabled standard equipment to be used for frequency

analysis besides reducing the analysis time of an 8 hour record to
about 8 minutes for a single frequency band.   Magnetic tape recording has the further important advantage of providing a much greater
dynamic range (about 50 dB) than can reasonably be expected from
visual analysis of the standard paper chart.   Furthermore, a contracted amplitude vs. time chart, either for the broad band or for
certain selected frequency bands, gives a good bird's eye view of
the progress of sleep for the night.   The advantages are somewhat
similar to those of a road map as compared to an aerial photograph.
While the former does not contain all the information of the latter
it is easier to use and quite adequate for most purposes.

    One result that is not normally obvious is shown in Figure 1
where the amplitude of the alpha rhythm of one subject is shown for
one whole night on a contracted scale.   Contrary to common belief
the alpha level, at least for frontal electrodes, is higher during
stages 2, 3 and 4 than during state 1 or dreaming or when awake.
It serves in fact, to classify the sleep stages into these two
groups for practically all subjects studied, while this is not
easily done by visual study of the paper record.

    The amplitude of the delta rhythm is also shown in Figure 1;
it is usually the frequency band used to sort out the sleep levels
2, 3 and 4, although the correlation with the definitions of these
levels as described by Dement and Kleitman (2) is by no means unambiguous.

    The subject is required to perform one task, viz., to press a
button once when he awakens, regardless of the cause, and to press
it twice if he has awakened from a dream.   With this information
the stages of sleep belonging to the group with low alpha rhythm
amplitude are sorted out.

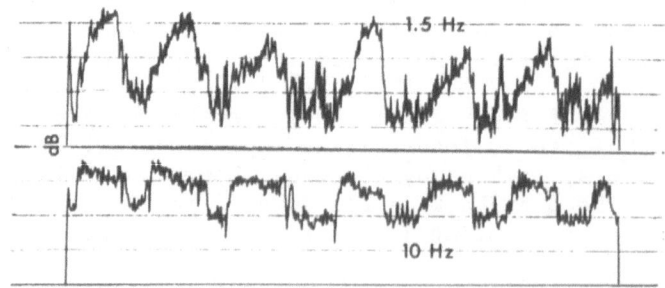

Figure 1.   The upper graph is a typical record of the variation,
during one night's sleep, of the amplitude of the EEG signal in
the 1/3-octave band centered about a frequency of 1.5 Hz.   The
lower graph shows the amplitude variation in the 1/3-octave band
which includes the ∝-rhythm.   The vertical scale is one decibel
per division.   The time scale is 15.2 mm/hr.

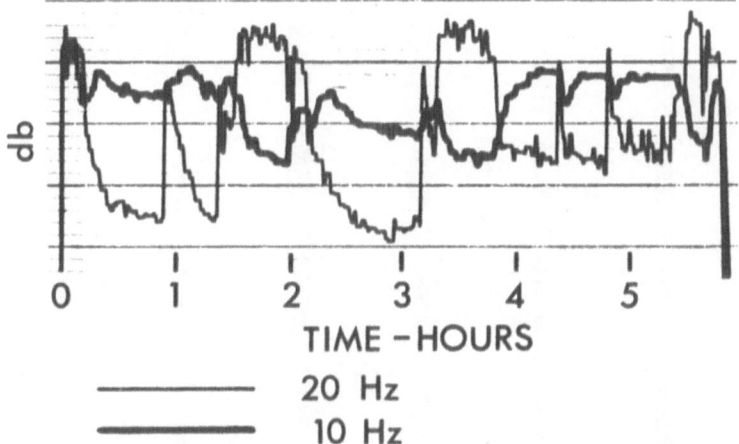

Figure 2.   Two 1/3-octave band frequencies are here superimposed
to show how, for this particular subject, the dreaming periods are
readily recognized.   They are those intervals during which the
amplitudes of the 20 Hz band is above that of the 10 Hz band.

Figure 2 shows an example of the records of a small minority
of subjects for whom the 10 Hz and 20 Hz frequency bands are the
most informative.   These two curves are superimposed to show that
generally the depth of the 20 Hz curve is a measure of the depth
of sleep (a negative correlation of the amplitude with the 1.5 Hz
band), there is a slight positive correlation between the two curves
except for some periods when the correlation is strongly negative.
These last periods always precede reports of dreaming and appear
to be unambiguous indicators.

For most subjects the sleep stages can be reasonably well
identified by ear when the electrical signal from the speeded up
tape is fed to a loudspeaker.   The simultaneous paper chart recording of overall amplitude, or just the 1.5 Hz band, is helpful in
providing a stable reference point.

The complete identification of the various stages of sleep is,
of course, not necessary in this study since the essential data are
the changes induced by noise.   Four levels of response are defined,
viz.:

(1) a brief change in brain wave pattern lasting only a few seconds
    and detectable only on the paper chart;
(2) a change lasting up to a minute and usually only detectable on
    the paper chart;
(3) a change in sleep level which is readily observed by rapid
    analysis of the magnetic tape record;
(4) awakening.

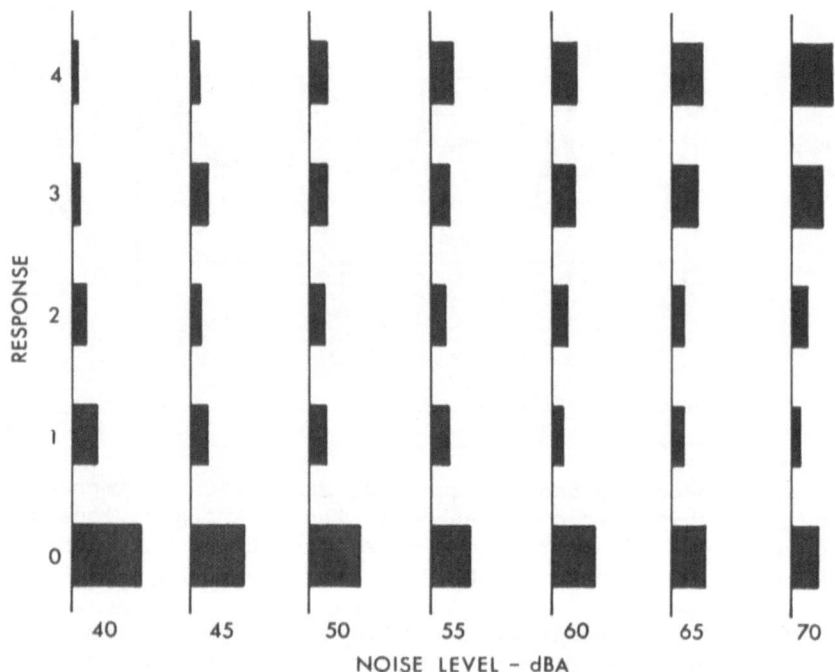

Figure 3. A series of bar graphs showing the relative probabilities of the four different responses for seven different values of the peak levels of the truck noise used as a stimulus.

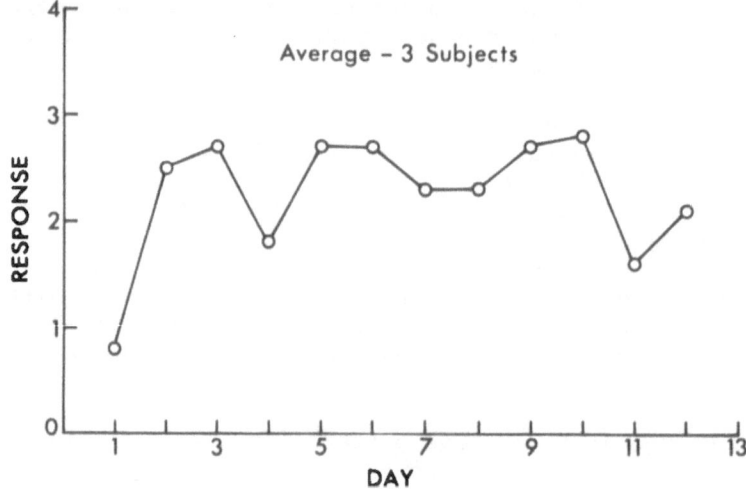

Figure 4. Average responses of three subjects when exposed to seven truck passages per night at the peak level of 65 dBA for twelve successive nights.

Figure 3 shows, in the form of a bar graph, a summary of the results for levels of 40 to 70 dBA. At 70 dBA the most probable reaction is to awaken while the next most probable reaction is to shift the level of sleep. At 50 dBA the probability is about 50% that there will be no reaction while the remaining 50% is divided about equally between the four levels of response. Even at 40 to 45 dBA there is a probability of more than 10% that the response will involve a change in sleep level or cause awakening.

Three subjects were subjected to truck passage noises at constant levels of 65 dBA on 12 successive nights (the total number of noises being 84 for each subject). Figure 4 shows the averaged responses for all three subjects and to within statistical error there appeared to be no adaptation except possibly for the tendency to awaken.

REFERENCES

1. Wilson, S., 1963. Noise, Final Report. Cmnd 2056, Her Majesty's Secretarial Office, London. pp. 10-180.

2. Dement, W. and N. Kleitman, 1957. Cyclic variations in EEG during sleep and their relation to eye movement, body motility and dreaming. Electroencephalography and Clinical Neurophysiology 9:673-690.

AUDITORY STIMULATION, SLEEP LOSS AND THE EEG STAGES OF SLEEP

HAROLD L. WILLIAMS*

Department of Psychiatry and Behavioral Sciences
University of Oklahoma Medical Center
Oklahoma City, Oklahoma 78104

Chronic loss of sleep may impair performance and cause psychological distress. In fact, severe disturbances of sleep precede
and accompany most acute psychiatric syndromes, and complaints of
sleeplessness are among the most frequent symptoms presented to
the general medical practitioner.

The psychological and social consequences of sleep-disturbing
stimuli are greater for middle-aged and older persons, for day-
time sleepers, for the physically and mentally ill, and for other
special groups than they are for the young male volunteers usually
studied in sleep laboratories. Investigation of the effects of
sonic boom on sleep, performance and health should be extended
to these groups.

To date most investigations of sleep and arousal, in addition
to having used young male volunteers as subjects, have exposed the
subjects to acoustic stimuli of fairly low intensity. The effects
of intense, complex stimuli such as the sonic boom have not been
thoroughly investigated.

_____
*Dr. Williams was unable to provide a manuscript in time for publi-
cation. This paper was adapted by the Editor from Dr. Williams
symposium presentation and from his review in Human Response to
Sonic Booms: A Research Program Plan. Final Report FAA No. 70-2,
Contract No. DOT-FA69WA-2103, Project No. 550-007-00, February, 1970,
Department of Transportation, Federal Aviation Administration Office
of Noise Abatement, Washington, D.C. (Prepared by Bolt, Beranek and
Newman, Inc., Cambridge, Massachusetts).

WAKING THRESHOLDS

All studies cited below of responses to stimulation during sleep found evidence of specificity for both stimulus and response systems. When a standard acoustic stimulus of constant intensity is used, different physiological systems have different threshold properties, patterns of response, and relations to EEG stage of sleep. Also, for a given physiological system, different classes of stimuli cause different patterns of response. Thus, the specific effects of sonic booms on the waking response should be examined.

Studies by Oswald (5), Zung and Wilson (15), Williams (11-14), Goodenough (i) and Rechtschaffen (6) show that auditory thresholds of awakening during sleep are functions of several variables. These include: stimulus intensity, EEG stage of sleep, subject differences, accumulated sleep time, time of night, amount of prior sleep deprivation, and the subject's past experience with the stimuli. In general, the probability of waking increases with stimulus intensity, with background EEG frequency, with time of night, and with accumulated sleep. For low-voltage, fast EEG stages of sleep that occur in the last half of the night, stimulus intensities no higher than 35 dB above waking-sensation thresholds can induce full arousal.

With normal aging sleep becomes light and increasingly fragmented. Specifically, by age 45 the quantity of high-voltage stage-4 sleep has diminished, and by age 60 this stage is almost entirely absent from the EEG sleep record. Thus, normal aging is accompanied by a specific kind of insomnia where sleep is broken several times a night by lacunae of waking. The only reported studies of stimulation during the sleep of aging subjects are those in the sonic-boom simulator at Stanford Research Institute. Dr. Jerome Lukas* has stated that two elderly men showed very low waking thresholds for simulated sonic boom and that, once awakened, they remained awake for very long periods. Two children between 5 and 7, studied in the same laboratory, were easily awakened by boom stimuli but were able to go back to sleep quickly. Almost nothing is known about the effects of intense stimuli on the sleep of infants.

Stimuli that have personal significance for the subject, that carry connotations of threat, or that are identified prior to sleep as "critical" stimuli are considerably more potent inducers of arousal than neutral stimuli (see studies cited above). Pilot studies by H.L. Williams' group at the University of Oklahoma showed that among the several physical parameters of neutral acoustic stimuli, rise-decay time was most significant for determining the potency of a signal for arousal.

---

*This volume.

ADAPTATION

Of major importance is the degree of long-term adaptation associated with repeated exposures to sonic booms. Information on this matter is fragmentary and inconclusive. The studies of acoustic stimulation during sleep that were cited above, as well as those by Williams (10) and Johnson (2), found that most physiological responses (such as heart rate, finger pulse volume, and EEG) show little or no adaptation during sleep. Apparently, the central-nervous-system mechanisms that permit rapid habituation during waking are not available to the sleeping subject. However, Williams (14) found evidence that waking responses did show adaptation over four nights of acoustic stimulation.

On the basis of anecdotal evidence, most sleep investigators would expect to find adaptation of the waking response, even to intense stimuli. Everyone is familiar with stories of soldiers able to sleep in the vicinity of artillery fire or people able to sleep in the vicinity of railroad trains, jet planes, or pile drivers; however, Dr. Jerome Lukas* reports that two young males exposed to many nights of stimulation in the Stanford Research Institute sonic-boom simulator showed no evidence of adaptation. Apparently, thresholds for waking remained relatively constant over the entire series. Obviously, this work needs to be replicated and extended.

SLEEP LOSS AND PERFORMANCE

Acute sleep deprivation impairs performance of certain kinds of tasks. In general, jobs that require short-term memory and high-speed processing of information are extremely sensitive to small amounts of sleep deprivation. Chronic sleep loss, where the subject is partially deprived of sleep every day for many days, has not been systematically studied. However, anecdotal reports from military and industrial sources suggest that subjects who lose some sleep each night eventually develop the same impairments found with acute sleep loss (8-14).

Prolonged sleep deprivation can lead to transient but severe psychological disturbance in which the subject experiences vivid visual hallucinations, delusions of persecution, and disorientation for time and place (4,7).

There have been no systematic studies of sleep deprivation in aging subjects, but the investigation by Williams and Williams (14) showed that normal young men who lacked stage-4 sleep were especially vulnerable to small amounts of sleep deprivation. Stage-4

---

*This volume.

sleep, as mentioned before, is substantially reduced with normal aging. Thus, the performance efficiency and psychological health of middle-aged and older persons will most likely be particularly susceptible to chronic sleep disturbance.

Further study of the effect of sonic boom on sleep should be given rather high priority. Without doubt, interruption of sleep is particularly annoying to everyone, and people are inordinately concerned about their sleep. In fact, doctors administer barbiturates to some 20 million patients each year, and our drug companies produce about 6 to 10 billion barbiturate capsules a year (3). These sleep compounds, tranquilizers, and stimulants are prescribed for patients who complain of sleep disturbance. Thus, administrative decisions that would permit a substantial increase in sleep-disturbing stimuli in the environment could affect the health and well-being of a significant portion of the population.

REFERENCES

1. Goodenough, D.R., H.B. Lewis, A. Shapiro, L. Jaret, and I. Sleser, 1965. Dream reporting following abrupt and gradual awakenings from different types of sleep. Journal of Personality and Social Psychology 2:170-179.

2. Johnson, L.C., E. Slye and A. Lubin, 1965. Autonomic response patterns during sleep. Abstract of paper presented at meeting of the Association for the Psychophysiological Study of Sleep, Bethesda, Maryland.

3. Luce, G.G. and J. Segal, 1966. Sleep. Coward-McCann, Inc., New York.

4. Morris, G.W., H.L. Williams and A. Lubin, 1960. Misperception and disorientation during sleep deprivation. American Medical Association Archives of General Psychiatry 2:247-2 54.

5. Oswald, I., A.M. Taylor and M. Treisman, 1960. Discriminative responses to stimulation during human sleep. Brain 83:440-453.

6. Rechtschaffen, A., P. Hauri and M. Zeitlin, 1966. Auditory awakening thresholds in REM and NREM sleep stages. Perceptual and Motor Skills 22:927-942.

7. West, L.J., 1967. Psychopathology produced by sleep deprivation, In: Proceedings of the Association for Research in Nervous and Mental Disease, Sleep and Altered States of Consciousness, S.S. Kety (Ed.). Williams & Wilkins Company, Baltimore.

8. Wilkinson, R.T., R.S. Edwards and E. Haines, 1966. Performance following a night of reduced sleep. Psychonomic Science 5: 471-472.

9. Williams, H.L., A. Lubin and J.J. Goodnow, 1959. Impaired performance with acute sleep loss. Psychological Monographs 73 number 14:1-26.

10. Williams, H.L., A.M. Granda, R.C. Jones, A. Lubin and J.C. Armington, 1962. EEG frequency and finger pulse volume as predictors of reaction time during sleep loss. Electroencephalography and Clinical Neurophysiology 14:64-70.

11. Williams, H.L., J.T. Hammack, R.L. Daly, W.C. Dement and A. Lubin, 1964. Responses to auditory stimulation, sleep loss and the EEG stages of sleep. Electroencephalography and Clinical Neurophysiology 16:269-279.

12. Williams, H.L., C.F. Gieseking and A. Lubin, 1966. Some effects of sleep loss on memory. Perceptual and Motor Skills 23:1287-1293.

13. Williams, H.L., H.C. Morlock and J.V. Morlock, 1965. Instrumental behavior during sleep. Psychophysiology 2: 208-216.

14. Williams, H.L. and C.L. Williams, 1966. Nocturnal EEG profiles and performance. Psychophysiology 3: 164-175.

15. Zung, W.W.K. and W.P. Wilson, 1961. Response to auditory stimulation during sleep. American Medical Association Archives of General Psychiatry 4: 548-552.

AWAKENING EFFECTS OF SIMULATED SONIC BOOMS AND

SUBSONIC AIRCRAFT NOISE

JEROME S. LUKAS and KARL D. KRYTER

Sensory Sciences Research Center
Stanford Research Institute, 333 Ravenswood Ave.
Menlo Park, California 94025

This paper will briefly describe a simulator of sonic booms as the booms would be experienced in houses under the flight path of a supersonic aircraft. It will also describe research on the effects of these simulated booms upon sleeping humans.

THE SIMULATOR

When a sonic boom strikes a typical frame dwelling three some-what distinguishable stimulations of the inhabitants are possible. These are (i) the sound of the sonic boom noise as transmitted through the walls and windows, (ii) felt vibrations from the walls and floors, and (iii) the sound resulting from the vibrations and shaking of pictures, lamps, plates, bric-a-brac and other items of furniture and decoration found in the house. The exact char-acteristics of these effects in any house are somewhat unique to the house, being dependent upon construction and furnishings. However, closely similar responses can be expected for frame houses built under common construction codes (1).

The device built at Stanford Research Institute under contract with the National Aeronautics and Space Administration simulated those three factors in the manner illustrated in Figure 1. The forward movement of a motor-driven diaphragm rapidly compresses the air in a small, sealed pressure chamber contiguous with one wall of the test room proper; withdrawal of the diaphragm beyond the neutral point reduces the pressure below atmospheric level, and another rapid movement of the diaphragm forward returns chamber pressure to near ambient level. These diaphragm movements effect-ively produce a sonic boom in the chamber which generates sound

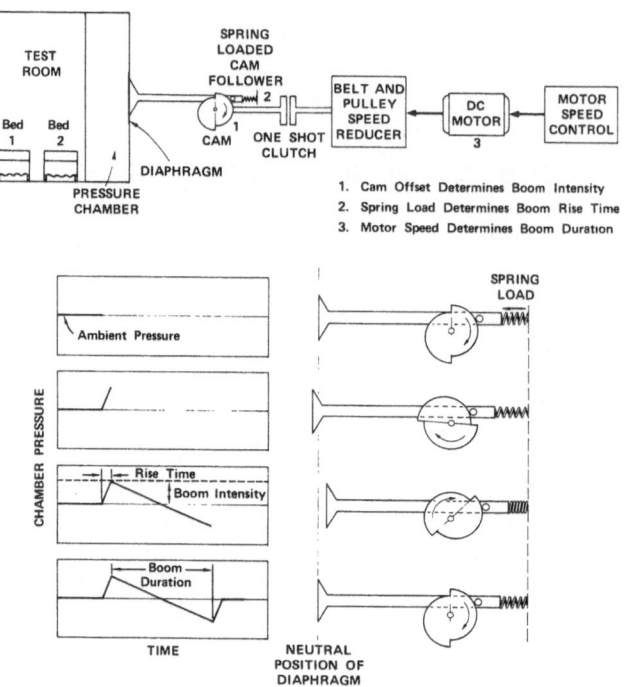

Figure 1.   Schematic of sonic boom simulator.

stimuli and affects the walls and floor of the test room in a manner similar to that possible in a room in houses struck by aircraft-
generated booms.

    Peak spectral intensity for actual and simulated sonic booms
occurs at about five to twenty Hertz, depending on the duration
(which could be systematically varied in the simulator) of the
boom.   At frequencies below and above the peak, intensity drops
off at the rate of about six decibels per octave.   At about 200-
500 Hertz, depending on the rise time of the boom, the intensity
starts dropping off at the rate of twelve decibels per octave.
Our simulated booms had effective rise times of about ten milliseconds resulting in a drop-off rate change at about 200 Hertz.
The rise time of the simulated booms fell well within the range of
rise times found for actual sonic booms, but was somewhat above
the median.   For example, booms from B-58 aircraft range from one
to as high as fifty milliseconds with a median of about five milliseconds (2).

    Figure 2 shows overall sound and vibration tracings found in
the test chamber.   These patterns are very similar to those found
in houses exposed to actual booms of comparable rise times and
durations (2).

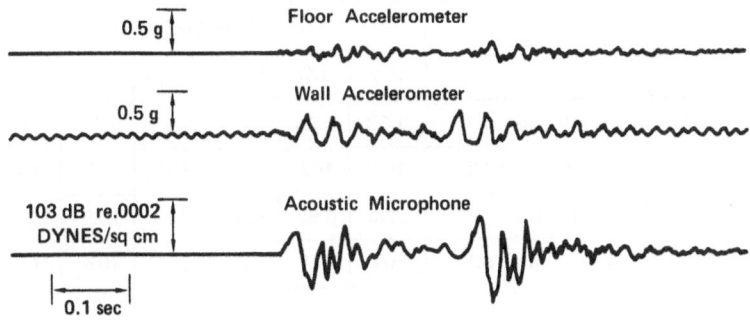

Figure 2. Accelerometer and indoor acoustic microphone tracings in test room during a 2.1 psf simulated sonic boom.

EXPERIMENTAL SUBJECTS

A. Procedure

The study to be reported here involved three age groups of subjects. They are to be designated as young, about eight years old, middle-aged, about forty-five years, and old, about seventy years of age. Two subjects were included in each age group. The subjects had normal hearing relative to their age groups except the old group whose losses were twenty to forty decibels greater than the fifty decibels loss expected at 4000 Hertz and above.

After an initial briefing regarding the purposes of the experiment, what was required of them, and a demonstration of a single one psf (pounds per square foot) boom, the subjects were permitted four to five nights for accommodation to sleeping in the laboratory with electrodes attached. Each of the subjects were tested twice a week, usually on non-consecutive nights, over a span of about twelve weeks. Pairs of subjects in the same age group were tested simultaneously.

On any given night the intensity of the subsonic jet flyover noise or the simulated sonic booms was not varied. On different nights, however, sonic booms at intensities representative of those expected at various places on the ground from proposed supersonic transports were presented. The flyover noise of the subsonic jet aircraft was selected from recordings made in a bedroom of a typical dwelling when the aircraft was passing overhead at an altitude of about 500 ft.; it had a duration of about five seconds. This recording was presented on different nights at intensities that represented the noise from the aircraft at higher flight altitudes.

Table 1. Intensity (as would be measured outdoors)
of stimuli on different experimental nights.

SUBJECT AGES (In Years)	STIMULUS	EXPERIMENTAL NIGHTS						
		1&2	3&4	5 to 8	9&10	11&12	13&14	15&16
7 & 8	Sonic Boom (in psf)	0.63	1.25	2.50	1.25	0.63	1.25	2.50
41 & 54	Subsonic Flyover (in PNdB)	101	107	113	107	101	107	113
69 & 72	Sonic Boom (in psf)	0.63	0.63	0.63	0.63	1.25	1.25	0.63
	Subsonic Flyover (in PNdB)	101	101	101	101	107	107	101

The intensities of booms and aircraft noise presented on each
test night are shown in Table 1. The young and middle-aged groups
received the same intensities of stimuli, but the young received,
usually, ten booms and ten subsonic aircraft noises per night, and
the middle-aged an average of but seven of each per night. Because
the older subjects complained about excessive fatigue in the morning
after the first two or three nights of tests, the number of stimulations was decreased for this age group to three booms and three
subsonic aircraft noises, and the most intense booms and flyover
noises were not presented.

Three nights of testing of each group included control trials
in which the stimulus triggering switches would be pushed, but no
stimuli occurred. The specific trials for these control tests
were selected at random before the subjects arrived at the laboratory, and were superimposed upon the usual test procedure. Of the
thirty-seven control trials, in only one instance was an electroencephalographic (EEG) change found to occur within ten seconds of
the test, when about two seconds of "spindling" was observed. No
behavioral awakening occurred in any of these trials.

B. Results

1. Response to sonic booms. That the subjects in the three
age groups responded differently to the simulated booms is shown in
Figure 3. Clearly, old people were awakened much more frequently
than were the middle-aged subjects who, in turn, were awakened more
frequently than were the children.

It should be noted here that an "Awake Response" was scored
only if the subject activated a switch hanging from the headboard
on his bed. An "EEG Change" was scored if the EEG indicated that
the subject had shifted from one sleep stage to the next "lighter"
stage. The occurrence of so-called K complexes shortly following

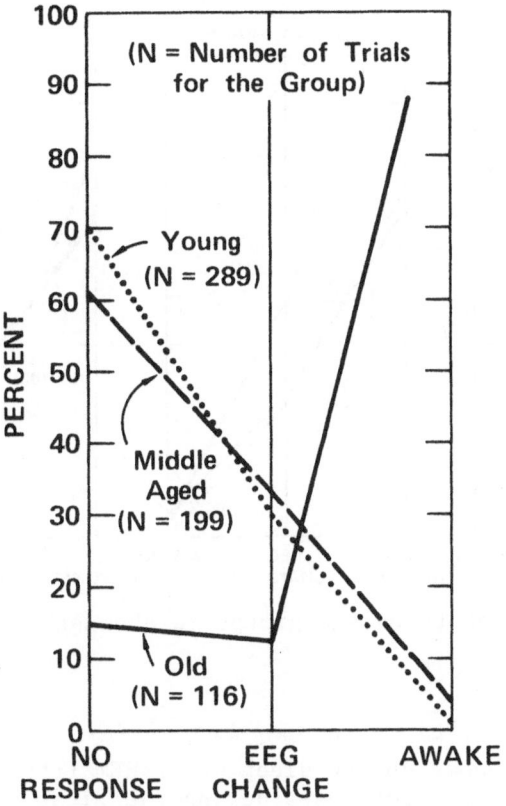

Figure 3. Relative sensitivity during sleep of three age groups to simulated sonic booms.

a stimulus was scored as "No Response" since K complexes are not specific to auditory stimuli (3) and tend not to be found in stages REM or 4 (4,5). Observed difference in responsiveness to sonic booms of the middle-aged and young groups was found to be statistic- ally significant (x^2 = 7.706, 2df; 0.025 $p < 0.02$) due, in the main, to the relatively large number of awake responses in the middle- aged group.

That the responses were about the same to the different in- tensities of booms presented to each group of subjects is shown in Figure 4. It must be noted, however, that because of the general increase in boom level during the course of the study and the relatively few nights of testing, effects of adaptation are possibly counteracting, or at least confounded with, the effects of boom intensity.

Sleep stage is clearly a determinant of the effect of sonic booms. As shown in Figure 5, the groups uniformly were awakened

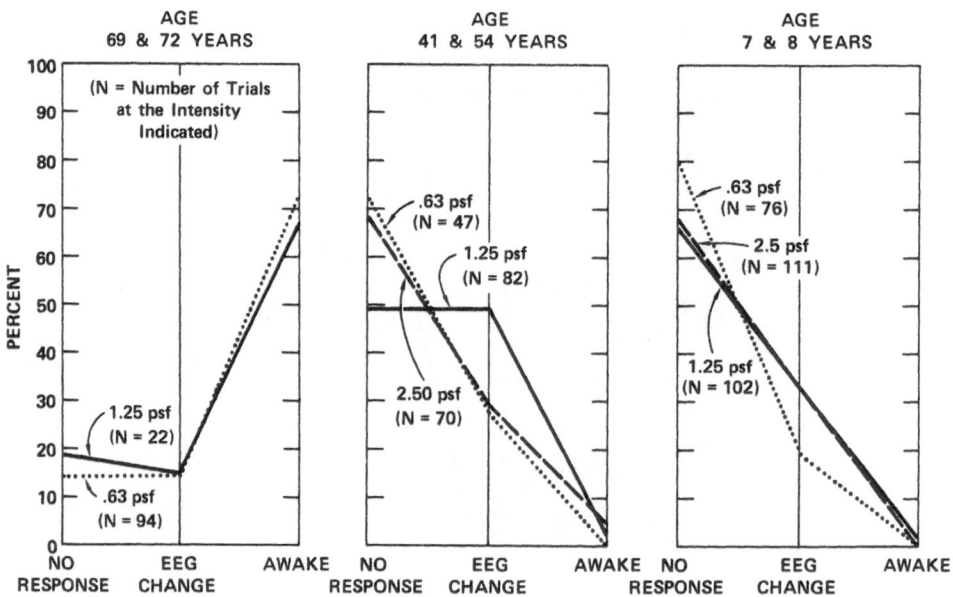

Figure 4. Response of three age groups to changes in simulated sonic boom intensity.

most frequently by booms during sleep stage REM (rapid eye movement). For each sleep stage the responses during the first six nights of testing in which boom intensity was increased from 0.63 psf to 2.5 psf (except for the old group who were tested with 0.63 psf and 1.25 psf only) were compared with the last six nights during which the intensity was increased in an identical manner. Statistically significant adaptation was found for the old group during sleep stage 2 only. Adaptation to booms was not found in either the young or middle-aged groups during any of the three sleep stages studied. However, as noted before, the separation of possible effects of adaptation from intensity level changes is questionable for this study.

2. <u>Response to aircraft noise</u>. As was found with sonic booms, the age groups responded differently to aircraft noise. However, in contrast to the lack of response differences to changes in boom intensity, as the intensity of the flyover noise increased the groups rather uniformly showed increased rates of awakening, increased rates of EEG changes, and a decrease in the percentage of no responses. These effects are illustrated in Figure 6.

The old group, as was true with respect to booms, was more

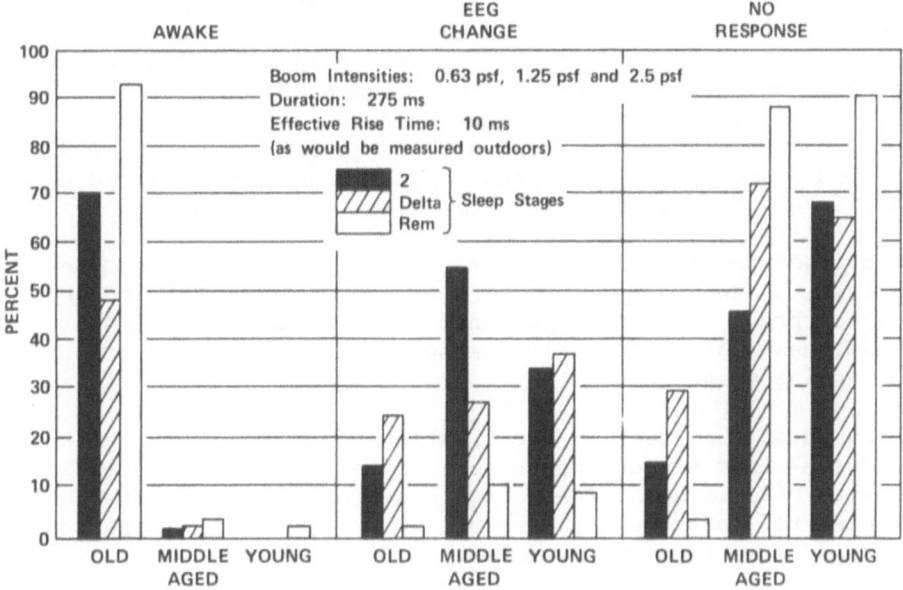

Figure 5. Response frequencies during the sleep stages to simulated sonic booms.

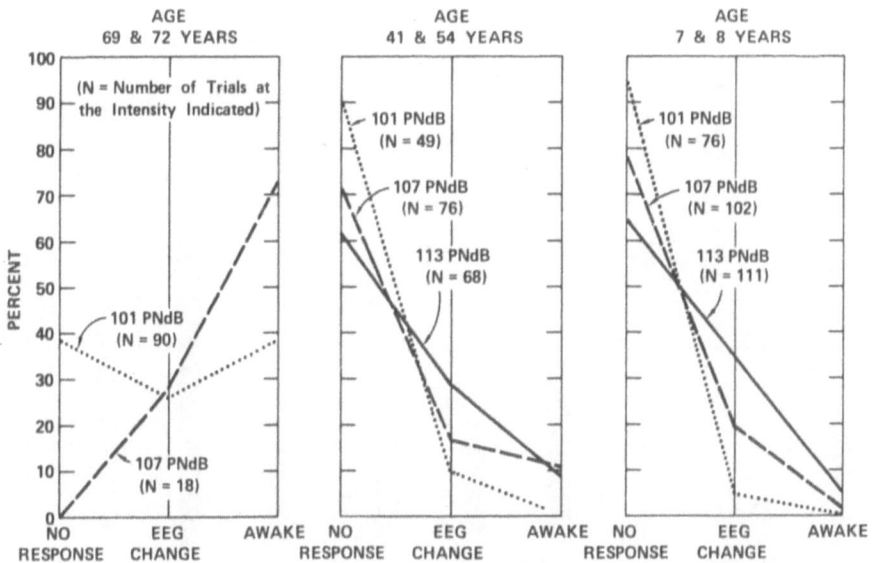

Figure 6. Response of three age groups to subsonic jet flyover noise of different intensities.

likely to be awakened by the flyover noise than were two younger
groups but, in addition, showed the most dramatic change in rate
of awakening with a change of intensity.   An increment of 6 PNdB
(from 101 to 107 PNdB) of intensity resulted in about double (from
37.8 to 72.2 percent) the percentage of awake responses in the old
group, whereas the middle-aged group shifted from 0 to about 10
percent, and the young group from 0 to but a few percent.   (PNdB
is a commonly used unit for expressing the frequency-weighted sound
pressure level of aircraft noise.)(6-8).

     Responses to flyover noise were in part dependent upon the
sleep stage, as shown in Figure 7.   As was the case with sonic
booms, all three groups were awakened most frequently while in
stage REM, next most frequently during stage 2, and least frequently during stage Delta.

     Because of the small number of stimuli occurring during each
of the three sleep stages, tests for adaptation at each intensity
of flyover noise in each sleep stage was precluded.   It was necessary, therefore, to average the results over all sleep stages to
assess adaptation to aircraft noise.

     Adaptation, as indicated by a decrease in the rate of behavior-
al awakening and an increase in the rate of no response, was found

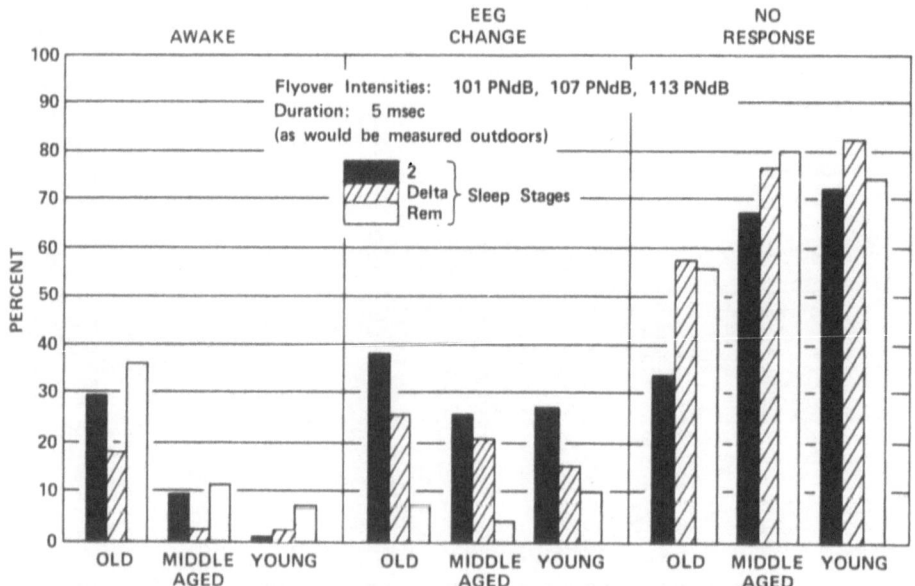

Figure 7. Response frequencies during the sleep stages to subsonic jet flyover noise.

in the old group to flyovers of 101 PNdB, and in the middle-aged
group to flyovers of 113 PNdB, as shown in Figure 8.   In the
children, adaptation was found to any intensity of flyover noise.

   3. Comparison of rates of awakening to sonic booms and aircraft noise. It is perhaps of interest to compare the frequency
of awakening of the three age groups to booms and subsonic aircraft flyover noise.   These data are presented in Table II.
Although the groups did not respond differently to changes in boom
intensity, the average percentage of awake responses was positively

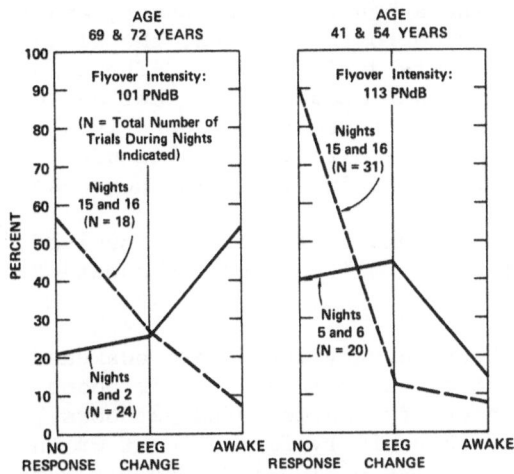

Figure 8.   Adaptation in the old and middle aged groups to subsonic jet flyover noise.

Table II.   Awake response of three age groups to simulated sonic
   booms and subsonic jet flyover noise of several intensities.

STIMULUS INTENSITY	SUBJECT AGES (in years)		
	7 & 8	41 & 54	69 & 72
Boom Intensity*			
0.63 psf	0/76 = 0 %†	0/47 = 0 %	69/94 = 73.4%
1.25 psf	1/102 = 0.9	2/82 = 2.4	15/22 = 68.2
2.50 psf	0/111 = 0	3/70 = 4.3	
Flyover Intensity*			
101 PNdB	0/76 = 0 %	0/49 = 0 %	34/90 = 37.8%
107 PNdB	1/102 = 0.9	8/76 = 10.5	13/18 = 72.2
113 PNdB	6/111 = 5.4	6/68 = 8.8	

* As Would be Measured Outdoors

† (Number of Awake Responses/Number of Stimulations
   at Intensity) x 100 = Percent

correlated with age. A similar correlation was found with respect
to age and the average percentage of awake responses to subsonic
aircraft noise, but in this case increasing flyover intensity resulted in an increase in the percentage of awake responses for each
age group.

SUMMARY

The boom intensities used in this study (0.63 to 2.5 psf,
measured six feet above the ground outdoors) are comparable to
those to be found inside a house located under, and to a distance
of twenty-five miles or so to the side of, the flight path of proposed supersonic transports during cruise. The subsonic jet
noises used (101 to 113 PNdB measured six feet above the ground
outdoors) are comparable to those to be found in a house located
under a large commercial jet aircraft at altitudes ranging from
about 500 to 2000 feet, depending on engine power settings.

Because of the small number of subjects involved, the results,
which can be summarized as follows, must be considered with some
caution at this time:

1. People aged about seventy years were awakened by about
   seventy percent of the simulated sonic booms of 0.63
   to 1.25 psf intensity (as measured outdoors), and on
   the average by about fifty-five percent of the subsonic
   jet aircraft flyovers of 103-107 PNdB (as measured outdoors).

2. People aged about eight years and 41-54 years were
   awakened about seven percent of the time by subsonic
   jet aircraft flyovers of intensities from 107-113 PNdB
   and about only two percent of the time by sonic booms
   in the intensity range of .63 to 2.5 psf.

3. Stage of sleep, as indicated by EEG pattern, was correlated with sensitivity to behavioral awakening.
   Subjects were most likely awakened in stage REM, less
   likely in stage 2, and least likely in stage Delta.

4. As indicated by a comparison between responses on the
   first and last several nights of testing, there appeared to be little adaptation to the booms, but some
   adaptation to subsonic aircraft noise.

ACKNOWLEDGEMENT

Supported by the National Aeronautics and Space Administration
under Contracts NAS 1-6193 and NAS 1-7592.

REFERENCES

1. Anon., 1967. Sonic Boom Experiments at Edwards Air Force Base, Interim Report. National Sonic Boom Evaluation Office Report; NSBEO-1-67. Annex G. (Prepared by Stanford Research Institute, Menlo Park, Calif., under Contract AF 49(638)-1758.)

2. Lukas, J.S. and K.D. Kryter, 1968. Preliminary Study of the Awakening and Startle Effects of Simulated Sonic Booms. National Aeronautics and Space Administration Report No. CF-1193.

3. Davis, H., P.A. Davis, A.L. Loomis, E.N. Harvey and G. Hobart, 1939. Electrical reactions of the human brain to auditory stimulation during sleep. Journal of Neurophysiology 2:500-514.

4. Rechtschaffen, A., P. Hauri and M. Zeitlin, 1966. Auditory awakening thresholds in REM and NREM sleep stages. Perceptual and Motor Skills 22:927-942.

5. Williams, H.L., J.T. Hammack, R.L. Daly, W.C. Dement and A.Lubin, 1964. Responses to auditory stimulation sleep loss and the EEG stages of sleep. Electroencephalography and Clinical Neurophysiology 16:269-279.

6. Kryter, K.D., 1968. Concepts of perceived noisiness, their implementation and application. Journal Acoustical Society America 43:344-361.

7. Kryter, K.D. and K.S. Pearsons, 1963. Some effects of spectral content and duration on perceived noise level. Journal Acoustical Society America 35:866-883.

8. Society of Automotive Engineers, 1965. Definitions and Procedure for Computing the Perceived Noise Level of Aircraft Noise, ARP 865, Society of Automotive Engineers, New York (1965).

EFFECTS OF NOISE ON THE PHYSIOLOGY AND BEHAVIOR

OF FARM-RAISED ANIMALS

JAMES BOND

Animal Husbandry Research Division
Agricultural Research Service
Beltsville, Maryland 20705

With the constantly increasing intensity of ambient sound in
the modern world, questions arise as to how well the present noise
level is tolerated by livestock, and whether noises of still higher
intensities and greater frequency will interfere with the efficiency of production.   While it has been assumed that most farm
animals would seldom be exposed to sounds of extremely high intensities, some animals, nevertheless, are frequently exposed to
loud aircraft, truck and motorcycle sounds.   Any environmental
factor that places farm animals under severe stress can decrease
efficiency and profits.   One question that needs to be answered
is whether sound places livestock under stress and, if it does,
the extent and mode of interference, i.e., how does it interfere
with production.   Although little research has been conducted on
the noise problem with farm animals, extensive studies have been
made of the responses of man and laboratory animals to various sounds.

This manuscript will present a general review of literature on
sound effects on farm animals and of specific studies conducted
by the personnel of the Animal Husbandry Research Division, Agricultural Research Service, U.S. Department of Agriculture.

Ely and Petersen (5) presented data from a series of experiments in which Jersey cows were subjected to fright stimuli caused
by exploding paper bags every few seconds for two minutes which
resulted in an immediate cessation of milk ejection.   Thirty
minutes after the fright stimuli, seventy percent of the normal
amount of milk was removed from the udder by hand milking.   Oda (7)
reported that motorboat racing noises caused a decrease in milk
production of dairy cows but growth of calves and heifers was not
affected.

Parker and Bayley (9) of the Animal Husbandry Research Division (AHRD) made a survey to determine if there was any measurable effect of jet aircraft noise and flyovers on the production of dairy herds located in the vicinity of existing air bases. Data covering a period of twelve months were obtained on the daily milk deliveries from 182 herds located within three miles of eight Air Force Bases using jet aircraft.

An analysis of data from forty-two herds surrounding the Lockbourne Air Force Base, Ohio, and from complete data on flight activity at that base did not show any evidence that flyovers or proximity to the ends of the active runways had an effect on the milk production of the herds. Comparisons at this base were made between days of flight activity and no activity as well as between the areas mentioned above. Analyses of the less complete data available at the other seven bases confirmed the results obtained regarding the Lockbourne Base. The results of this survey showed no evidence of an effect on milk production of dairy cattle resulting from flyovers by jet aircraft or proximity to the air base.

Casady and Lehmann (3) selected animal installations for observations on animal behavior under sonic boom conditions in areas near Edwards Air Force Base, California. Numbers of animals observed* in this study approximated 10,000 commercial feedlot beef cattle; 100 horses; 150 sheep; and 320 lactating dairy cattle. Booms during the test period were scheduled at varying intervals during the morning hours, Monday through Friday of each week. The land area around Edwards Air Force Base had been exposed to about four to eight sonic booms per day for several years prior to this study but the intensity of the previous booms was somewhat less than those presented during the sonic boom tests of this study.

Observers were stationed to watch specified groups of animals and to note behavior patterns of the animals just prior to, during, and immediately following each boom. They also noted disturbances caused by low flying aircraft used in noise tests.

Results of the study showed that the reactions of the sheep and

---

*Footnote by Editor: Also observed were 125,000 turkeys, 35,000 chicken broilers and 50,000 pheasants. The observers were fourteen senior-high school students working part-time, supervised by a high school science teacher. Observations were made during morning hours Monday through Friday, June 6-23, 1966. The poultry exhibited evidence of fright and/or pandemonium, especially during the early stages of the program. Reactions consisted of flying, running, crowding, just cowering. The owners of the pheasant breeding flock filed a claim with the U.S. Air Force, stating that there had been a severe drop in egg production (3).

horses to sonic booms were slight.   The dairy cattle reactions
were affected little by sonic booms (2.6 - .75 pounds per square
foot (psf).   Only 19 of 104 booms produced even a mild reaction,
evidenced by a temporary cessation of eating, raising of heads, or
slight startle effects in a few of those being milked.   Milk production was not affected during the test period, as evidenced by
total and individual milk yield.

Indications were that in beef cattle the daily frequency of
total changes in activity was somewhat lower after sonic booms at
a farm which was much closer to the flight track than another farm.
This may simply reflect observer differences in judging activity
changes in animals.   At all farms there was an apparent decrease
in activity after sixteen days of sonic booming which might be
attributed to adaptation to noise.   However, it is also possible
that this was due to animal adaptation to the presence of observers.

Casady and Lehmann (3) developed a summary by species and farms
indicating that the few abnormal behavioral changes observed were
well within the range of activity variation within a group of animals.
They defined these changes as horses jumping up and galloping around
the paddock, bellowing of dairy cattle, and increased activity by
beef cattle.

Since so few abnormal behavioral changes are evident in the
Edwards Air Force Base test results, it seemed advisable to con-
duct some closely controlled observations on normal changes in
behavior of animals.   Therefore, a series of tests were conducted
at the Agricultural Research Center, Beltsville, Maryland, utilizing
groups of beef cattle, dairy cattle, and sheep to supplement the
Edwards Air Force Base studies.   These groups were observed by
two individuals per group, working independently, from 9-11:30 a.m.
on three consecutive days.

From these data, differences among classifiers, days, and
time periods of a day were observed.   Each of these effects was
evaluated statistically.   With respect to the Edwards Air Force
Base data, the pertinent figures are simply the percentages of normal changes for each of the species.   When the values obtained in
the Beltsville study were compared to percentage changes in activity
among animals at the test farms at Edwards Air Force Base, it was
concluded that the booms had very little effect on changing the
behavioral pattern of farm animals.   The assumption is that behavioral changes in activity among animals at Beltsville would reflect
normal activity changes among animals at the test farms at Edwards
Air Force Base.   It was therefore concluded that the observed
behavior reactions of large animals to the sonic booms were minimal.
It was, however, noted that the reactions by animals were more pronounced to low flying subsonic aircraft noise than to booms.   The
reactions were of similar magnitude and nature to those resulting

from flying paper, the presence of strange persons, or other moving
objects which may indicate that stress may be more pronounced
when an object is seen.

Dawson and Revens (4) of the AHRD studied the behavior of
forty-two female pigs by using an electrical sparking device which
caused a bluish white spark and made a distinct, though not loud
hissing and crackling sound. They found that while many of the
pigs were appreciably bothered by the spark at first, almost all
the pigs showed very rapid adjustment to it. No association was
found between the average time required by the pigs to return to
feeding and their average daily gain. In those females which
raised litters the next spring, no relation could be shown between
previous treatment and whether one or more of her suckling pigs
were or were not crushed by her during early lactation because the
death loss was similar to that of the control. No relation could
be shown between the weekly weight of the suckling pigs raised by
the sow and the time required by the sow to return to feeding
during the previous test.

To determine possible harmful effects of aircraft noise, pigs,
boars, and sows were exposed at Beltsville, Maryland, (1) to reproduced aircraft and other loud sounds at various stages of the
life cycle. The swine unit, animals, and diets used were typical
of those found at most swine production operations. The sound reproduction system was made up of components consisting of a tape
reproducer, amplifiers, horn-type loud-speakers, and control equipment. The tape recording used in these studies was of propeller
driver and jet aircraft in flight and airfield background noises.
The schedule of jet and B-36 flyovers was random so that the intervals of sound and no-sound periods were not uniform.

Observations were made of animals during exposure to the following sounds: fixed frequency (104 to 120 decibels (db); varying
frequency (from 200 to 1,200 cycles per second (c.p.s.) at 100 to
115 db); alternating sound and quietness (450 to 4,500 c.p.s. at
115 to 120 db); and the recorded squeal of a baby pig (at 110 db).

The typical reaction of a nursing sow to those sounds was
initial alarm during which she arose to her feet and appeared to
search for the source of sound, followed by resumption of suckling
by the baby pig and apparent indifference to the sound. When suckling pigs were exposed to the sounds in the absence of the dam,
they appeared to be alarmed and crowded together. No differences
were detected in the responses to the various sounds used; sounds
of frequencies ranging from 200 to 5,000 c.p.s. at 100 to 120 db
intensity elicited like responses, while the effect of a recorded
squeal of a baby pig reproduced at 100 db was similar to that of
the other sounds used. A detailed report of behavior during exposure to loud sounds is given elsewhere (12).

Boars and sows were almost entirely indifferent to loud sounds
during mating.   Conception rate of exposed sows was similar to
that of unexposed sows.   Neither the number of pigs farrowed (born)
nor the number of survivors were influenced by exposure of the
parents to sounds during mating.   Sows were exposed to reproduced
sounds, from 6 a.m. to 6 p.m. at 120 decibels, three days before
farrowing until their pigs were weaned.   The birth weight and
weaning weight of the experimental pigs were higher than the control
pigs.

To simulate conditions in proximity to an airfield, weaned pigs
were exposed to jet and propeller aircraft sounds reproduced at 120
to 135 db daily from 6 a.m. to 6 p.m. from weaning time, or earlier,
until slaughter at 200 pounds body weight in five trials.   The
trials were carried out with three to five groups of four to six
weaned pigs each.   Aside from ambient sound levels, environmental
conditions were similar for all groups.   An analysis of variance
failed to show a significant difference between the rates of gain
of treated and untreated groups.   Feed intake, efficiency of feed
utilization, and rate of gain of treated and untreated pigs were
similar. (Table 1.)

Measurements of heart rate before, during, and after sound exposure were made of a large number of weaned pigs to supplement the
prior production results.   A telemetering electrocardiograph that
recorded heart action without necessitating the presence of atten-
dants in the acoustical chamber was employed to eliminate possible
errors due to unfavorable responses of the animals to the presence
of humans.   These studies showed that the heart rate was significantly increased during exposure but that it decelerated rapidly
after the sound was discontinued while the pattern of the electro-
cardiogram appeared to be unchanged.   In trials in which previously
unexposed pigs were exposed to loud sound, differences in response
between intensities ranging from 100 to 130 db were just below the
level of significance.   A significant intensity effect was found
when previously exposed animals were subjected to sounds of 120,
130, and 135 db.   No significant difference was found in responses
of unexposed pigs to frequencies ranging from 50 to 2,000 c.p.s. at
110 to 120 db.

Ramm and Boord (10) studied the effects of aircraft noise on
the cochlea of the ear of experimental pigs from the AHRD study (1).
They found no evidence of injury caused by exposure to jet and propeller aircraft sound.   A morphological and histological study of
the organ of Corti of 39 experimental animals showed no evidence of
severe or extensive injury as compared with the normal ear.   Minor
variations noted in the organ of Corti were of such a nature that
in this study it could not be determined if they were due to delayed fixation, tissue artifacts, or minor noise injury.   Histological examinations were also made of the thyroids and adrenal

Table 1. Influence on Swine of Chronic Exposure to Aircraft Sound*

Trial No.		No. of animals	Gain daily (1b)	Feed per pig daily (1b)	Feed per 1b gain (1b)
1	Control	6	1.7	6.6	3.9
	Exposed	12	1.5	6.1	3.9
2+	Control	5	1.8	6.0	3.3
	Exposed	14	1.7	6.4	3.8
3	Control	5	1.7	7.2	4.2
	Exposed	15	1.7	6.9	4.1
4	Control	10	1.8	7.4	4.1
	Exposed	15	1.9	7.8	4.1
5	Control	10	1.7	6.7	3.9
	Exposed	15	1.6	6.2	3.9
Mean**	Control	31	1.7***	7.0	4.0
	Exposed	57	1.7	6.8	4.0

+ The results of trial No 2 are not included in the average given
   above because some distinct differences existed between this
   trial and the other ones.   In trial No 2, some sows were exposed to chronic sound during late pregnancy and the litters
   were farrowed in acoustical chambers and remained there until
   the end of the trial.   Because the animals in each pen were
   littermates, it is probable that observed differences between
   the controls and experimental animals were due, in part, to inheritance.   In all trials except No 2, littermates were divided
   among the various pens so that the latter were approximately
   equal so far as inherited characteristics are concerned.

* Intensity: 120 to 135 db re 0.0002 dyne/cm^2.
** These mean values are weighted in regard to the number of individuals involved.
***Probability is 80% that actual differences of 10 to 15% in gains
   of sound exposed as compared with control pigs would have been
   detected by these experiments.

glands of pigs that had been exposed to recorded aircraft sounds
of intensities up to 135 db.   No evidence of injury or changes
suggesting impaired function was found.   None of these lines of
investigation (1, 4, 10) has yielded evidence indicating that swine
are influenced significantly by noise.

     Bugard et al. (2) studied the neuro-endocrine system of young
castrated male pigs.   They found that noise of 93 db for several
days caused aldosteronism and severe retention of water and sodium.
They also noted that sounds ("alarm signals") recorded from pigs
in the slaughter-house were more disturbing than noise from mechanical sources.

     Pallen (8) reported that if a female mink becomes nervous she
is very liable to unintentionally kill some, if not all, of her
kits in an attempt to protect or hide them.   Such worry and excitement may also cause hardening of the mammary glands and, being unable to nurse, the kits will die.   Low flying planes and severe
electrical storms have caused losses of very young kits.   Hartsough
(6) reported that noises or disturbances to which the mink are un-
accustomed may result in kit losses.

     During the spring of 1967, a study (11) was conducted by the
Agricultural Research Service, U.S. Department of Agriculture, on
2 commercial mink farms in Virginia to determine the effects of
simulated sonic booms upon pregnancy, parturition and kit production
of farm-raised mink.   The sonic booms were simulated by a device
which was a large exponential horn, into which 2 charges of compressed air were released sequentially by time-controlled ruptures
of the two diaphragms.   This produced overpressures in the general
range of magnitude of 0.5 to 2.0 pounds per square foot.   Three
hundred breeding pastel females and over 1,250 kits were used.

     The experimental design was as follows: Mink were randomly
divided into 7 groups according to age and subjected to the following treatments:

     (1) 60 females, maintained at Control farm; not boomed or moved.
     (2) 20 females, maintained at Control farm; moved and returned
         to new pens at start of test (April 8).   Not boomed.
     (3) 20 females, maintained at Control farm; moved and returned
         to new pens just before whelping period (April 18).   Not
         boomed.
     (4) 20 females, maintained at Control farm; moved and returned
         to new pens at start of test and again just before whelping.
         Not boomed.
     (5) 60 females, moved from Control farm to Experimental farm
         (April 8), boomed 8 times per day during pregnancy; re-
         turned to Control farm (April 18).

  (6) 60 females, moved from Control farm to Experimental farm
      (April 18); boomed 8 times per day during the late gesta-
      tion and whelping period.
  (7) 60 females, moved from Control farm to Experimental farm
      at start of tests and subjected to 8 booms per day through-
      out pregnancy and whelping.

All females were observed daily, starting April 25, to determine whether or not they had whelped. Each female was also observed the day after birth of the kits and at 5 and 10 days after
whelping to determine the number of kits born alive and dead, and
mortality after whelping. Observations were discontinued 11 days
following the whelping of each litter. Representative kits that
died from the boomed mink groups were autopsied to determine the
cause of death.*

The daily booming of the mink was initiated on April 8, when
the first mink were taken to the Experimental Farm, and was discontinued June 1. Mink were boomed four times between 0930 and
1130 and four times between 1300 and 1530. The times were picked
at random with no less than 10 minutes between booms.

The results were as follows:

Since all groups of mink were bred in a similar manner before
the moving or booming was initiated, the production of kits per
female whelping is considered to be the most valid criterion of
reproductive performance in this experiment.

Kit production per female on experiment for the mink receiving
the sonic boom treatment was statistically higher than that of the
control (724 live kits at 10 days from 180 females for an average
per female kept of 4, compared to 427 live kits from 120 females
for an average of 3.6). This was primarily because of a higher
percentage of females whelping. The percentage of females whelping was 90 percent for the boomed mink compared to 78 percent for
mink that did not receive the boom.

On a basis of kits per female whelping, the mink receiving the
sonic boom treatment had essentially the same litter size at 10 days
(4.4 kits per female whelping compared to 4.5 kits per female whelping in the groups not boomed).

The highest percent of kit mortality was for mink boomed the
entire period. This contributed to slightly higher mortalities
in the boomed group and the group whelped at the farm where mink
were boomed. However, the overall production (kits per female on
experiment) was higher for boomed mink and for mink whelping at the
farm where the mink were boomed.

*Footnote by Editor: See Appendix I to this paper for amplification.

There was no effect that could be attributed to differences in intensity of the boom.   There were no visible indications that the repeated booming caused increased excitability or nervous reactions in the mink that were boomed.╪ There was no evidence observed from autopsies of the dead kits that indicated mortality because of the effects of the sonic boom.*

Under the conditions of this study, no harmful effects to mink were observed that could be attributed to exposure to the simulated sonic booms.   Reproduction in both the boomed and not boomed groups could be considered normal.*

SUMMARY

The effects of noise on the physiology and behavior of farm-raised animals results in the following:

(1) Milk production in dairy cows has been shown to be adversely affected by sudden noises but in studies with noise resulting from jet aircraft or sonic booms there was no evidence of an effect on milk production.

(2) In an observed behavior study, the reactions of dairy and beef cattle, race horses and sheep to sonic booms were minimal.

(3) Swine were exposed to reproduced aircraft and other loud sounds (usual intensities ranged from 100 to 120 decibels, but intensities as high as 135 db.) at various stages of the life cycle.   No evidence was found that the rate of growth, feed intake, efficiency of feed utilization, or reproduction was influenced by these loud sounds or that ear, adrenal, or thyroid injury detectable by gross examination or microscopic methods took place.

(4) It has been observed that unusual noises and disturbances may cause kit losses in mink.   It was found that under the conditions of a USDA study, no harmful effects to mink were observed that could be attributed to exposure of the mink to the simulated sonic booms.   Reproduction in both the boomed and not boomed groups could be considered normal.

╪Footnote by Editor: No systematic behavioral studies were conducted.
*Footnote by Editor: see Appendix I for amplification.

REFERENCES

1. Bond, James, C.F. Winchester, L.E. Campbell and J.C. Webb. 1963.
   Effects of loud sounds on the physiology and behavior of swine.
   U.S. Department of Agriculture, Agricultural Research Service
   Technical Bulletin No. 1280.

2. Bugard, P., M. Henry, C. Bernard and C. Labie. 1960. Aspects
   neuro-endocriniens et metaboliques de l'agression sonore.
   Revue de Pathologie Generale et de Physiologie Clinique
   60:1683-1707.

3. Casady, R.B. and R.P. Lehmann. 1967. Responses of farm animals
   to sonic booms. Sonic Boom Experiment at Edwards Air Force Base.
   National Sonic Boom Evaluation Office Interim Report NS BE-1-67
   of 28 July, 1967. Annex H.

4. Dawson, W.M. and R.L. Revens. 1946. Varying susceptibility
   in pigs in alarm. Journal of Comparative Psychology 39:297-305.

5. Ely, F. and W.E. Peterson. 1941. Factors involved in the
   ejection of milk. Journal of Dairy Science 24(3):211-23.

6. Hartsough, G.R. 1968. American Fur Breeder. September 1968:21.

7. Oda, R. 1960. Noise and Farm Animals. Animal Industry
   Japan 14:888.

8. Pallen, D. 1944. Practical Mink Breeding Methods. Fur Trade
   Journal of Canada. November 1944:8.

9. Parker, J.B. and N.D. Bayley. 1960. Investigations on effects
   of aircraft sound on milk production of dairy cattle. 1957-58.
   U.S. Department of Agriculture, Agricultural Research Service.
   44-60.

10. Ramm, Gordon M., and L. Boord. 1957. Anatomy of the pig's ear
    and some effects of the noise on swine. 36 pp., illus. Univ.
    of Maryland. (Unpublished Report, U.S. Department of Agriculture
    Contract No. 12-14-100-282(53).

11. Travis, H.F., G.V. Richardson, J.R. Menear, and James Bond.
    1968. The effects of simulated sonic booms on reproduction and
    behavior of farm-raised mink. U.S. Department of Agriculture,
    Agricultural Research Service 44-200.

12. Winchester, C.F., L.E. Campbell, James Bond, J.C. Webb. 1959.
    Effects of aircraft sound on swine. WADC Technical Report 59-200.

APPENDIX I to DR. BOND'S PAPER
(By the Editors)

In the mink study by Travis et al. (11), 15.5% of the kits "boomed-entire-period" died, whereas 7.2% from the "not-boomed" controls died.   During the period between the first and tenth day after birth, 12% of the kits "boomed-entire-period" died, whereas 3.2% of the kits "not-boomed" died.

Page 17 of the published report (11) stated: "Autopsies on the representative group of boomed kits found dead (42 of 60) were conducted by the Veterinary Science Labora tory of Virginia Polytechnic Institute.   Nothing was observed that could be attributed to the effects of the sonic boom."

The senior editor wrote the following letter of query:

December 27, 1968

Director
Veterinary Science Laboratory
Virginia Polytechnic Institute
Blacksburg, Virginia

Dear Sir:

Personnel in your laboratory conducted autopsies on the mink studied in the simulated sonic boom studies reported in ARS 44-200, June 1968, "The effects of simulated sonic booms on reproduction and behavior of farm-raised mink," by H. F. Travis et al.

I am involved in a long-term effort to evaluate the possible effects of sonic boom and similar impulsive sounds upon experimental animals and humans.

It would be very helpful to me if you could tell me how detailed your inspection was and whether you think it adequate to discount possible important changes, either "normal" or pathological, in cardiovascular or endocrine function.

Sincerely yours,

Bruce L. Welch, Ph.D.

BLW:pda

The following reply was received:

VIRGINIA POLYTECHNIC INSTITUTE
COLLEGE OF AGRICULTURE
BLACKSBURG, VIRGINIA 24061

DEPARTMENT OF VETERINARY SCIENCE   January 7, 1969

Dr. Bruce L. Welch
Memorial Research Center and Hospital
1924 Alcoa Highway
Knoxville, Tennessee

Dear Dr. Welch:

I have your letter of December 27 in which you refer to our work with mink
subjected to simulated sonic booms.

I would like to emphasize that we did not have any part in drawing up the
protocol or determining what was to be investigated or interpretations. We
were only asked to determine by ordinary objective laboratory means whether
the fetus or young mink were stillborn or had breathed after birth. This was
all that we did. I do not even know what interpretation was placed on the
data which we submitted. I certainly would agree with you that there are
many other things that need to be done.

I regret that I cannot be of further help.

                                            Sincerely,

gw   D. F. Watson
                                            Head, Veterinary Science

EFFECT OF SONIC BOOMS ON THE HATCHABILITY OF CHICKEN EGGS AND OTHER

STUDIES OF AIRCRAFT-GENERATED NOISE EFFECTS ON ANIMALS

JACK M. HEINEMANN*

USAF Environmental Health Laboratory
Kelly Air Force Base, Texas

ABSTRACT ⨍

White leghorn chicken eggs were incubated for 21 days in an area where 30 or more sonic booms per day were produced by Lockheed F-104 Starfighter Aircraft. The overpressures generated were equal to or much greater than those normally produced by operational military aircraft or estimated to be produced by the proposed supersonic transport (SST). The hatchability of the experimental eggs (percent of eggs set which hatched) was not reduced by exposure to over 600 sonic booms.

In other studies, cattle were subjected to extremely high intensity sonic booms and mink to ones of normal intensity. In both cases, the animals either did not react to the sonic booms or reacted in a much less exaggerated manner than had been reported. In still another study, the reactions of poultry broiler and layer flocks to low-flying sub- and trans-sonic jet aircraft were observed. Panic, crowding, and smothering--commonly blamed upon aircraft operations--were not seen.

*Present address: Hazelton Laboratories, Inc., TRW Life Sciences Center, P.O. Box 30, Falls Church, Virginia 22046.

⨍Reference for the experiments on hatching of eggs: Heinemann, J.M. and E.F. LeBrocq, Jr., 1965. Effect of sonic booms on the hatchability of chicken eggs, Technical Report, Regional Environmental Health Laboratory, United States Air Force, Kelly AFB, Texas, Project No. 65-2, February 1965. The other studies have not been published. They consist of observations of gross behavioral reaction to sonic boom, which the author said were not sufficiently systematic or definitive to warrant preparation of a manuscript. (Editor).

MAN AND SONIC BOOM: ENVIRONMENTAL CHANGE

CHARLES W. NIXON

Aerospace Medical Research Laboratory
Wright-Patterson Air Force Base, Ohio

On the eve of the commercial supersonic transport era wide-spread concern exists about the possible undesirable effects on people of repeated sonic boom exposures. Many experimental programs and observations of human response to sonic booms have been conducted in an attempt to estimate the nature and extent of these projected exposures. Based upon the integrated body of results from physiological, psychological and sociological response programs conducted in various nations in recent years estimates have been made of the acceptability of frequent, regular commercial supersonic flights over populated areas. In this report, human response studies conducted in France, the United Kingdom and the United States are reviewed and considered in terms of present foundations for criteria for sonic boom acceptability.

Human response to sonic boom is exceedingly complex, involving the physical stimulus, the immediate environment, the ambient noise conditions, the experiences, attitudes and opinions of those exposed as well as various factors not related directly to the stimulus. Consequently the possibility of formulating a completely satisfactory method for reliably estimating the effects on individuals and communities of operational sonic boom exposures on the basis of the physical stimulus alone is recognized as ambitious and perhaps unattainable at this time. Nevertheless, knowledge gained from various observations, overflight programs, experimental laboratory and field studies of noise and sonic boom effects on people does form a basis for present day estimates of the acceptability of sonic booms by man.

Although large gaps exist in technical knowledge regarding this matter, guidance is needed today, and is in fact overdue, for

authorities responsible for certification, regulation and operation
of supersonic vehicles, for rendering legal judgments and for general planning purposes.   The urgency of the need for measures to
allow populations to be protected from possible adverse effects of
overexposure to sonic booms, dictates that provisional recommendations
and guidelines be established now and continually refined as our
technical understanding is increased by future research efforts and
experience.   In the following the methods and results of laboratory,
controlled field studies and uncontrolled overflight field studies
on individual subjects, groups and communities are briefly reviewed
(Figure 1).   The last section gives the best present day yardstick
for estimating sonic boom acceptability taking all experimental
evidence and experience into account.

LABORATORY STUDIES

In the laboratory study of sonic boom the physical stimulus
as well as other important parameters may be controlled to a degree
unattainable in any other investigative situation considered.
Within these constraints knowledge has been provided regarding
effects of stimulus characteristics on subjective psychological
responses such as loudness or noisiness, comparative judgments of
sonic boom and aircraft noise, and some physiological responses
including startle.

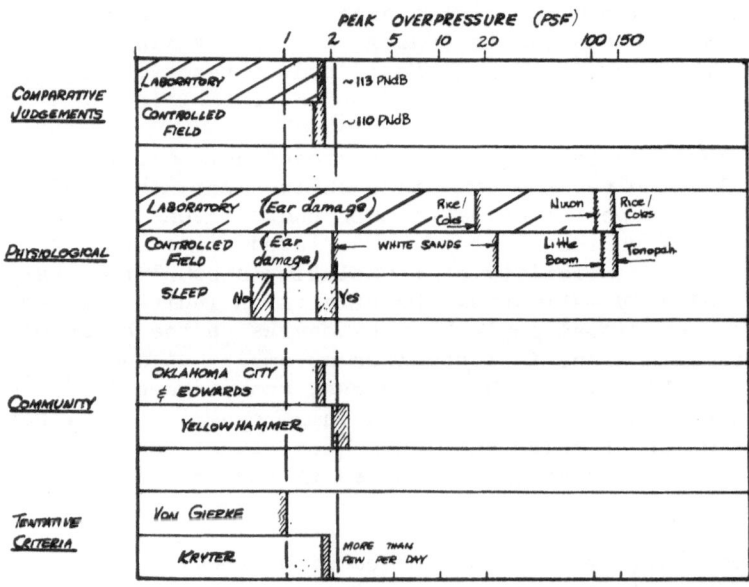

Figure 1.   Summary of laboratory, field and community studies of
human exposures in terms of type of response behavior and overpressure.

Loudness and Annoyance Judgment

It is well known that the loudness and annoyance of an impulse type noise such as the sonic boom are directly related to its energy spectral density function and qualitatively successful attempts have been made to calculate these attributes from the boom pressure time function (19).   However, since no generally accepted method for these calculations exists laboratory judgment tests were conducted to confirm in detail variations in loudness and annoyance with the boom signature.

The perceived loudness or annoyance of sonic booms is primarily related to components of the pressure-time-history of the event, i.e., the peak pressure and rise time (15,20).   Peak overpressure is the most obvious contributor to loudness for in general, the higher the peak pressure the greater is the judged loudness of the boom.   Annoyance or loudness is increased as rise time is decreased. A decrease in rise time from 10 ms to 1 ms (Figure 2) corresponds to a 13 dB increase in loudness.

The presence of a peak factor which extends the rise time of a specific signature, with the remainder of the signature unchanged, increases the positive peak overpressure and the corresponding judged relative loudness.   As shown in Figure 3 the peak factor

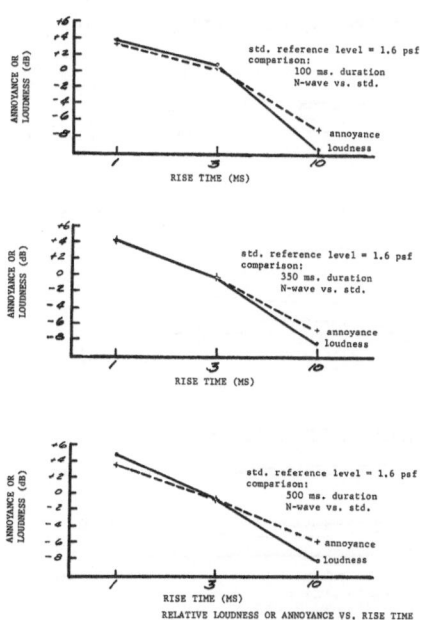

Figure 2.   Relative loudness and annoyance vs rise time of sonic booms.   Laboratory free-field judgments.   From Reference 15.

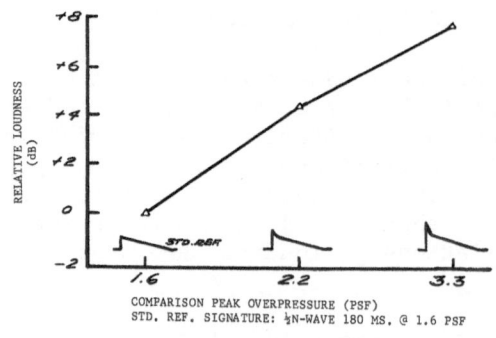

Figure 3. Contribution of peak component to relative loudness.
Laboratory free-field judgments. From Reference 15.

alone increased the loudness of 4 dB and 8 dB relative to the
standard reference signature. Relative annoyance and loudness
are generally unaffected by N-wave signature durations. Figure
4 displays signature durations ranging from 100 ms to 500 ms for

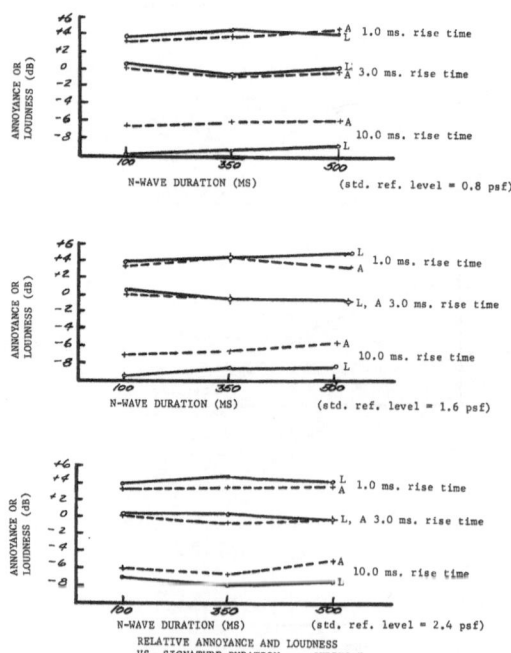

Figure 4. Relative annoyance and loudness vs signature duration.
Laboratory free-field judgments. From Reference 15.

which the annoyance or loudness was reported as essentially the same. The relationships between sonic boom signature characteristics and subjective judgments of loudness or annoyance are reasonably well established.

Comparative Judgments of Booms vs. Aircraft Noise

Methods for estimating the community acceptability of aircraft noise exposure are in use today, therefore, to study the acceptability of sonic booms relative to these established aircraft noise control guidelines is a logical approach to the sonic boom question. Although it is recognized that the experience of a sonic boom exposure is vastly different from that of aircraft noise, both from the standpoint of the stimulus characteristics and the nature of the exposure situation, quite good agreement in judgment tests has been found in the laboratory as well as in field studies to be discussed.

Pearson & Kryter (13) found (Figure 5) that a simulated indoor sonic boom of 1.7 ps was judged equally acceptable as a subsonic aircraft noise of 113 PNdB. Broadbent and Robinson (3) report findings that are almost identical to these 1.7 psf and 113 PNdB values. The present level of acceptability of noise around airports in terms of perceived noisiness is about 100 to 110 CNR* (outdoors) and this includes the wide range accounting for variations in time of day, number of exposures and the like, which are not considered in the laboratory sonic boom judgments. If a direct

```
               RESULTS  TO DATE
       (NOTE - ALL PHYSICAL MEASURES AS THOUGH MADE OUTDOORS)
                       I.         SUBSONIC AIRCRAFT
       P&K (N=20)  BOOM  |   NOISE EQUAL TO BOOM
               1.7 psf    |  INDOORS  |  OUTDOORS
                          |  113 PNdB |  94 PNdB

       B&R (N=87)  BOOM  |  NOISE EQUAL TO BOOM
               1.7 psf    |  INDOORS  |  OUTDOORS
                          | 107-113 PNdB |

               2. LOUDNESS AND ANNOYANCE OF OUTDOOR BOOM
                  PREDICTABLE FROM SPECTRUM AND CALCULATED
                  LOUDNESS OR PERCEIVED NOISE LEVELS
```

Figure 5. Relative loudness or annoyance of sonic booms vs subsonic aircraft noise. Controlled field studies (P & K = Pearson and Kryter, B and R = Boardbent and Robinson) From Reference 8.

---

*Composite Noise Rating (CNR) is a procedure whereby physical measures of noise are adjusted to take into account number of occurrences, time of day of occurrences, spectra and intensity of the noise to provide a rating number which allows a meaningful interpretation of effects of the noise on human behavior.

extrapolation was appropriate, the level of the 1.7 psf boom (113 PNdB) would exceed the range of acceptability as defined by the aircraft noise control guidelines.

The two major sonic boom exposure situations experienced by residents in a community are the indoor and outdoor exposures. The outdoor exposure is nominally a clean signal or N-wave with a rapid initial rise time. The indoor sonic boom exposure is quite different because it is affected by the response characteristics of the building, i.e., cavity resonances of the rooms and mechanical excitation of the main building modes. Indoor booms are lower in magnitude, longer in duration, have slower rise times and visably vibrate objects within the building which generate rattling and other noises. Indoor sonic boom exposures, with the additional visual, vibratory and acoustic cues, are clearly less acceptable than those experienced out of doors. This well recognized difference between the acceptability of exposures to the same sonic boom stimulus indoor vs outdoor is demonstrated in Figure 5. The 1.7 psf sonic boom heard outdoors was judged equally as acceptable as a 94 PNdB aircraft noise. The large difference of 19 PndB between the same sonic boom heard indoors and outdoors emphasizes the importance of recognizing the significantly greater unacceptability of indoor boom exposures.

Physiological Responses

Physiological responses of humans to sonic boom and other impulsive acoustic stimuli have been considered in the laboratory in the form of startle, effects on the auditory system and sleep interference. Although data are meagre in many areas and additional research is required some conclusions may be reached relative to the human auditory system and other immediate direct effects of sonic boom exposure.

1. Startle was investigated during a psychomotor tracking task in which the subjects were exposed to simulated sonic booms (8). Myographic responses (EMG's) taken from the contralateral (to tracking arm) trapezius muscle were the criterion measures. The experimental design for the study of startle response to simulated sonic booms is summarized in Figure 6. Group 1 began the tracking task in the initial test phase with simultaneous boom exposure while Group 4 tracked only during the final test session having experienced three boom sessions earlier. Startle responses to sonic booms, as measured by an increase in skeletal muscle tension, did occur. Figure 7 displays the mean differences in the electromyograms between the measures recorded before and during the boom exposures. The mean-difference myographic responses appear directly related to the boom exposures as evidenced by their relative absence from the control group data. Some adaptation in this muscular

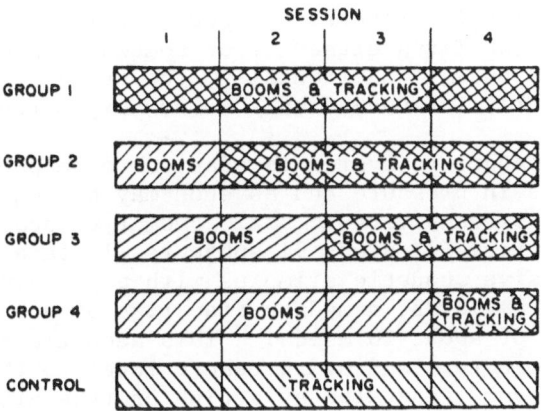

Figure 6. Experimental design for study of "startle" to simulated sonic booms. From Reference 9.

startle response may be observed as a decrease in the mean difference values with successive sessions; however, after the four sessions (36 booms) the EMGs were still greater than those recorded by the control group. An extrapolation of these data, based on the assumption of a continued linear growth rate with successive sessions, would suggest that Groups 1, 2 and 3 would reach the Control level mean difference values around sessions 7 to 9.

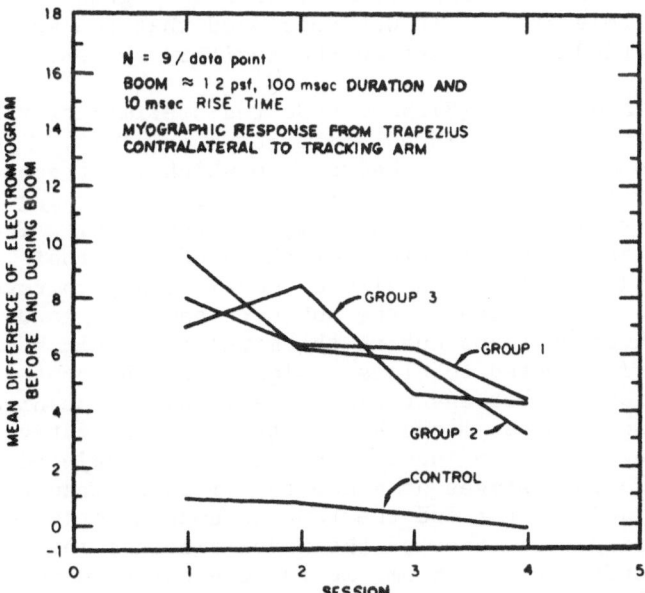

Figure 7. Mean normalized muscular startle response to simulated sonic boom as a function of test session. From Reference 9.

It would be of particular interest to pursue this response behavior
until adaptation stabilizes either at or somewhere above the Control
Group level of response.   Incidentally, the tracking performance
was not significantly influenced by the boom exposures.

2. The human auditory system is adapted to respond to very
small fluctuations in pressure and as such may be considered the
human mechanism most sensitive to sonic boom type acoustic impulses.
In a recent laboratory (1) study 91 subjects (12 females) were exposed to an impulsive acoustic stimulus with a positive peak pressure level of 168 to 170 dB, a median rise time of about 5 msec and
a median duration of about 20 msecs.   This acoustic exposure was
very similar in signature and overpressure level to the intense
near field sonic booms generated by a fighter type aircraft at
very low altitude (12).   Otological and audiometric examination
of the subjects exposed to this laboratory stimulus revealed no
adverse effects on the tympanic membrane or on auditory acuity.
The sonic boom type acoustic impulses of 168-169 dB, were shown to
be safe for the participating subjects, thus indicating a very wide
margin of safety for community populations which experience nominal
sonic booms of about 130 dB or so.

Rice and Coles (14), measured temporary threshold shift (TTS)
for subjects exposed to simulated sonic booms produced by special
explosive charges.   Resulting TTS suggested that exposure to a
sonic boom type N-Wave of 17 psf (152 dB) would not constitute an
acoustic hazard and that exposures considerably greater than this
can be safely tolerated.   It was concluded that the sonic boom
can be disregarded as a threat to the auditory system.

3. Lucas and Kryter (8) have reported preliminary results of
a study of effects of simulated sonic booms on sleep.   Their laboratory consists of a furnished bedroom to which is attached a sonic
boom simulator (Figure 8).   Indoor booms of varying magnitudes
and with good realism are generated inside the room.   Two subjects
slept in the test room each night of the study and their electro-
encephalograms (EEG's) were monitored continuously to reveal the
sleep patterns experienced by the subjects and changes which
occurred in these patterns due to the simulated sonic booms.   The
subjects were instructed to close a signal switch located on their
bed if they awakened for any reason at all during the night.
This "awakening response measure" was the only subjective behavior
observed during the sleep interference study.   Subjects were exposed during various stages of sleep to simulated indoor booms of
0.6, 0.8, 1.6 and 2.1 psf and the EEG and awakening response behavior were analyzed.   Some of the major findings indicate (a)
that sleep interference by booms is to some extent dependent on
the individual, (b) that significantly more awakening occurred with
booms of 1.6 and 2.1 psf magnitudes (Stage 2 sleep) than from booms
of lesser intensity, (c) that some adaptation to booms of 0.6 and 0.8

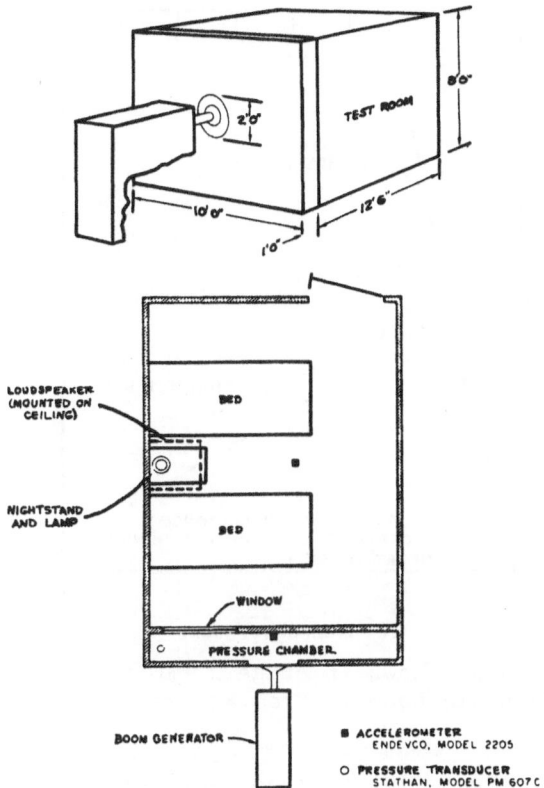

Figure 8.   Schematic of sonic boom study facility.   From Reference 9.

psf occurred during stage 2 sleep (Figure 9), but adaptation to the 1.6 and 2.1 psf booms was not found and (d) that awakening was about the same for all booms during the REM sleep stage.   This last find-ing is significant for it suggests that booms of very low intensity level will be sufficient to awaken sleepers in the REM stage of sleep.

Laboratory studies have demonstrated (a) that loudness or annoyance is directly related to particular characteristics of the sonic boom signature, (b) that the acceptability of sonic booms can be related to the acceptability of subsonic aircraft noise with some degree of success, (c) that muscle startle response to sonic boom did occur as did partial adaptation in successive exposures, (d) that the human auditory system was not adversely affected by single impulsive stimuli of much greater intensities than are ever expected for sonic booms in communities and (e) that sleep interference adaptation were observed for booms below 1.0 psf but not for booms in excess of 1.0 psf.

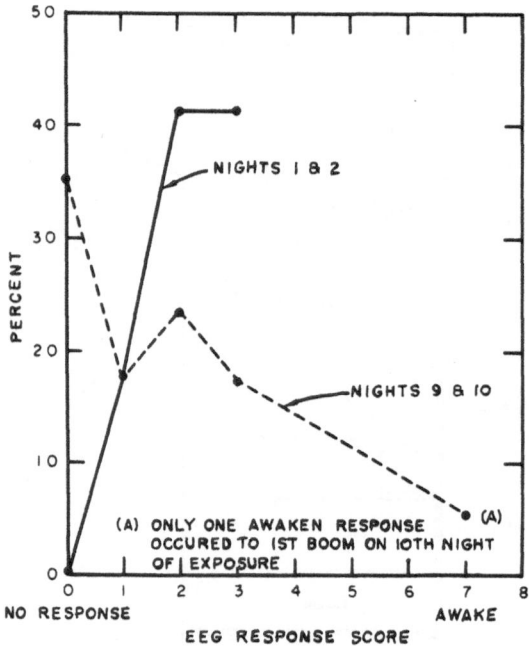

Figure 9. Adaptation as shown in stage 2 EEG response to low
intensity simulated sonic booms. From Reference 9.

FIELD STUDIES (CONTROLLED)

     Experimental field studies have been conducted in the United
States and the United Kingdom to further demonstrate the reactions
of people to the sonic boom. Generally, aircraft in supersonic
flight generate programmed sonic booms exposing personnel participating in the programs as well as panels of observers selected to provide subjective judgments of the exposures on some sort of a statistical basis. One type of field study has investigated comparative
magnitude judgments of the sonic booms relative to subsonic jet aircraft flyover noise while another type, involving very high level
sonic booms, has examined observable direct physiological response
behavior or possible injury of man.

Loudness - Annoyance

     The comparative loudness or annoyance of sonic booms vs sub-
sonic jet aircraft flyover noise work is well represented by the
UK Project Westminster (19) and the US Edwards AFB Program (5).
In Project Westminster two juries of observers were exposed, both
indoors and outdoors, to sonic booms from military aircraft, explosive

bangs, a subsonic jet aircraft flyover noise and indoors to an
occasional door slam. Of particular interest is the comparative
judgment findings shown in Figure 10. A sonic boom of 1.7 psf
was judged equally annoying as a flyover noise of 110 PNdB heard
indoors. This finding is surprisingly close to that reported in
the laboratory comparative judgment studies wherein 1.7 psf was
judged equivalent to 113 PNdB aircraft noise. As a matter of
added interest, it was reported that the door slams were judged
29% more annoying than any of the sonic booms experienced in the
study.

The Edwards AFB study, utilizing the same general jury type
approach as that used in Westminster, measured the relative acceptability of sonic booms and noise from various types of aircraft
employing the psychophysical technique of paired-comparison judgments. Observations were made both indoors and outdoors at the
specially constructed test site. Individuals with prior experience with aircraft noise and sonic boom, those with little and
those with no prior experience were included among the observers.

A major finding of these experiments was the clearcut difference in subjective judgments of annoyance as a function of the prior
experience of the subjects with noise and sonic boom exposure.
As shown in Figure 11, for indoor booms, individuals with prior
experience (Edwards) judged the B-58 nominal sonic boom of 1.69 psf
equally acceptable with a subsonic jet aircraft noise of 109 PNdB.
Those with little experience (Redlands) judged the same boom equal
to 118 PNdB aircraft noise and those with no experience (Fontana)
equal to 119 PNdB noise. The same relationships hold for the outdoor listening conditions. In addition, the 1.69 psf sonic boom
was rated as "just acceptable" to "unacceptable" by 27% of the
Edwards subjects and by 40% of the others.

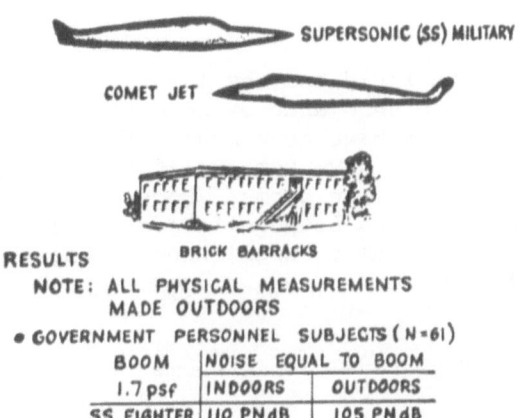

RESULTS

NOTE: ALL PHYSICAL MEASUREMENTS
MADE OUTDOORS

• GOVERNMENT PERSONNEL SUBJECTS (N=61)

BOOM	NOISE EQUAL TO BOOM	
1.7 psf	INDOORS	OUTDOORS
SS FIGHTER	110 PNdB	105 PNdB

Figure 10. U.K. Project Westminster, 1966. From Reference 7.

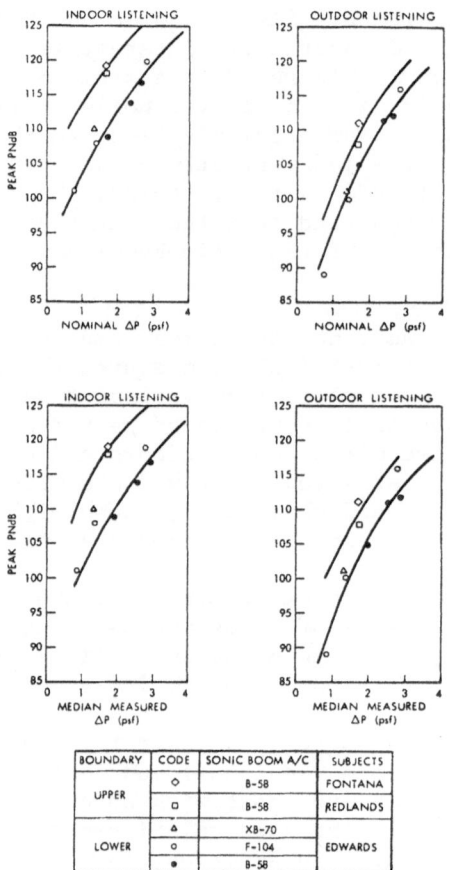

Figure 11. Results of paired-comparison judgments for subjects
from different communities.   From Reference 6.

It is clear from the field and the laboratory study that comparative judgments of the annoyance or acceptability of sonic boom
and aircraft flyover noise have been made with reasonably good reliability and in this sense have some validity for estimating
relative acceptability of sonic booms.   However, it must be re-
emphasized that relationships between laboratory judgments of annoyance and the annoyance actually experienced during everyday living
have not been established.

High Level Sonic Booms

In the United States various programs have been conducted
primarily for military purposes during which personnel were exposed
to very intense sonic booms several orders of magnitude greater than

what would ever occur in a community. In one study, personnel experienced sonic booms at a maximum peak overpressure level of 120
psf (9), while in another the maximum sonic boom exposure of the
experimenters was 144 psf (11). Although no biomedical monitor-
ing by electronic means* was accomplished, the subjective observations of individuals experienced in noise exposure were carefully
reviewed. Transient tinnitus and fullness in the ears were reported
by the observers exposed to the very intense booms. No aural pain
was reported for sonic booms of up to 144 psf. Neither temporary
nor permanent effects on auditory acuity were subjectively reported
or observed.

During a study at White Sands (16), no significant temporary
shift in hearing levels was measured after days of exposure to an
average of 30 sonic booms daily ranging in overpressure from 2 to
24 psf. In general, no adverse effects of the very intense sonic
boom exposures on any human function were observed.* As stated
earlier relative to the laboratory study of impulsive (sonic boom)
type noises, the margin of safety for community exposures to sonic
booms is very wide and the probability of direct damage to the
human auditory or other physiological system under those conditions
is essentially non-existent.

Controlled field study results are in good agreement with the
laboratory findings enumerated above. In addition, the significance
of prior exposure to sonic boom and noise in affecting judgments of
annoyance or loudness was clearly demonstrated.

FIELD STUDIES (UNCONTROLLED)

The community survey approach to the sonic boom problem in-
vestigates the attitudes, opinions and reactions of individuals
exposed to sonic booms during their day-to-day living activities.
This approach has been taken in specific programs in France, in
the United Kingdom and in the United States. The paradigm of
these studies consists of exposures of whole communities to sonic
booms generated by aircraft flying supersonic over the designated
populated areas, of the measurement or calculation of sonic boom
magnitude occurring within the community and the assessment of the
community response by various means.

United States

The first extensive survey of community reactions to sonic
booms in the United States was conducted in St. Louis (10) where
the population was repeatedly exposed to sonic booms ranging in

---

*Footnote by Editors: No studies of physiological function were
made.

overpressures up to 3.1 psf.   Results were obtained from personal
interviews in more than 1,000 households, analyses of complaint
files and of alleged damage evaluations.   About 90% of the community experienced some interference with ordinary living activities,
35% were annoyed and 10% had considered complaint action.   This
study served as a basis for and was followed approximately 2 years
later by the Oklahoma City study.

    The Oklahoma City (2) program is considered by many to be the
most comprehensive experiment to date on sonic boom exposure of a
large community over a relatively long period of time.   Eight sonic
booms per day (total of 1,253) were programmed at overpressure levels
of from 1.5 to 2.0 psf and personal interviews were completed with
3,000 families three times each, during the six month program.
Results of this program were quite extensive and were obtained
from personal interview, from complaint files, and claim actions,
all of which were related in the analysis to the sonic boom pressure measurements.

    Responses of residents of the Oklahoma City area were quite
similar and in good agreement with the findings in the St. Louis
program.   Interference, annoyance and complaint activity were
generally the same.   Community adaptation to the boom could not
be determined because the intensity of the stimulus was gradually
increased at various times during the program.   Annoyance was
strongly related to believed or expected damage to personal property.   At the end of six months, about one fourth of the residents
felt they could not learn to live with or accept eight sonic booms
daily.

    Of the many conclusions to be drawn from this study, one re-
lating to complaint activity is considered vitally significant.
This may be observed in Figure 12, which is a chronological history

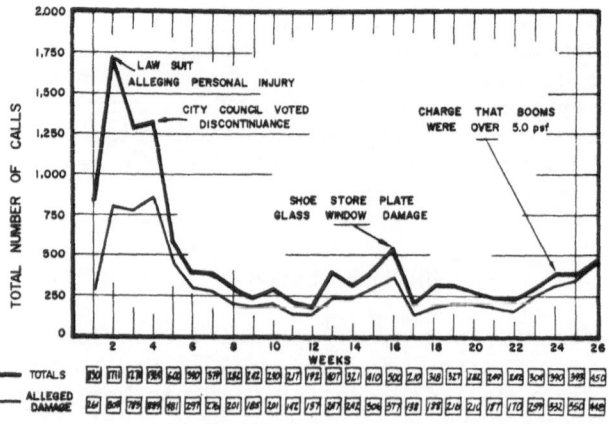

Figure 12.   Chronology of complaints in Oklahoma City study.

of the complaints and alleged incidents of damage to property received at the Complaint Center during the program.   In general,
the curves represent the high initial response expected at the introduction of overflights to a community and some degree of adaptation with time and continuation of the same stimulus exposure.
One of the most significant aspects of these curves consists of
the irregularities or peaks which appear thereon and the factors
which caused or contributed to them.

Each of the peaks can be related to some event or incident
which received widespread coverage via the local news media and
which in almost all instances was independent of the stimulus.
This means that the community reactions as reflected by the peaks
on these curves were not directly related to the sonic boom but
were strongly influenced by factors other than the stimulus.   This
finding is not encouraging for one who wishes to estimate reactions
to sonic boom based solely on the physical stimulus.

The numerous facets of this study serve (a) to emphasize that
the problem of community exposure to sonic boom is very complicated
and (b) to offer a considerable basis for interpretation.

A community survey was conducted at Edwards AFB in 1966 which
permitted the comparison of attitudes from 793 persons toward sonic
boom exposures prior to and during a special test period of boom
activity which was increased over that usually experienced by the
residents.   Prior to the test the exposure was 4 to 8 booms per
day at a mean overpressure level of 1.2 psf.   During the tests
(a one month period) 289 booms were generated at a mean overpressure
level of 1.7 psf.

More than 50% of the respondents had experienced the 4-8 boom
exposure schedule for over a year.   Some adaptation to sonic boom
exposure is believed possible when one is exposed regularly on a
daily basis to sonic booms since 60% found the boom more acceptable
after being regularly exposed to it before the tests.   Ten booms
per day at an overpressure level of 1.7 psf was less than acceptable for these respondents as it was for Oklahoma City residents.

United Kingdom

A number of community programs have been conducted in the
United Kingdom (UK) in which the impulsive stimuli were in some
cases sonic booms from supersonic aircraft and in others pressure
waves from explosive charges.   Exercise Crackerjack (17) exposed
people to sonic booms and explosive bangs at intensities of from
1.0 to 2.0 psf.   Among the findings it was concluded that explosive
bangs closely resembled sonic booms but were more annoying than the
booms after subjects become familiar with them.   Explosive bangs

were judged to be adequate stimuli for future studies of sonic boom
effects on people.

Project Yellow Hammer (18) was a more extensive community program, which used explosive charges as the basic impulsive stimuli.
A small community of less than 300 residents were exposed to explosive bangs ranging from 0.5 psf to 7.5 psf on an exposure schedule
of from 8 to 72 booms per day (Figure 13). The standard program
called for 24 bangs daily during the working hours at an overpressure level of 2.5 psf. Residents were interviewed weekly to determine the extent of their annoyance with the booms heard during the
previous week. In general, it was found that annoyance decreased
with familiarity with the bangs and this appears particularly significant because the overpressure level of the bangs remained at
about 2 to 2.5 psf during this time.

As clearly shown in Figure 14, the percentages of both those
persons considerably annoyed and those persons less annoyed (at
all) decreased as the program progressed from the first through
the fourteenth week. The considerably annoyed group appeared to
decrease only about five percent from about 17% to 12% whereas the
other group decreased about 33% from about 53% to 20%. The two
sharp peaks of approximately equal magnitude appearing for each
group at the 10th and 14th weeks correspond to a 3-fold increase
in the number of booms and an increase in the overpressure by a
factor of 2 in that order. This is interpreted as indicating that
the annoyance generated by increasing the number of booms per day
by a factor of three was essentially equivalent to that resulting
from a doubling of the standard overpressure of 2.6 psf. The
community reflected some adaptation to the repeated sonic boom

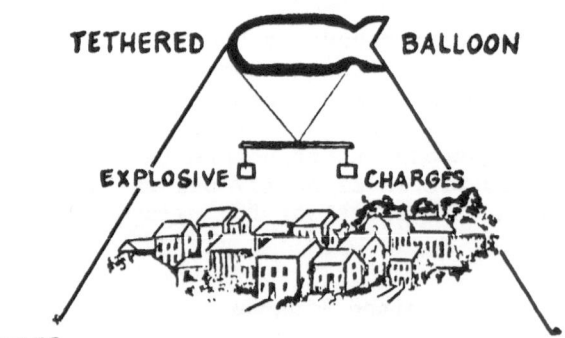

RESULTS
 • ANNOYANCE DECREASED WITH FAMILIARITY
   OF BANGS
 • INCREASE IN ANNOYANCE WITH DOUBLING OF
   INTENSITY OF BANGS SAME AS WITH INCREASE
   OF 2½ TIMES NUMBER OF BANGS

Figure 13. U.K. Project Yellow Hammer, 1966. From Reference 18.

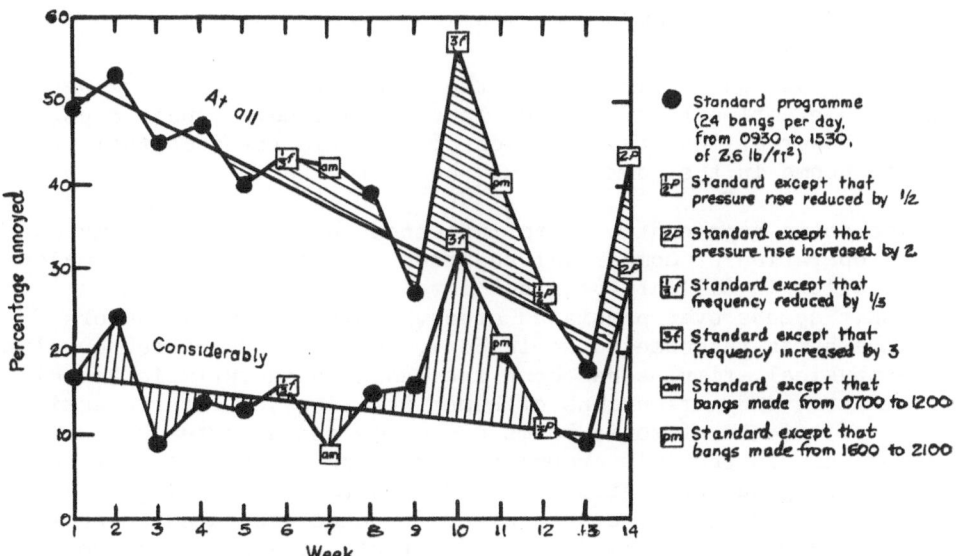

Figure 14.   Percentages of Annoyed Respondents (Project Yellow
Hammer).   From Reference 18.

exposures; however, the community was considered not to be a noise
sensitive community.

France

In 1965, approximately 2,300 personal interviews were conducted in eastern (Strasbourg area) and southwestern (Bordeaux
area) France to assess community responses to military aircraft
flying supersonic over those regions and to estimate projected
acceptability of future booms to the citizens (4).   No data on
the physical stimulus, number of booms, prior exposure of individuals and the like, were available.   Among the results it was
found that (a) interference with daily living activities was about
the same as in St. Louis and Oklahoma City, but a higher percentage
of respondents felt they could not live with 10 booms per day,
(b) expectation of property damage and personal injury were significant factors in annoyance, (c) reaction to general noise was un-
related to reaction to the sonic boom, and (d) the response "no"
was given to the proposal that an Oklahoma City type study be conducted in France.   Observations of the behavior of residents and
attitude surveys during community studies have contributed to an
increased understanding of the sonic boom problem.

PRESENT SUGGESTED CRITERIA FOR ACCEPTABILITY

Based upon an integration of the community overflight experiments with experimental field and laboratory studies, past experience and observations, the question of criteria for acceptability of sonic booms in the community may be approached.

The sonic boom continues to influence the design, economy and planned operation of commercial supersonic aircraft and is a major consideration in military programs involving aircraft flying at supersonic speeds over populated areas. Although many factors influence human response to sonic boom, the peak overpressure value of the physical stimulus has been the basic characteristic associated with measures and estimations of human reactions. Interim estimations of effects of sonic booms on people and expected community reactions considered only ground overpressure as the relevent factor. A present proposed extension of the Composite Noise Rating (CNR) method of estimating effects of aircraft noise exposures on communities(6) to include estimates of effects of sonic booms considers in addition to peak overpressure, the frequency of the boom occurrences and other variables. Major objections to such an extrapolation are recognized. However, some support is derived from the fact that data from laboratory, field and community response studies are consistent with each other (7).

Composite noise rating (CNR) is a method of relating the undesirable aspects of noise exposure to response behavior of people by means of calculations based on the characteristics of the noise exposure. The calculation provides a number (CNR) for the noise which is related to the expected behavior of a population exposed to the noise. These methods are well established for various habitual noise environs and especially for noise associated with aircraft operations. This rating method is based primarily upon the magnitude of the noise exposure, the duration of the individual exposure and the number of occurrences and the time of day. Another method, Noise Exposure Forecast (NEF) procedures, define the undesirable aspects of aircraft noise in the same general manner as the CNR, however, the NEF includes additional corrections for duration and for the presence of discrete frequencies in the exposure (2). Reactions of people in communities exposed to aircraft noise environments of different CNR's do correspond to the predicted behavior, confirming the validity of the procedure for aircraft noise evaluations.(Figure 15).

The experimental field and laboratory studies revealed that even though the sonic boom is a vastly different acoustic stimulus from aircraft noise, its loudness or annoyance can be reliably judged relative to the noise. Equally as important is the fact that annoyance with sonic boom exposures is strongly influenced

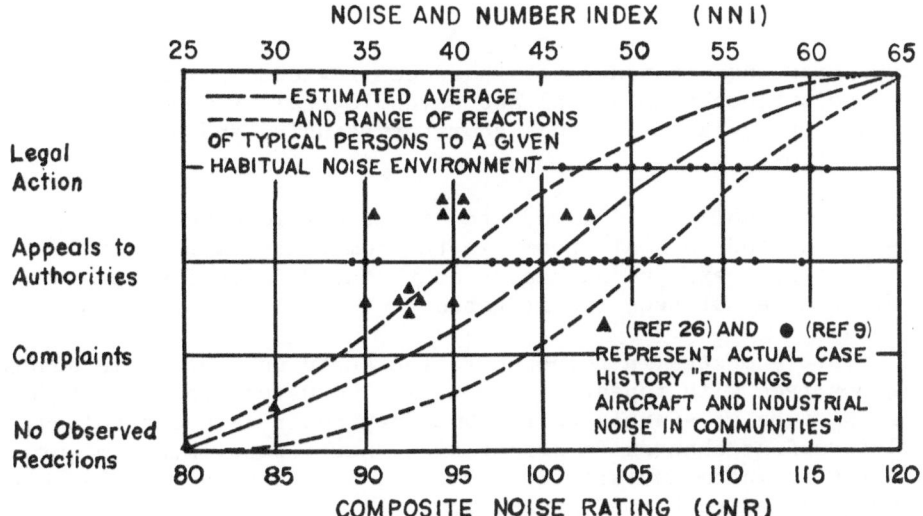

Figure 15. General relation between community response to aircraft or other noises and composite noise rating or noise and number index. From Reference 8.

if not determined by the magnitude and frequency of occurrence of the boom, as it is for noise. Intuitively, since basic factors contributing to annoyance with the signal are the same for noise and sonic boom, the CNR might be considered as a tentative present criteria for acceptability of sonic booms as well.

Kryter (7) extrapolates composite noise ratings for sonic booms, accepting the equation that a sonic boom of 1.9 psf (population has adapted over the years) will be subjectively equal to a subsonic aircraft noise of 110 PNdB. On this basis, one boom per day would provide a CNR of 98 and the CNR would increase with greater frequencies of occurrence of the boom. In general, sonic booms of 1.9 psf occurring at a rate of 2-3 or more daily would result in CNR's of 100 and more which would indicate sporadic to vigorous, widespread complaints and appeals to authorities by exposed communities.

Immediately, objections to this consideration may be stated in terms of the differences in durations of the various types of signals, the absence of startle with aircraft noise, significantly greater presence of vibrations and rattlings of dwelling contents with sonic booms, the presence of believed damage to property and of the fact that sufficient evidence is not presently available to fully substantiate such a proposal. These are rather formidable arguments which cannot be ignored and a more positive conclusion will require additional information either for or against such a consideration.

SUMMARY

Human response to sonic boom has been reviewed primarily in terms of possible physiological and psychological responses as found in experimental field and laboratory studies and community survey programs.

1. The probability of immediate direct physiological injury to persons exposed to sonic booms in the community is essentially zero. Long term effects on health of repeated daily exposures to sonic booms have not been investigated.

2. Startle occurs in response to the sonic boom and some adaptation is observed with repeated exposures. However, the extent to which adaptation of startle to the boom may occur is undetermined. Typical transient changes in respiration, heart rate, etc., might be expected to accompany startle; however, this does not imply that the exposed is being harmed in any way.

3. Sleep interference from night time booms, which may be a major determinant of public reaction, was observed for simulated sonic booms in excess of 1.0 psf for which adaptation did not occur during the test period. All sonic booms in that study were adequate stimuli for awakening subjects during their REM state of sleep. Possible long term effects on sleep of repeated nightly exposure to sonic booms are not known.

4. Comparative judgments of the relative annoyance of sonic booms and aircraft noise are in good agreement and form a basis for considering the acceptability of sonic boom exposures (primarily during day time) in terms of Composite Noise Rating.

5. A level of acceptability of sonic boom exposures in the community has not been determined. Determination of the exposure level below which no adverse response occurs would be desirable.

6. Sonic booms from fully operational SST schedules of present configuration vehicles flying over the US several flights per route per day, in terms of current estimates of the overpressures they would generate, would likely result in widespread action against the boom and its source. Nighttime operation would make this action considerably stronger. There is evidence that reactions in European countries would not be very different.

7. Although physical parameters of the sonic boom signature important to annoyance or loudness have been identified, a standard procedure for measuring and describing the sonic boom is not agreed upon and in use. Positive peak overpressure is widely described as a less than satisfactory measure but still remains the primary measure found almost universally in the scientific and technical

community to physically describe the exposure experience.   Energy spectral density likely includes more of the relevant information than peak pressure alone.

Although basic human behavior and community response to the sonic boom is not expected to drastically change with time, different factors motivating the attitudes and opinions that result in response behavior may develop in the future.   It is possible that human responses measured today may differ somewhat from those re-corded in the next eight to ten years.   It would be expected that acceptability might increase slightly with adaptation over a few years; however, it cannot be determined at this time whether such a change would have any appreciable effect on the overall community response.

Finally, it must be considered that introduction of the sonic boom from commercial SST operation constitutes a new phenomenon in aircraft noise and environmental pollution in general: probably never before has a technological development exposed with one step such a large number and large percentage of the general population to such an increase in disturbing acoustic stimuli.   The new phenom-enon would be such that it would be virtually impossible for people to escape the sonic boom and avoid boom exposed areas similar to the way some people avoid noise polluted areas in our major cities today.   The sonic boom from SST operation would probably not make the environment more noisy than the environment many people have to tolerate today in the vicinity of airports; the main difference will be that instead of a small minority a large percentage of the popu-lation will be exposed almost without escape to this new noise.

In spite of all research and predictive capability the final decision as to what boom environment people will be willing to tolerate cannot be made until the real environment will surround their daily living.   Therefore, a decision to fly over water and not over land is wise for it will introduce the new environment as slowly and controlled as possible gathering experience which can be used to modify the situation without undue harm to technological progress, economy and, first of all, human health and happiness.

REFERENCES

1. Bishop, D.E. and R.D. Horonjeff, 1967. Procedures for develop-ing noise exposure forecast areas for aircraft flight operations. Aircraft Development Service, Federal Aviation Agency, DS-67-10.

2. *Borsky, P.N. 1965. Community reactions to sonic booms in the
   Oklahoma City area. National Opinion Research Centre, Aerospace
   Medical Research Laboratory Technical Report 65-37.

3. Broadbent, D.E. and D.W. Robinson, 1964. Subjective assessment
   of the relative annoyance of simulated sonic bangs and aircraft
   noise. Journal of Sound and Vibration 1-2, 162.

4. de Brisson, Med. Lt. Col., 1966. Opinion study on the sonic
   bang, Centre d'etudes et d'instruction psychologiques de l'armee
   de l'air, Study No. 22 (1966) France. Royal Aircraft Establishment Translation 1159, ARC 27993.

5. *Kryter, K.D., 1967. Sonic boom experiments at Edwards Air
   Force Base, Stanford Research Institute, Interim and Final Report,
   National Sonic Boom Evaluation Office Report 1-67.

6. Kryter, K.D., 1968. Concepts of perceived noisiness, their
   implementation and application, Journal of the Acoustical Society
   of America 43:344-361.

7. Kryter, K.D., 1969. Sonic booms from supersonic transport,
   Science 163:359-367.

8. Lukas, J.S. and K.D. Kryter, 1968. A preliminary study of the
   awakening and startle effects of simulated sonic booms, Stanford
   Research Institute Final Report, National Aeronautics and Space
   Administration Report 1-6193, Langley Station, Hampton, Virginia.

9. Maglieri, D.J., V. Huckel and T.L. Parrott, 1966. Ground measurements of shock-wave pressure for fighter airplanes flying
   at very low altitudes and comments on associated response phenomena, National Aeronautics and Space Administration Technical
   Note D-3443, Langley Station, Hampton, Virginia.

10. Nixon, C.W. and H.H. Hubbard, 1965. Results of USAF-NASA-FAA
    Flight Program to study community responses to sonic booms in
    the Greater St. Louis area, National Aeronautics and Space Ad-
    ministration Technical Note D-2705, Langley Station, Hampton,
    Virginia.

11. *Nixon, C.W., H.K. Hille, H.C. Sommer and E. Guild, Lt. Col.,
    USAF, 1968. Sonic booms resulting from extremely low-altitude
    supersonic flight: Measurement and observations on houses,

---

*Available at nominal cost from: Clearinghouse for Federal Scient-
ific and Technical Information, 5285 Port Royal Road, Springfield,
Virginia 22151.

livestock and people, Aerospace Medical Research Laboratory
Technical Report 68-52.

12. *Nixon, C.W., 1969.   Human auditory response to an air bag
    inflation noise, Report Number PB-184-837, Clearinghouse for
    Federal Scientific and Technical Information, 5285 Port Royal
    Road, Springfield, Va., 22151.

13. Pearsons, K.S. and K.D. Kryter, 1965.   Laboratory tests of
    subjective reactions to sonic boom, National Aeronautics and
    Space Administration Contractors Report 187, Langley Station,
    Hampton, Virginia.

14. Rice, C.G. and R.R.A. Coles, 1968.   Auditory hazards from sonic
    booms? International Audiology, V11, No. 1.

15. Shepherd, L.J. and W.W. Sutherland, 1968.   Relative annoyance
    and loudness judgments of various simulated sonic boom waveforms,
    National Aeronautics and Space Administration Contractors Report
    1-6193, Langley Station, Hampton, Virginia.

16. Sonic Boom Structural Response Test Program, 1965, White Sands
    Missile Range, New Mexico, Supersonic Transport Report 65-4,
    Federal Aviation Agency, Washington, D.C.

17. Warren, C.H.E., A preliminary analysis of the results of Exercise
    Crackerjack and their relevance to supersonic transport aircraft,
    Royal Aircraft Establishment Technical Note Number Aero 2789.

18. Webb, D.R.B., and C.H.E. Warren, 1967.   An investigation of the
    effects of bangs on the subjective reaction of a community,
    Journal of Sound and Vibration, 6:375-384.

19. Webb, O.R.B., and C.H.E. Warren, 1965.   Physical characteristics
    of the sonic bangs and other events at Exercise Westminster,
    Royal Aircraft Establishment Technical Report 65248.

20. Zepler, E., and J.R.P. Harel, 1965.   The loudness of sonic booms
    and other impulsive sounds, Journal of Sound and Vibration, 2:
    249-256.

APPENDIX: CURRENT SONIC BOOM RESEARCH IN FRANCE

In keeping with our objective of making this volume a comprehensive
presentation of the current state of knowledge of the physiological
effects of noise, we include photographs of a letter recently received by Dr. Welch from Professor J. E. Dubois.

MINISTERE D'ETAT
CHARGE DE LA DEFENSE NATIONALE

DÉLÉGATION MINISTÉRIELLE POUR L'ARMEMENT

**DIRECTION DES RECHERCHES
ET MOYENS D'ESSAIS**
5 bis, avenue de la Porte-de-Sèvres
PARIS-15ᵉ

Tél. : 828 70-90 - 79-80 — Poste 632-47
SOUS-DIRECTION DES RECHERCHES

GROUPE 6
"MECANIQUE ET PHYSIQUE DES FLUIDES"

PARIS, le
009709

Nº   /D.R.M.E./ G.6.

30 JUIN 1970

              Le Professeur J.E. DUBOIS
         Directeur des Recherches et Moyens d'Essais
                          à

       Monsieur le docteur Bruce L.Welch
            Department of mental hygiene
       Maryland state psychiatric research center
                     Box 3235
              Baltimore maryland 21228

                 - ETATS-UNIS -
                 -=-oOo-=-

   O B J E T : Recherches françaises sur les effets physiologiques de
               l'exposition prolongée au bruit chez l'animal.

   REFERENCE : Votre lettre du 28 avril 1970.

   P.JOINTE : Programme de recherche sur le bang sonique.

                 -=-oOo-=-

          Monsieur,
             Les recherches actuellement menées en France sur les
   effets physiologiques de l'exposition prolongée au bruit ont lieu à ma
   connaissance dans les laboratoires suivants :

                 - Centre de biophysique sensorielle (Professeur BURGEAT)
                   Hopital Lariboisière - rue Ambroise Paré - PARIS -

                 - Institut franco allemand de St Louis (Dr RIGAUD)
                   12 rue de l'Industrie - 68 - SAINT-LOUIS -

                 - Laboratoire national des Arts et Métiers (Professeur
                   WISNER)
                   45 rue Gay Lussac - PARIS - Vème.

                                        .../...

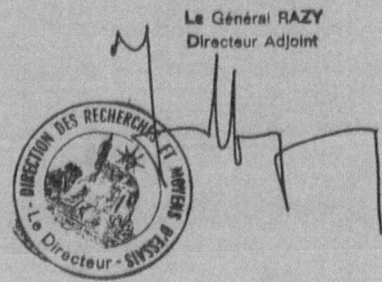

- 2 -

      - Faculté de pharmacie de Paris (Professeur RAYNAUD et Professeur ROSSIGNOL)
avenue de l'Observatoire - PARIS - VIème.

      Des recherches appliquées sur l'homme sont en outre menées par le Professeur WISNER, le Professeur HENRY (ARPAT Centre d'études bioclimatiques 21 rue Bequerel - 67 - STRASBOURG)

      Le Professeur MEYER - SCHWERTZ (Laboratoire de physiologie appliquée de la faculté de médecine, rue Kischleger 67 - STRASBOURG) et les docteurs GROGNOT et BORREDON au C.E.R.M.A. (5 bis avenue de la porte de Sèvres - PARIS XVème) .

      Les recherches sur l'effet du bang sonique sur les animaux ont été confiées au Professeur COTTEREAU (Ecole nationale vétérinaire 8 quai Chauveau - 69 - LYON 9) au Professeur GELOSO (Laboratoire de physiologie animale - Bâtiment L - 11 quai St Bernard - PARIS Vème), au docteur vétérinaire BOUTELIER (Laboratoire de médecine aérospatiale C.E.V. - 91 - BRETIGNY-SUR-ORGE) et au Professeur BURGEAT. Vous trouverez en annexe le programme détaillé des recherches entreprises sur le bang.

      En restant à votre disposition pour vous fournir de plus amples renseignements je vous prie de croire, Monsieur, à l'assurance de ma considération distinguée.

                                 Le Général RAZY
                                 Directeur Adjoint

PROGRAMME DE RECHERCHES SUR LE BANG SONIQUE

S U J E T	LIEU, ORGANISME RESPONSABLE	OBSERVATIONS
FRANCE :		
C.- REPONSE HUMAINE ET ANIMALE		
C.1.- Etude des effets sur les animaux domestiques et le gibier.	- Ecole Vétérinaire de LYON Professeur COTTEREAU - V.E.T. Cdt BOUTEILLER (C.E.V. - BRETIGNY)	Les études comportementales porteront sur les oiseaux gibier, les volailles domestiques, le comportement des troupeaux, et la gestation des porcins et bovins. Les études à l'aide du simulateur de bang analyseront les effets des bangs répétés sur le système cardiovasculaire (chien) sur les abeilles en hivernage, et sur l'incubation d'oeufs de poule domestique et de poule faisane.
C.2.- Etude des effets sur l'audition et l'équilibration de l'animal et de l'homme.	- Hôpital Lariboisière PARIS Professeur BURGEAT - C.E.R.M.A.	Sur l'animal seront étudiés l'audition et l'équilibre en recherchant les effets de stimulation ou de perturbation avec étude histologique fine. Sur l'homme on recherchera les effets de masque et de fatigue, subjective ou objective ; l'équilibre sera étudié par statokinésimétrie. Ces études seront menées sur simulateur.
C.3.- Etude expérimentale des effets hormonaux.	- Faculté des Sciences PARIS - Mr GELOSO	Il s'agira de préciser et de définir les paramètres endocriniens susceptibles d'être influencés par le bang, et de déterminer les effets sur ces paramètres d'agressions répétitives, en vue de mettre en évidence une éventuelle adaptation.
C.4.- Etude des effets de la vigilance (réaction de sursaut).	- L.E.M.P. (MONT DE MARSAN) - Méd. Chef CHATELIER	L'étude comprendra deux parties : des enregistrements électrophysiologiques (rythme cardiaque, rythme respiratoire, électromyographe, électrodermographe, électroencéphalographe), et l'utilisation d'une batterie de tests tendant à apprécier sur le plan psychophysiologique, le degré d'inhibition sensorielle et l'incoordination motrice.

- 2 -

S U J E T	LIEU, ORGANISME RESPONSABLE	OBSERVATIONS
C.5.- Etude des effets sur le système cardio-vasculaire de l'homme	LAMAS (BRETIGNY) Méd. Chef DEMANGE	L'étude des réactions circulatoires des sujets humains se fera par plethysmographie électrique permettant la mesure en continu des variations du volume d'éjection systolique, de la fréquence cardiaque et donc du débit cardiaque.
C.6.- Etude du retentissement psychosociologique	CERPAIR Méd. Chef BREMOND	Il s'agira de reprendre, si possible sur une base plus large l'enquête faite antérieurement, en tenant compte du fait qu'il s'agira maintenant de bangs produits par des avions civils ; les bangs nocturnes, troublant le sommeil devront faire également l'objet d'une recherche spéciale.

SUMMARY OF THE SYMPOSIUM

CHAUNCEY D. LEAKE

School of Medicine
University of California
San Francisco, California 94122

This has been an exciting session. We've learned a lot. Dr.
and Mrs. Bruce Welch have truly arranged an admirable symposium on
the physiological effects of audible sound. They have been able
to bring together competent scientists from all parts of the world
to report on noise as a polluting factor in our environment and on
the often unrecognized effects which noise may produce within us.

As was indicated in the announcement for this conference, we
are very sensitive to our environment and are indeed a part of it.
We are now in the process of changing that environment more rapidly than has ever occurred before. We are polluting it. Noise is
a form of pollution that isn't often recognized as such. But actually we are all aware that we are frequently annoyed by noise,
and it is the overflow from this emotional disturbance that we
are now recognizing as being threatening to our good health.

We have considered noise as a form of stress. In general
we've agreed that it is unwanted or intolerable sound. We have
tried to reach agreement on characterization of sound and there
is nothing that we can do beyond agreeing with the ordinary
measurement of sound by decibels. This is logarithmic, running
from the 20 or 30 decibels in a bedroom in a quiet dwelling at
midnight in the city or in the quiet of a piece of wilderness.
In an average residence, during the day, the noise may reach 40
decibels, and this may be doubling as people get new appliances,
including vacuum cleaners, dish washers and other kitchen apparatus.
Heavy trucks give us something around 90 decibels at 25 feet.
Train whistles at 500 feet, no matter how pleasant they may sound
on a cold winter night, nevertheless may yield around 90 decibels.
Within 20 feet or so of a subway train the decibelar level is 95.

A loud outboard motor or a power lawnmower may yield 100 decibels. A rock and roll band at peak can give 125 decibels, like a thunder clap.

While we have become conditioned in part to increasing noise in our environment, we still do not realize the extent to which noise can activate subcortical neuronal systems in our brain. Whether continuous or intermittent, noise will modify the pacing by the brain of our cardiovascular, endocrine, metabolic and reproductive functions. While it is true that the cortex may be inhibitory upon lower portions of the brain stem, this is only mildly modulated, and may be related to the extent of conditioning to which we have been subject. Even though we may have learned both behaviourly and physiologically to ignore noise and thus to reduce the intensity of its emotional response, some of the neurological stimulus may spill into the autonomic nervous system, producing cardiovascular-renal disturbance, together with endocrine, metabolic and reproductive abnormalities.

Unexpected noise is usually an alarm signal and can produce the alarm reaction which has been so thoroughly explored by Hans Selye. It is to be remembered here that usually there is overcompensation for stress as expressed in the alarm reaction. While stimulation may be necessary for the development of our brains, as well as for other physiological functions in our body, overstimulation may be harmful. Certainly we have been increasing the amount of stimulus from noise to which we are exposed and we may be approaching a point where it becomes intolerable and harmful.

Our program was well begun by Dr. Samuel Rosen, who described the cardiovascular effects of noise. His extensive experience and that of Dr. Gerd Jansen of Dortmund, Germany, who reported supporting observations, gave great strength to the important observations which he reported. His comments were further explored with regard to cardiovascular and biochemical effects by Dr. Joseph Buckley of the University of Pittsburgh and by Dr. William Geber. Some of these cardiovascular effects may spill over from a pregnant mother to the fetus and produce developmental abnormalities. This is a field that deserves extensive study.

There were discussions on sleep with respect to the effects of noise. This was well explored by Dr. Harold Williams of the University of Oklahoma and by Dr. C.J. Thiessen of the National Research Council of Canada. Reproductive disturbances brought about by noise were discussed by Dr. A. Arvay of the University of Debrechen, Hungary, by Dr. R. I. Tamari, Hebrew University of Jerusalem and by Dr. Lester Sontag of Yellow Springs, Ohio. Extensive publicity resulted from Dr. Sontag's presentation. He pointed out that noise during pregnancy may not only influence fetal development, but may also condition the resulting developing adult, and modify behaviour often in an unsatisfactory manner.

Audiogenic seizure, a remarkable result of noise in certain susceptible animals, perhaps also in humans who are prone to an epileptic-type of convulsion, was explored by Dr. Alice Lehmann of the Acoustic Physiological Laboratory of Jouy-En-Josas, France, and further by Dr. Paul Sze of the University of Connecticut. There were further reports along the same line by Dr. John Fuller of the Jackson Laboratory in Maine, by Dr. Kenneth Henry and Dr. Robert Bowman of the University of Wisconsin and by Dr. Gregory Fink and Dr. Ben Iturrian of Oregon State University and the University of Georgia.   Dr. Benson Ginsburg discussed the developmental factors involved in seizure susceptibility, and Dr. Francis Forster of the University of Wisconsin extended these reflections to human studies on epileptic seizures induced by sound.   A considerable report from Moscow by L. B. Krushinsky and his colleagues related to the variations in the functional state of the brain during audiogenic stimulation.

There were interesting reports on the effects of audiogenic stress on susceptibility to infection, as given by Dr. M.M. Jensen of Brigham-Young University and by Dr. A. Rasmussen of the University of California at Los Angeles.   Dr. Mary F. Lockett of the University of Western Australia gave a well received discussion on the effects of sound on endocrine function and upon electrolyte excretion.   It does seem to be remarkable that noise will produce significant and measurable effects on ion balance.   This was further explored by Dr. A. E. Arguelles of Buenos Aires with respect to endocrine and metabolic effects of noise in normal and psychotic patients.

The conference was closed by a panel discussion in which various aspects of physiological disturbances to noise was considered.   In this there was participation by C. E. Forkner of Cornell, W. I. Gay of the National Institute of General Medical Sciences, P. Henning Von Gierke of Wright-Patterson Air Force Base, R. E. Hodgson of the U. S. Department of Agriculture and A. S. Miller of George Washington University.   The latter considered some of the legal responsibilities that may be involved in connection with untoward effects from noise.

In prospect, there was consensus that education among the people on the health hazards of noise may produce the individual motivation needed for effective social and political action to reduce the noise contamination of our environment.   We may be helped by recognition of the fact that increasing noise with its associated dangers to our health is another aspect of the pollution crisis resulting from too many of us.   If we can learn how satisfyingly to lower our numbers we may find that we also will be

reducing the toxic levels of noise, and of environmental pollution
in general.   The United Nations may aid in world wide agreement on
noise abatement.

     We all crave satisfaction in living: we all seek satisfaction,
even as the Hebrew Prophet Elijah sought the joyful satisfaction
of being with his Lord.   The great music of Mendelsohn dramatized
the story from the 19th chapter of 1 Kings: There came a great
storm, but the Lord was not in the tumult of the storm; there came
a crashing earthquake, but the Lord was not in the earthquake; there
came a soaring fire, but the Lord was not in the fire, and then
there came a still small voice, and in that still voice, onward
came the Lord.   Perhaps we need to listen to the still small
voices if we wish to get the satisfactions in living which we all
crave.

SUPPLEMENTARY REFERENCES
(Compiled by the Senior Editor)

CARDIOVASCULAR

Shatolov, N.N., A.O. Saitanov and K.V. Glotov, 1962. On the state
of the cardiovascular system under conditions involving exposures
to the action of permanent noise. Labor Hygiene and Occupational
Disease 6: 10-14 (Russian with English summary).

> Studied 300 workers regularly exposed to industrial noise,
> medium to high-frequency, 85-120 dB. ECG showed bradycardia
> and tendency to reduced intraventricular conductivity. After
> physical stress and at the end of the work day, the ECG showed
> an abnormal falling off, leveling out and diphasic appearance
> of the T- wave, these observations being paralleled by corres-
> ponding ballistographic changes. Hypertension,abnormal lower-
> ing of vascular tone were common. In workers exposed to the
> higher noise levels of 114 - 120 dB these changes were more
> common and more extreme.

Taccola, A., G. Straneo and G.C. Bobbio, 1963. Changes of the cardiac dynamics induced by noise. Lavoro Umano 15: 571-579 (Italian
with English summary).

> Auditory stimulation causes changes in cardiac deformation time,
> true isometric contraction and isotonic systole. Authors attri-
> bute changes to a double stimulation by noise: one neuro-vegeta-
> tive, the other an indirect consequence of altered peripheral
> hemodynamics.

Darner, C.L., 1966. Sound pulses and the heart. Journal of the
Acoustical Society of America 39: 414-416.

> Heart rate tends to synchronize with repetitive audible sound
> pulses. Prolonged daily exposure to sound pulses having a
> repetition rate higher than the normal heart rate (140-160
> pulses/min for several hours each working day for 3 mos)
> appears to fix the heart rate at an abnormally high rate

(120-140/min vs. normal 85/min). The heart rate can be reduced again by listening to sound pulses at lower rates, but
the time required to return heart rate to normal is considerably longer than is required to cause the abnormality.

Pretorius, P.J. and J.J. Van Der Walt, 1967. Influence of a loud
acoustic stimulus on the ultra-low frequency acceleration ballisto-
cardiogram in man. Acta Cardiology (Brux) 22: 238-246.

> Young men and women were startled by blank pistol shots.
> Maximum cardiovascular change was observed 12-30 sec after
> the stimulus (peripheral vasoconstriction, increased heart
> rate, decreased duration of systole). The low-frequency
> acceleration ballistocardiogram was a more sensitive indicator
> of rapidly altered cardiac function than the electocardiogram,
> phonocardiogram or finger plethysmogram.

Alekseen, S.V. and G.A. Suvorov, 1965. Effects produced by continuous spectrum noise on some physiological functions of the organism.
Labor Hygiene and Industrial Disease 9: 8-11 (Russian).

Caporale, R. and M. DePalma, 1963. Effect of high intensity noise
on the human body. Rivista Di Medicina Aeronautica E Spaziale
(Roma) 26: 273-290 (Italian with English summary).

> Young humans, supine and relaxed, experienced 110 dB white noise
> for 3 min. Rheographic measures during and 5 min post-stimulation
> showed moderate increases in heart rate, but large increases in
> pulse amplitude and arteriolar flow index during the 10 sec post-
> stimulus period. Values trended to normalize in the following
> minutes.

Pokrovskii, N.N., 1966. On the effect of industrial noise on the
blood pressure in workers in machine building plants. Labor Hygiene
and Industrial Disease 10: 44-46 (Russian).

Jerkova, H. and B. Kremarova, 1965. Observation of the effect of
noise on the general health of workers in large engineering factor-
ies; attempt at evaluation. Pracovni Lekarstvi 17: 147-148
(Russian with English summary).

> Health records of 969 workers in factories having noise levels
> above 85 dB were studied. Observations indicate a higher
> incidence of subjective complaints, hypertension and peptic
> ulcer in people exposed to noise.

Ponomarenko, I.I., 1966. The effect of constant high frequency
industrial noise on certain physiological functions in adolescents.
Hygiene and Sanitation 31: 188-193.

Constant 1000 - 2000 Hz, 85 dB industrial noise had the following effects on 30 normal adolescents: In the course of the
working day there was an increase in the threshold for auditory
response; a decrease in the systolic and an increase in the
diastolic blood pressures; a decrease in vascular tone pulse
pressure, in pulse rate, and in the systolic index; an increase
in the diastolic and cardiac cycle time; an increase in the
time of reflex response to light and sound stimuli; and a
deterioration in the capacity for mental work. Such changes
did not occur on control days of classroom instruction.

Yeakel, E.H., H.A. Shenkin, A.B. Rothballer and S.M. McCann, 1948.
Blood pressures of rats subjected to auditory stimulation. American
Journal of Physiology 155: 118-127.

After a year's exposure to the sound of a blast of compressed
air, 5 min each day for 5 days each week, the average systolic
pressure of 25 gray Norway rats rose from an initial value of
113 mm Hg to 154 mm Hg in the last 2 months. Controls, mostly
littermates, showed an increase from 124 to 127 mm Hg in the
same period.

Tsysar, A.I., 1966. Effects of combined noise and vibration on the
cardiovascular system of working adolescents. Hygiene and Sanitation 31: 193.

Farris, E.J.; E.H. Yeakel and H.S. Medoff, 1945. Development of
hypertension in emotional gray Norway rats after airblasting.
American Journal of Physiology 144: 331.

Strakhov, A.B., 1964. The influence of intense noise on certain
functions of the organism. Hygiene and Sanitation 29: 29-36.

Rabbits and dogs were exposed to 1,500 - 3,000 cps, 90-93 dB
noise for periods from a few days to 152 days (rabbits) or 82
days (dogs). Humans experienced similar noise for 30 - 45 min.

In animals, the blood pressure became markedly labile. The
common tendency was for elevation of blood pressure, going in
some animals from initial values of 100 - 120 mm Hg to 170 mm
Hg. Hypertension was persistent for several months after
cessation of noise. Both in rabbits and dogs, there was a
pronounced increase in the ratio of the QRST interval to the
R-R interval in the ECG, indicative of impaired functional
properties of cardiac muscle. During exposure to noise, there
was marked lengthening of the P - Q segment of the ECG, lengthening of the QRS complex, and changes in wave forms, particularly
R and T. In dogs, there was a sharp strengthening of respiratory cardiac arrhythmia.

EEG in animals showed generalized cortical and subcortical low-
amplitude, high-frequency rhythms during the first days of exposure, subsequently replaced by slow synchronized rhythms.
Conditioned reflex salivary secretion was initially inhibited;
released from inhibition and exceeded normal levels; and finally
subjected to a new, deeper and longer-lasting inhibition.
In humans, noise caused increasing depression of the alpha
rhythm with passing time, and low-amplitude beta rhythm appeared. After 30 - 45 min noise exposure, these changes persisted for 30 - 45 min.

Observations were interpreted in terms of increased activation
of the reticular formation of the brainstem and of tonic in-
hibitions upon it.

NERVOUS SYSTEM

Connors, K. and D. Greenfield, 1966. Habituation of motor startle
in anxious and restless children. Journal of Child Psychology and
Psychiatry 7: 125.

Hyperactive children had poorer than normal control over startle
responses.

Mohr, G.C., J. N. Cole, E. Guild and H. E. von Gierke, 1965. Effects
of low frequency and infrasonic noise on man. Aerospace Medicine
36: 817-824.

Limits of voluntary tolerance of humans for exposure to low
frequency noise of 50 - 100 cps was determined in a special
Air Force facility as being 150 - 153 dB. The decision to
stop exposures was based upon the following subjective responses:
mild nausea, giddiness, subcostal discomfort, cutaneous flushing
and tingling, coughing, severe substernal pressure, respiratory
difficulty, salivation, pain on swallowing, choking, headache,
visual blurring, testicular pain. All subjects complained of
marked post-exposure fatigue. One subject continued to cough
for 20 min, and one retained some cutaneous flushing for approximately four hours. Earplug and earmuff combinations were worn,
and no shifts in hearing threshold were measurable two min post
exposure.

Halstead, W.C., 1953. Neuropsychological effects of chronic intermittent exposure to noise. Benox Report: An Exploratory Study of
the Biological Effects of Noise. Office of Naval Research Project
NR 144079, Contract N6 ori-020 Task Order 44, The University of
Chicago, December 1, 1953. pp. 96-111.

Civilian maintenance workers on jet aircraft at Wright Field are intermittently exposed to sounds of about 140 dB. They had abnormally high skin resistance on GSR tests. They complained of tiredness, insomnia, difficulty in relaxing at home, gradual reductions in libido, and irritability and short-temper with their families. Five of ten men tested who had been on jet maintenance for two years or more showed impairment of higher brain functions, most notable of which was loss of information through the tactile route. Of the ten men, two had mild and two had severe neurotic symptoms. Similar findings were made for maintenance personnel on the aircraft carriet Wasp.

Granati, A., F. Angeleri and R. Lenzi, 1959. L'influenza dei rumori sul sistema nervoso; Studio clinico ed elettroencefalografico sui tessitori artigiani. Folia Medica 42: 1313-1325.

Effects of intense noise on the central nervous system were studied by clinical examination and EEG. Psychoneurotic syndromes were diagnosed in a high percentage of individuals examined. In some, the EEG was similar to those of psychoneurotic patients or patients undergoing personality conversions.

Hermelin, B. and N. O'Connor, 1968. Measures of the occipital alpha rhythm in normal, subnormal and autistic children. British Journal of Psychiatry 114: 603-610.

Children sitting quietly were exposed to 800 cps, 75 dB, for 2 min intermittent, followed by 2 min continuous. No adaptation was seen in any group. Autistic children startled less, but were more aroused by continuous auditory stimulation than normal children.

Sternbach, R.A., 1960. A comparative analysis of autonomic responses in startle. Psychosomatic Medicine 12: 204-210.

Young human males, reclining on a cot, were startled by firing a blank pistol. Ambient noise was 46 dB, and the pistol shot 106 dB. Measured: systolic and diastolic blood pressures; finger, face and axillary temperatures; palmar and volar forearm skin conductances; heart and respiration rates; finger pulse volume; and stomach motility. Startle was different in effect from cold pressor stress, exercise and norepinephrine, but similar in direction of response to epinephrine, except that contraction period of the stomach was not decreased.

Szabo, I. and P. Kolta, 1967. Transitory increase of the acoustic startle reaction during its habituation. Acta Physiologica Academiae Scientiarum Hungaricae 31: 51-56.

Rats were startled by simulated gunshot 13 times in 15 sec intervals, each day for 10 consecutive days. The threshold for
startle was lowered during the first three days of exposure,
prior to being elevated at longer periods of time.

Minckley, B.B., 1968. A study of noise and its relationship to
patient discomfort in the recovery room. Nursing Research 17: 247–
250.

It was hypothesized that patients' subjective sensation of pain
in the immediate postoperative period would be increased at
times when noise levels were high (60–70 dB with spikes to
80+ dB vs. lowest level of 40–50 dB and median of 50–60 dB).
Findings were that more pain medication was given per patient
at times of high noise levels than at times of low noise levels.

Dumkina, G.Z., 1966. Some clinico-physiological investigations
made in workers exposed to the effects of stable noise.

Workers on regular turret lathes were exposed to noise of 250 –
4000 cps, 82–87 dB, and workers on automatic turret lathes to
92 – 99 dB. Functional nervous disorders in the form of astheno-
vegetative syndromes affected 29.4 percent of regular turret
lathe operators and 43 percent of automatic turret lathe operators. There were also paralleling functional changes in
hemodynamics.

SLEEP

Zarcone, V., G. Gulevich, T. Pivik and W. Dement, 1968. Partial
REM phase deprivation and schizophrenia. Archives of General
Psychiatry 18: 194–202.

Schizophrenics are abnormally sensitive to REM sleep deprivation. Suggest that the central nervous system changes associated with REM sleep deprivation may directly underlie psychotic
disintegration.

Mendels, J. and D.R. Hawkins, 1967. Sleep and depression. Archives
of General Psychiatry 16: 536–542.

Depressed patients studied had significantly less than normal
deep (Stage 4) sleep, more time awake and drowsy and more
spontaneous awakenings. Observations discussed in terms of
residual heightened activity of central arousal mechanisms
and/or a relative inefficiency in the functioning of these
mechanisms involved in achieving Stage 4 sleep.

REPRODUCTIVE

Ishii, H., T. Kamei and S. Omae, 1962. Effects of concurrent ad-
ministrations of chondroitin sulfate with cortisone, vitamin A or
noise stimulation on fetal development of the mouse. Gunma Journal
of Medical Science 11: 259–264.

Sovia, K., M. Gronroos and A.J. Aho, 1959. Effect of audiogenic-
visual stimuli on pregnant rats. Annales Medicinae Experimentalis
et Biologiae Fenniae (Helsinki) 37: 464–470.

> Audiovisual stimulation throughout pregnancy caused abortion
> in 25 percent of rats, but surviving litters were normal in
> number, average weight and morphology.

Ishii, H. and K. Yokobori, 1960. Experimental studies on teratogenic activity of noise stimulation. Gunma Journal of Medical
Sciences 9: 153–167.

> Female mice were subjected to white noise for 6 hours daily at
> 90, 100 or 110 phon, from the 11th through the 14th day of
> pregnancy. Noise increased the number of malformed young,
> increased the rate of still-birth, and reduced the weight of
> embryos. Noise interacted with trypan blue injections to in-
> tensify these effects.

Shidara, T., 1963. Experimental studies on the mechanism of teratogenic activity of noise; the role of the maternal adrenals. Japanese Journal of Otolaryngology 66: 532–547. (Japanese with English
summary).

> In pregnant rats, noise plus trypan blue injections caused a
> higher level of congenital malformation than did the dye alone;
> this effect was markedly reduced by adrenalectomy.

Takahashi, I. and S. Kyo, 1968. Studies on the difference of adapt-
abilities to the noisy environment in sexes and the growing process.
Journal of Anthropology Society Nippon 76: 34.

> In families living near noisy airfields, there was an increased
> rate of premature births and a depressed weight gain in the
> younger children exposed to noise.

ENDOCRINE and METABOLIC

Grognot, P., H. Boiteau and A. Gibert, 1961. L'influence de
L'Exposition aux Bruits sur le Taux du Potassium Plasmatique.
Presse Thermale et Climatique 98: 201–203.

Human subjects were exposed to 50 - 3,500 Hz (dominant 3000 Hz).
90 dB lowered plasma potassium, 105 dB either raised or lowered,
and 115 dB tended to raise plasma potassium.

Serra, C., A. Barone, C. De Vita and L. Laurini, 1964. Risposta
neuroumorali alla stimolazione acustica intermittente. Acta
Neurologica (Naples) 19: 1018-1035. (Italian with English summary).

> 132 adult humans experienced 60 sec exposures to noise varying
> from 64 - 8192 cps. Physiological measures were made 1 min
> and 30 min after cessation of noise. Serum potassium and sodium
> were markedly elevated or lowered, depending upon frequency of
> stimulation. Heart rate decreased at low frequencies and in-
> creased at high frequencies. Blood sugar and urinary 17-
> ketosteroids were increased. Serum cholinesterase was increased
> by some frequencies and lowered by others.

Gregorczyk, J., 1966. Activity of certain enzymes of intermediate
metabolism and behavior of serum proteins and their fractions in
workers exposed to intense noise. Acta Physiologica Polonica 17:
107-118.

> In one or more age groups of men tested, long-term exposure to
> noise of 90 - 108 dB of various broad-band frequencies was assoc-
> iated with lowered blood serum, total protein, beta globulins,
> gamma globulins, gamma-glutamyl transpeptidase and malic dehy-
> drogenase.

Hale, H.B., 1952. Adrenalcortical activity associated with exposure
to low frequency sounds. American Journal of Physiology 171: 732
(abstract).

> Workmen exposed to, and accustomed to, sound fields around
> aircraft engines showed eosinophil changes indicative of mild
> stress. Rats were exposed for 30 - 60 min to pure tones ranging
> from 50 - 6000 cps at 120 and 140 dB. Ascorbic acid content of
> adrenals was determined one hour later. Adrenalcortical acti-
> vation was greatest at low frequencies, and it was similar in
> magnitude for 30 min and 60 min exposures. At some frequencies,
> the activation caused by 120 dB could be increased by 140 dB.

Miline, R. and O. Kochak, 1951. L'Influence du bruit et des vibrations sur les glandes surrenales. Compte Rendu Association Des
Anatomistes Communications (Paris) 38e: 692-703.

> Rabbits were subjected to noise 16 hours per day for one to
> sixty days. The various zones of the adrenal cortex and also
> the adrenal medulla showed histological changes indicative of
> sustained increases in hormone secretion.

Miline, R., 1952. Effet du bruit et des vibrations sur la glande thyroïde. Compte Rendu Association Des Anatomistes Communications (Paris) 39: 649-656.

    Rabbits were subjected to noise 16 hours per day for 1 to fifty days. The thyroid gland showed histological changes indicative of altered secretatory activity.

RESPIRATION

Maugeri, U., 1965. Respiratory effects of industrial noise. Lavoro Umano 17: 331-338. (Italian with English summary).

    Noise of 90-95 dB in industrial situations altered various respiratory functions: frequency, tidal volume, outer ventilation, oxygen uptake, carbon dioxide exhaled.

AUTHOR INDEX

Abell, L. 55
Abelson, D. 37
Aceto, M.D.G. 84
Ackerman, E. 17, 88, 176
Adams, I.S. 176
Ader, R. 19, 140
Ades, H.W. 149, 150
Aho, A.J. 343
Akelseen, S.V. 338
Alexander, F. 83
Al-Hachim, G.M. 225
Amoroso, E.C. 127
Andersen, R.N. 36
Anderson, T.A. 88, 89
Angeleri, F. 341
Anthony, A. 17, 88, 176
Anticaglia, J.R. _143_, 147
Aprison, M.H. 269
Aresin, N. 114
Arguelles, A.E. _43_, 54
Armendariz, H. 254
Armington, J.C. 281
Arvay, A. _91_, 111, 112, 113,
   114, 127
Asakura, T. 41
Aschheim, S. 127, 130
Aulsebrook, L.H. 40
Axelrod, J. 84

Babcock, S. 88
Bainbridge, F.A. 128
Bajusz, E. 54, 113
Balazsy, L. 112
Ballantina, E.E. 176
Ballantine, E. 254
Balzer, A. 253
Barbeau, A. 268

Bargman, W. 128
Barka, T. 90
Baron, D.N. 37
Barone, A. 344
Barry, H. 83
Basowitz, H. 36
Baughman, F. 157
Baxter, C.F. 253, 269
Bayley, N.D. 304
Bazso, J. 113, 115
Beach, F.A. 176
Bekesy, G. 149
Beletsky, V. 176
Bellerby, C.W. 129
Benedetti, G. 114
Bengzon, A.R.A. 157
Benjamin, D.C. 18
Benko, E. 148
Bennett, D.R. 157
Bennholdt-Thomsen, C. 111
Beretta, L. 148
Bergman, M. 65
Bernard, C. 304
Bernard, J. 141
Bernstein, D.F. 84
Besser, G.M. 149
Bevan, W. 225
Bielec, S. 255
Bioteau, H. 55
Biro, J. 55, 88
Bishop, D.E. 329
Bisset, G.W. 39, 40
Blechman, H. 84
Bliss, E.L. 200, 257
Bobbio, G.C. 337
Bocci, G. 149
Boesiger, E. 210, 252